AMERICAN ORNITHOLOGY.

PRINTED BY BALLANTYNE, HANSON AND CO.
EDINBURGH AND LONDON

Drawn from Nature by A. Wilson.

Engraved by W.H.Lizars.

69.

1. Hooded or Crested Merganser. 2. Red-breasted M. 3. Blue Bill or Scaup Duck. 4. American Widgeon Male. 5. Female Snow Goose. 6. Pied Duck.

AMERICAN ORNITHOLOGY;

OR,

THE NATURAL HISTORY

OF THE

BIRDS OF THE UNITED STATES.

BY

ALEXANDER WILSON

AND

PRINCE CHARLES LUCIEN BONAPARTE.

The Illustrative Notes and Life of Wilson

BY SIR WILLIAM JARDINE, BART., F.R.S.E., F.L.S.

IN THREE VOLUMES.—VOL. III.

CASSELL PETTER & GALPIN:
LONDON, PARIS & NEW YORK.

CONTENTS OF THE THIRD VOLUME.

The names printed in italics are species not contained in the original, which have been introduced into the notes.

CONTENTS.

CONTENTS.

WILSON'S

AMERICAN ORNITHOLOGY.

HOODED MERGANSER.   (*Mergus cucullatus.*)

PLATE LXIX.—Fig. 1.

L'Harle Huppé de Virginie, *Briss.* vii. p. 258, 8.—*Pl. enl.* 935.—L'Harle Couronné, *Buff.* viii. p. 280.—Round-crested Duck, *Edw.* pl. 360.—*Catesby.* i. pl. 94.—*Arct. Zool.* No. 467.—*Lath. Syn.* 10, p. 426.—*Peale's Museum*, No. 2930.

MERGUS CUCULLATUS.—Linnæus.

Mergus cucullatus, *Cuv. Regn. Anim.* i. p. 540.—*Bonap. Synop.* p. 397.—*Selby, Illust. Brit. Ornith.* pl. 58.

This species on the sea-coast is usually called the *hairy-head*. They are more common, however, along our lakes and fresh-water rivers than near the sea ; tracing up creeks, and visiting millponds, diving perpetually for their food. In the creeks and rivers of the southern States they are very frequently seen during the winter. Like the red-breasted, they are migratory, the manners, food, and places of resort of both being very much alike.

The hooded merganser is eighteen inches in length, and two feet in extent ; bill, blackish red, narrow, thickly toothed, and furnished with a projecting nail at the extremity ; the head is ornamented with a large circular crest, which the bird has the faculty of raising or depressing at pleasure ; the fore part of this, as far as the eye, is black, thence to the hind head, white, and elegantly tipt with black; it is composed of two separate rows of feathers, radiating from each side of the head, and which may be easily divided by the hand ; irides, golden ; eye, very small ; neck, black, which spreads to and over the back ; part of the lesser wing-coverts, very pale ash, under which the greater coverts and secondaries form four alternate bars of black and white ; tertials, long, black, and streaked down the middle with white ; the black on the back curves handsomely round in two points on the breast, which, with the whole lower parts, are pure white ; sides, under the wings and flanks, reddish brown, beautifully crossed with parallel lines of black ; tail, pointed, consisting of twenty feathers of a sooty brown ; legs and feet flesh-coloured ; claws, large and stout. The windpipe has a small labyrinth.

The female is rather less, the crest smaller, and of a light rust or dull ferruginous colour, entirely destitute of the white ; the upper half of the neck, a dull drab, with semicircles of lighter, the white on the wings is the same as in the male, but the tertials are shorter and have less white ; the back is blackish brown ; the rest of the plumage corresponds very nearly with the male.

This species is peculiar to America ;* is said to arrive at Hudson's Bay about the end of May ; builds close to the lakes ; the nest is composed of grass, lined with feathers from the breast ; is said to lay six white eggs.   The young are yellow, nd fit to fly in July.†

* The female, or a young male of this bird, has lately been killed in England, and is figured in the last part of Mr Selby's " Illustrations." This, I believe, is the first instance of its occurrence in Europe.—ED.

† Hutchins, as quoted by Latham.

RED-BREASTED MERGANSER.   (*Mergus serrator.*)

PLATE LXIX.—FIG. 2.

L'Harle Huppé, *Briss.* vi. p. 237, 2, pl. 23.—*Buff.* viii. p. 273.—*Pl. enl.* 207.—
  Bewick, ii. p. 235.—*Edw.* pl. 95.—*Lath. Syn.* iii. p. 432.—*Peale's Museum*,
  No. 2936.
MERGUS SERRATOR.—Linnæus.*

Mergus serrator, *Linn. Syst.* i. 208.—*Bonap. Synop.* p. 397.—L'Harle Huppé,
  Temm. Man. ii. p. 884.—Red-breasted Merganser, *Mont. Ornith. Dict.* ii.
  and Supp.—*Flem. Brit. Anim.* p. 129.—*Selby, Illust. Br. Ornith.* pl. 58.

This is much more common in our fresh waters than either
of the preceding, and is frequently brought to the Philadelphia
market from the shores of the Delaware.   It is an inhabitant
of both continents.   In the United States it is generally
migratory, though a few are occasionally seen in autumn,
but none of their nests have as yet come under my notice.
They also frequent the sea-shore, keeping within the bays
and estuaries of rivers.   They swim low in the water, and,
when wounded in the wing, very dexterously contrive to elude
the sportsman or his dog by diving and coming up at a great
distance, raising the bill only above water, and dipping
down again with the greatest silence.   The young males of

---

* This beautiful species is also a native of both continents, and has
similar manners with its congeners.   In this country, during winter they
frequent the sea, but even in severe weather do not so frequently ascend
the rivers.   They breed throughout the whole of the north of Scotland,
by the edges or on the small islets of fresh-water lakes, both sexes
being seen in company only so long as the female continues to lay.
The nest is placed in some thicket of brushwood or rank herbage, and
is composed of the same materials which Wilson has mentioned.   The
eggs are a rich yellowish fawn colour.   Both Wilson and some of our
British writers mention them as white, or bluish white.   When they
have been sat upon for some time and approach to maturity, they receive
the latter tint from the transparency of the shell.

The female sits very close, and will allow an intruder to approach
within the distance of a yard.   All the nests which I have seen had two
runs in opposite directions, leading out of the cover, and when disturbed,
she followed one of these for a few yards before taking flight.—Ed.

a year old are often found in the plumage of the female; their food consists of small fry, and various kinds of shellfish.

The red-breasted merganser is said by Pennant to breed on Loch Maree, in the county of Ross, in North Britain, and also in the isle of Islay. Latham informs us that it inhabits most parts of the north of Europe on the Continent, and as high as Iceland; also in the Russian dominions about the great rivers of Siberia and the Lake Baikal. Is said to be frequent in Greenland, where it breeds on the shores. The inhabitants often take it by darts thrown at it, especially in August, being then in moult. At Hudson's Bay, according to Hutchins, they come in pairs about the middle of June, as soon as the ice breaks up, and build soon after their arrival, chiefly on dry spots of ground on the islands; lay from eight to thirteen white eggs the size of those of a duck; the nest is made of withered grass, and lined with the down of the breast. The young are of a dirty brown, like young goslings. In October they all depart southward to the lakes, where they may have open water.

This species is twenty-two inches in length, and thirty-two in extent; the bill is two inches and three-quarters in length, of the colour of bright sealing-wax, ridged above with dusky; the nail at the tip, large, blackish, and overhanging; both mandibles are thickly serrated; irides, red; head, furnished with a long hairy crest, which is often pendant, but occasionally erected, as represented in the plate; this and part of the neck is black glossed with green; the neck under this, for two or three inches, is pure white, ending in a broad space of reddish ochre spotted with black, which spreads over the lower part of the neck and sides of the breast; shoulders, back, and tertials, deep velvety black, the first marked with a number of singular roundish spots of white; scapulars, white; wing-coverts, mostly white, crossed by two narrow bands of black; primaries, black; secondaries, white; several of the latter edged with black; lower part of the back, the rump, and tail-coverts, grey, speckled with black; sides under the wings, elegantly crossed with numerous waving lines of black ; belly and vent, white ; legs and feet, red ; the tail, dusky ash ; the black of the back passes up the hind neck in a narrow band to the head.

The female is twenty-one inches in length, and thirty in extent ; the crested head and part of the neck are of a dull sorrel colour ; irides, yellow ; legs and bill red, upper parts, dusky slate ; wings, black ; greater coverts, largely tipt with white ; secondaries nearly all white ; sides of the breast, slightly dusky ; whole lower parts, pure white ; the tail is of a lighter slate than the back. The crest is much shorter than in the male, and sometimes there is a slight tinge of ferruginous on the breast.

The windpipe of the male of this species is very curious, and differs something from that of the goosander. About two inches from the mouth, it swells out to four times its common diameter, continuing of that size for about an inch and a half. This swelling is capable of being shortened or extended ; it then continues of its first diameter for two inches or more, when it becomes flattish, and almost transparent for other two inches ; it then swells into a bony labyrinth of more than two inches in length by one and a half in width, over the hollow sides of which is spread a yellowish skin like parchment. The left side of this, fronting the back of the bird, is a hard bone. The divarications come out very regularly from this at the lower end, and enter the lungs.

The intention of Nature in this extraordinary structure is probably to enable the bird to take down a supply of air to support respiration while diving ; yet why should the female, who takes the same submarine excursions as the male, be entirely destitute of this apparatus ?

SCAUP DUCK. (*Anas marilla.*)

PLATE LXIX.—Fig. 3.

Le Petit Morillon Rayé, *Briss.* vi. p. 416, 26, A.—*Arct. Zool.* No. 498.—*Lath. Syn.*
iii. p. 500.—*Peale's Museum,* No. 2668.

FULIGULA MARILLA.—Stephens.*

Fuligula marilla, *Steph. Cont. Sh. Zool.* xii. p. 108.—*Bonap. Syn.* p. 392.—
North. Zool. ii. p. 457.—Anas marilla, *Linn.* i. p. 19.—Scaup Duck, *Mont.*
Ornith. Dict. i. *and Supp.*—*Bew.* ii. p. 339.—Canard Milouinan, *Temm.*
Man. ii. p. 865.—Nyroca marilla, *Flem. Br. Anim.* p. 122.—Common Scaup
Pochard, *Selby, Illust. Br. Ornith.* pl. 66.

THIS duck is better known among us by the name of the blue-
bill. It is an excellent diver, and, according to Willoughby,
feeds on a certain small kind of shellfish called scaup, whence
it has derived its name. It is common both to our fresh-
water rivers and sea-shores in winter. Those that frequent
the latter are generally much the fattest, on account of the

---

* Common also to both continents, and in Britain a most abundant
sea-duck. Though generally to be found in the poultry markets during
winter, it is strong and ill flavoured, or what is called *fishy,* and of little
estimation for the table. In the " Northern Zoology," the American
specimens are said to be smaller, but no other distinctions could be perceived ; a single northern specimen which I possess agrees nearly with
the dimensions given of the smaller kind, and I can see no other important difference ; but there are also larger-sized birds, known to the
natives by the addition of " *Keetchee* " to the name, and I think it probable that two birds may be here confused, which future observations
will allow us to separate.

The young of both this bird and the tufted pochard have a white band
circling the base of the bill, which has caused them to be described as
distinct species.

To the description of Wilson, Mr Ord has added the following
observations :—" In the Delaware there are several favourite feeding
grounds of the blue-bill along the Jersey shore from Burlington to
Mantua Creek ; but the most noted spot appears to be the cove which
extends from Timber Creek to Eagle Point, and known by the name of
Ladd's Cove. Thither the blue-bills repair in the autumn, never quitting
it until they depart in the spring for the purpose of breeding, except
when driven away in the winter by the ice. It is no uncommon circum-

greater abundance of food along the coast. It is sometimes abundant in the Delaware, particularly in those places where small snails, its favourite shellfish, abound, feeding also, like most of its tribe, by moonlight. They generally leave us in April, though I have met with individuals of this species so late as the middle of May among the salt marshes of New Jersey. Their flesh is not of the most delicate kind, yet some persons esteem it. That of the young birds is generally the tenderest and most palatable.

The length of the blue-bill is nineteen inches; extent, twenty-nine inches ; bill, broad, generally of a light blue, sometimes of a dusky lead colour; irides, reddish ; head, tumid, covered with plumage of a dark glossy green, extending half way down the neck ; rest of the neck and breast, black, spreading round to the back ; back and scapulars, white, thickly crossed with waving lines of black ; lesser coverts, dusky, powdered with veins of whitish ; primaries and tertials, brownish black ; secondaries, white, tipped with black, forming the speculum ;

stance to see many hundreds of these birds at once, constantly diving for food ; but so shy are they, that even with the aid of a very small and well-constructed skiff, cautiously paddled, it is difficult to approach them within gunshot. So very sagacious are they, that they appear to know the precise distance wherein they are safe ; and, after the shooter has advanced within this point, they then begin to spread their lines in such a manner, that, in a flock of a hundred, not more than three or four can be selected in a group at any one view. They swim low in the water ; are strong feathered, and are not easily killed. When slightly wounded, and unable to fly, it is almost hopeless to follow them, in consequence of their great skill in diving. Their wings being short, they either cannot rise with the wind when it blows freshly, or they are unwilling to do so, for they are invariably seen to rise against the wind. In a calm they get up with considerable fluttering. Though often seen feeding in places where they can reach the bottom with their bills, yet they seldom venture on the shore, the labour of walking appearing repugnant to their inclinations. When wounded, they will never take to the land if they can possibly avoid it ; and when compelled to walk, they waddle along in the awkward manner of those birds whose legs, placed far behind, do not admit of a free and graceful progression."—ED.

rump and tail-coverts, black; tail, short, rounded, and of a dusky brown; belly, white, crossed near the vent with waving lines of ash; vent, black; legs and feet, dark slate.

Such is the colour of the bird in its perfect state. Young birds vary considerably, some having the head black, mixed with gray and purple, others the back dusky, with little or no white, and that irregularly dispersed.

The female has the front and sides of the same white; head and half of the neck, blackish brown; breast, spreading round to the back, a dark sooty brown, broadly skirted with whitish; back, black, thinly sprinkled with grains of white; vent, whitish; wings the same as in the male.

The windpipe of the male of this species is of a large diameter; the labyrinth similar to some others, though not of the largest kind; it has something of the shape of a single cockleshell; its open side, or circular rim, covered with a thin transparent skin. Just before the windpipe enters this, it lessens its diameter at least two-thirds, and assumes a flattish form.

The scaup duck is well known in England. It inhabits Iceland and the more northern parts of the continent of Europe, Lapland, Sweden, Norway, and Russia. It is also common on the northern shores of Siberia. It is very frequent on the river Ob. Breeds in the north, and migrates southward in winter. It inhabits America as high as Hudson's Bay, and retires from this place in October.*

* Latham.

AMERICAN WIDGEON. (*Anas Americana.*)

PLATE LXIX.—Fig. 4.

Le Canard Jenson, *Pl. enl.* 955.—*Buff.* ix. p. 174.—*Arct. Zool.* No. 502.—
Lath. iii. p. 520.—*Peale's Museum*, No. 2798.

MARECA AMERICANA.—Stephens.*

Mareca Americana, *Steph. Cont. Sh. Zool.* xii. p. 135.—*North. Zool.* ii. p. 445.—
Anas Americana, *Bonap. Synop.* p. 384.

This is a handsomely-marked and sprightly species, very common in winter along our whole coast from Florida to

* This species is closely allied to the European widgeon, and may be taken as the American analogue. They seem to meet each other about the arctic circle ; that of America extending beyond it, and that of Europe reaching to the European verge. They will form the types of Stephens's genus *Mareca*, which will probably stand in the rank of a more subordinate group only. The form is one of considerable interest, possessing many combinations, which may be found to connect some parts of the natural system. The bird of Europe, except in the breeding season, is mostly an inhabitant of the sea-shore ; during a severe winter, a few stray inland to the larger lakes and rivers, but as soon as a recurrence of moderate weather takes place, they return to their more favourite feeding grounds. In Britain they are mostly migratory, and at the first commencement of our harder weather, are found in vast flocks on the flatter coasts, particularly where there are beds of mussels and other shellfish. During day, they rest and plume themselves on the higher shelves, or doze buoyant on the waves, and only commence their activity with the approach of twilight. At this time they become clamorous, and rising in dense flocks from their day's resort, proceed to the feeding grounds, generally according to the wind in the same tract. At the commencement of winter, they are fat and delicate, much sought after by the sea-sportsmen, and are killed in numbers by persons lying in watch in the track of the known flight, or what, in some parts, is called *slaking*. The most propitious night for this sport is about half moon and strong wind ; the birds then fly low, and their approach is easily known by the whistling of their wings, and their own shrill cry ; whence their coast name of *hew*. They are subject to an annual change of plumage. Mr Ord mentions that a few of these birds breed annually in the marshes in the neighbourhood of Duck Creek, in the State of

Rhode Island, but most abundant in Carolina, where it frequents the rice plantations. In Martinico, great flocks take short flights from one rice-field to another during the rainy season, and are much complained of by the planters. The widgeon is the constant attendant of the celebrated canvas-back duck, so abundant in various parts of the Chesapeake Bay, by the aid of whose labour he has ingenuity enough to contrive to make a good subsistence. The widgeon is extremely fond of the tender roots of that particular species of aquatic plant on which the canvas-back feeds, and for which that duck is in the constant habit of diving. The widgeon, who never dives, watches the moment of the canvas-back's rising, and, before he has his eyes well opened, snatches the delicious morsel from his mouth and makes off. On this account the canvas-backs and widgeons, or, as they are called round the bay, *bald-pates,* live in a state of perpetual contention; the only chance the latter have is to retreat, and make their approaches at convenient opportunities. They are said to be in great plenty at St Domingo and Cayenne, where they are called vingeon, or gingeon. Are said sometimes to perch on trees, feed in company, and have a sentinel on the watch, like some other birds. They feed little during the day, but in the evenings come out from their hiding-places, and are then easily traced by their particular whistle or *whew-whew.* This soft note or whistle is frequently imitated with success, to entice them within gunshot. They are not known to breed in any part of the United States; are common, in the winter months, along the bays of Egg Harbour and Cape May, and also those of the Delaware. They leave these places in April, and appear upon the coasts of Hudson's Bay in May, as soon as the thaws come on, chiefly in pairs; lay there only from six to eight eggs, and feed on flies and worms in the swamps; depart in flocks in autumn.*

Delaware. An acquaintance of the editor's brought him thence, in the month of June, an egg which had been taken from a nest situated in a cluster of alders.—Ed.

* Hutchins.

These birds are frequently brought to the market of Baltimore, and generally bring a good price, their flesh being excellent. They are of a lively, frolicsome disposition, and, with proper attention, might easily be domesticated.

The widgeon, or bald-pate, measures twenty-two inches in length, and thirty inches in extent; the bill is of a slate-colour, the nail black; the front and crown, cream-coloured, sometimes nearly white, the feathers inflated; from the eye, backwards to the middle of the neck behind, extends a band of deep glossy green, gold, and purple; throat, chin, and sides of the neck before, as far as the green extends, dull yellowish white, thickly speckled with black; breast, and hind part of the neck, hoary bay, running in under the wings, where it is crossed with fine waving lines of black; whole belly, white; vent, black; back and scapulars, black, thickly and beautifully crossed with undulating lines of vinous bay; lower part of the back, more dusky; tail-coverts, long, pointed, whitish, crossed as the back; tail, pointed, brownish ash; the two middle feathers an inch longer than the rest, and tapering; shoulder of the wing, brownish ash; wing-coverts, immediately below, white, forming a large spot; primaries, brownish ash; middle secondaries, black, glossed with green, forming the speculum; tertials, black, edged with white, between which and the beauty-spot several of the secondaries are white.

The female has the whole head and neck yellowish white, thickly speckled with black, very little rufous on the breast; the back is dark brown. The young males, as usual, very much like the females during the first season, and do not receive their full plumage until the second year. They are also subject to a regular change every spring and autumn.

YOUNG OF THE SNOW GOOSE.   (*Anas hyperborea.*)

PLATE LXIX.—Fig. 5.

Bean Goose, *Lath. Syn.* iii. p. 464.—White-fronted Goose, *id.* iii. p. 463.—*Arct. Zool.* No. 476.—Blue-winged Goose, *Lath. Syn.* iii. p. 469.—*Peale's Museum,* No. 2636.

ANSER HYPERBOREAS.—Bonaparte.

The full-plumaged perfect male bird of this species has already been figured in the preceding plate, and I now hazard a conjecture, founded on the best examination I could make of the young bird here figured, comparing it with the descriptions of the different accounts above referred to, that the whole of them have been taken from the various individuals of the present, in a greater or lesser degree of approach to its true and perfect colours.

These birds pass along our coasts, and settle in our rivers, every autumn; among thirty or forty, there are seldom more than six or eight pure white or old birds. The rest vary so much, that no two are exactly alike; yet all bear the most evident marks, in the particular structure of their bills, &c., of being the same identical species. A gradual change so great as from a bird of this colour to one of pure white must necessarily produce a number of varieties or differences in the appearance of the plumage; but the form of the bill and legs remains the same, and any peculiarity in either is the surest means we have to detect a species under all its various appearances. It is therefore to be regretted that the authors above referred to in the synonyms have paid so little attention to the singular conformation of the bill; for even in the description of the snow goose, neither that nor the internal peculiarities are at all mentioned.

The length of the bird represented in our plate was twenty-eight inches; extent, four feet eight inches; bill, gibbous at the sides both above and below, exposing the teeth of the upper and lower mandibles, and furnished with a nail at the tip on both ; the whole being of a light-reddish purple or pale lake, except the gibbosity, which is black, and the two nails, which are of a pale light blue ; nostril, pervious, an oblong slit, placed nearly in the middle of the upper mandible ; irides, dark brown ; whole head, and half of the neck, white ; rest of the neck and breast, as well as upper part of the back, of a purplish brown, darkest where it joins the white ; all the feathers being finely tipt with pale brown ; whole wing-coverts, very pale ash, or light lead colour ; primaries and secondaries, black ; tertials, long, tapering, centred with black, edged with light blue, and usually fall over the wing ; scapulars, cinereous brown ; lower parts of the back and rump, of the same light ash as the wing-coverts ; tail, rounded, blackish, consisting of sixteen feathers, edged and tipt broadly with white ; tail-coverts white ; belly and vent, whitish, intermixed with cinereous ; feet and legs, of the same lake colour as the bill.

This specimen was a female ; the tongue was thick and fleshy, armed on each side with thirteen strong bony teeth, exactly similar in appearance, as well as in number, to those on the tongue of the snow goose ; the inner concavity of the upper mandible was also studded with rows of teeth. The stomach was extremely muscular, filled with some vegetable matter and clear gravel.

With this, another was shot, differing considerably in its markings, having little or no white on the head, and being smaller ; its general colour, dark brown, intermixed with pale ash, and darker below, but evidently of the same species with the other.

PIED DUCK.   (*Anas Labradora.*)

PLATE LXIX'—Fig. 6.

Arct. Zool. No. 488.—*Lath. Syn.* iii. p. 497.—*Peale's Museum*, No. 2858.
FULIGULA LABRADORA.—Bonaparte.*
Fuligula Labradora, *Bonap. Synop.* p. 391.

This is rather a scarce species on our coasts, and is never
met with on fresh-water lakes or rivers. It is called by some
gunners the *sand-shoal duck*, from its habit of frequenting
sandbars. Its principal food appears to be shellfish, which
it procures by diving. The flesh is dry, and partakes considerably of the nature of its food. It is only seen here during
winter; most commonly early in the month of March a few
are observed in our markets. Of their principal manners,
place, or mode of breeding, nothing more is known. Latham
observes that a pair in the possession of Sir Joseph Banks
were brought from Labrador. Having myself had frequent
opportunities of examining both sexes of these birds, I find
that, like most others, they are subject, when young, to a pro-
gressive change of colour. The full-plumaged male is as
follows :—Length, twenty inches ; extent, twenty-nine inches ;
the base of the bill, and edges of both mandibles for two-thirds
of their length, are of a pale orange colour; the rest, black ;
towards the extremity it widens a little in the manner of the
shovellers, the sides there having the singularity of being only
a soft, loose, pendulous skin ; irides, dark hazel ; head and
half of the neck, white, marked along the crown to the hind
head with a stripe of black ; the plumage of the cheeks is of
a peculiar bristly nature at the points, and round the neck
passes a collar of black which spreads over the back, rump,

---

* The Prince of Musignano places this bird among the *Fuligulæ*. I
have had no opportunity of seeing the bird itself, and cannot therefore
speak from examination as to its station. It seems a true sea-duck, and
agrees in general habits with the scaups and pochards.—Ed.

and tail-coverts ; below this collar the upper part of the breast is white, extending itself over the whole scapulars, wing-coverts, and secondaries ; the primaries, lower part of the breast, whole belly, and vent, are black ; tail, pointed, and of a black-ish hoary colour ; the fore part of the legs and ridges of the toes, pale whitish ash ; hind part, the same, bespattered with blackish ; webs, black ; the edges of both mandibles are largely pectinated. In young birds, the whole of the white plumage is generally strongly tinged with a yellowish cream colour ; in old males, these parts are pure white, with the exception sometimes of the bristly pointed plumage of the cheeks, which retains its cream tint the longest, and, with the skinny part of the bill, form two strong peculiarities of this species.

The female measures nineteen inches in length, and twenty-seven in extent ; bill, exactly as in the male ; sides of the front, white ; head, chin, and neck, ashy grey ; upper parts of the back and wings, brownish slate ; secondaries only white ; tertials, hoary ; the white secondaries form a spot on the wing, bounded by the black primaries, and four hoary tertials edged with black ; whole lower parts, a dull ash, skirted with brownish white or clay colour ; legs and feet, as in the male ; the bill in both is marked from the nostrils backwards by a singular heart-shaped outline.

The windpipe of the male measures ten inches in length, and has four enlargements, viz., one immediately below the mouth, and another at the interval of an inch ; it then bends largely down to the breast-bone, to which it adheres by two strong muscles, and has at that place a third expansion. It then becomes flattened, and, before it separates into the lungs, has a fourth enlargement much greater than any of the former, which is bony, and round, puffing out from the left side. The intestines measured six feet ; the stomach contained small clams and some glutinous matter ; the liver was remarkably large.

LONG-TAILED DUCK. (*Anas glacialis.*)

PLATE LXX.—FIG. 1, MALE.

Le Canard à longue queue de Terre Neuve, *Briss.* vi. p. 382, 18.—*Buff.* ix. p. 202, *Pl. enl.* 1008.—*Edw.* pl. 280.—*Arct. Zool.* No. 501.—*Lath. Syn.* iii. p. 528.- *Peale's Museum*, No. 2810.

HARELDA GLACIALIS.—LEACH.*

Anas glacialis and Anas hyemalis, *Linn. Syst.* i. p. 202 and 203.—*Lath. Ind.* ii. p. 864.—Fuligula glacialis, *Bonap. Synop.* p. 395.—Long-tailed Duck, *Mont. Ornith. Dict.* i. *and Supp.—Bew. Br. Birds,* ii. 363.—Long-tailed Hareld, *Selby's Illust. Br. Ornith.* pl. 61, *m.* and *f.*—Harelda glacialis, *North. Zool.* ii. p. 460.

THIS duck is very generally known along the shores of the Chesapeake Bay by the name of *south-southerly,* from the singularity of its cry, something imitative of the sound of those words, and also that, when very clamorous, they are supposed to betoken a southerly wind ; on the coast of New Jersey they are usually called *old wives.* They are chiefly salt-water

---

* This bird forms the type of Dr Leach's genus *Harelda.* It is remarkable for the decided change between the plumage of the breeding season and that of the winter, bearing analogy, in many particulars, to the Tringæ and their allies—for the prolongation of the scapulary feathers, and for the narrow lengthened tail. It is a native of both continents, but in Britain is only met with during winter, in the dress of that season, or in the plumage of the first year. It keeps to the open sea, and seldom ventures inland to rivers or lakes. The following is a description of a specimen killed on the 1st May, from the " Northern Zoology," and which agrees nearly with skins in my possession. " The whole upper plumage, the central pairs of tail feathers, and the under-plumage to the fore part of the belly, brownish black ; the lesser quills, paler. A triangular patch of feathers, between the shoulders and the scapulars, broadly bordered with orange brown." (In the winter plumage, the long scapulars are pure white, and form a beautiful contrast, hanging over the dark quills.) " Sides of head from the bill to the ears, ash grey ; eye stripe and posterior under-plumage, pure white ; flanks, sides of the rump, and lateral tail-feathers, white stained with brown ; axillaries and inner wing-coverts, clove brown ; bill, black, with an orange belt (*bright vermilion*) before the nostrils."—ED.

Drawn from Nature by A.Wilson.

Engraved by W.H Lizars.

1.Long-tailed Duck. 2.Female 3.Summer D. 4.Great-winged Teal. 5.Canvas back.D. 6.Red-headed D. 7.Mallard.

70.

ducks, and seldom ramble far from the sea. They inhabit our bays and coasts during the winter only; are rarely found in the marshes, but keep in the channel, diving for small shell-fish, which are their principal food. In passing to and from the bays, sometimes in vast flocks, particularly towards even-ing, their loud and confused noise may be heard in calm weather at the distance of several miles. They fly very swiftly, take short excursions, and are lively restless birds. Their native regions are in the north, where great numbers of them remain during the whole year, part only of the vast family migrating south to avoid the severest rigours of that climate. They are common to the whole northern hemisphere. In the Orkneys, they are met with in considerable flocks from Octo-ber to April; frequent in Sweden, Lapland, and Russia; are often found about St Petersburg, and also in Kamtschatka. Are said to breed at Hudson's Bay, making their nest among the grass near the sea, like the eider duck, and about the middle of June lay from ten to fourteen bluish white eggs, the size of those of a pullet. When the young are hatched, the mother carries them to the water in her bill. The nest is lined with the down of her breast, which is accounted equally valuable with that of the eider duck, were it to be had in the same quantity.* They are hardy birds and excellent divers. Are not very common in England, coming there only in very severe winters, and then but in small straggling parties; yet are found on the coast of America as far south at least as Charleston, in Carolina, during the winter. Their flesh is held in no great estimation, having a fishy taste. The down and plumage, particularly on the breast and lower parts of the body, are very abundant, and appear to be of the best quality.

The length of this species is twenty-two inches; extent, thirty inches; bill, black, crossed near the extremity by a band of orange; tongue, downy; iris, dark red; cheeks and front-let, dull dusky drab, passing over the eye, and joining a large

* Latham.

patch of black on the side of the neck, which ends in dark brown ; throat and rest of the neck, white ; crown, tufted, and of a pale cream colour ; lower part of the neck, breast, back, and wings, black ; scapulars and tertials, pale bluish white, long, and pointed, and falling gracefully over the wings ; the white of the lower part of the neck spreads over the back an inch or two ; the white of the belly spreads over the sides, and nearly meets at the rump ; secondaries, chestnut, forming a bar across the wing ; primaries, rump, and tail-coverts, black ; the tail consists of fourteen feathers, all remarkably pointed, the two middle ones nearly four inches longer than the others ; these, with the two adjoining ones, are black ; the rest, white ; legs and feet, dusky slate.

On dissection, the intestines were found to measure five feet six inches. The windpipe was very curiously formed ; besides the labyrinth, which is nearly as large as the end of the thumb, it has an expansion immediately above that of double its usual diameter, which continues for an inch and a half ; this is flattened on the side next the breast, with an oblong window-like vacancy in it, crossed with five narrow bars, and covered with a thin transparent skin, like the panes of a window ; another thin skin of the same kind is spread over the external side of the labyrinth, which is partly of a circular form. This singular conformation is, as usual, peculiar to the male, the female having the windpipe of nearly an uniform thickness throughout. She differs also so much in the colours and markings of her plumage, as to render a figure of her in the same plate necessary ; for a description of which see the following article.

FEMALE LONG-TAILED DUCK.

PLATE LXX.—Fɪɢ. 2.

Anas hyemalis, *Linn. Syst.* 202, 29.—*Lath. Syn.* iii. p. 529.—*Peale's Museum,* No. 2811.

HARELDA GLACIALIS.—Lᴇᴀᴄʜ.

Tʜᴇ female is distinguished from the male by wanting the lengthened tertials, and the two long pointed feathers of the tail, and also by her size and the rest of her plumage, which is as follows:—Length, sixteen inches; extent, twenty-eight inches; bill, dusky; middle of the crown, and spot on the side of the neck, blackish; a narrow dusky line runs along the throat for two inches; rest of the head, and upper half of the neck, white; lower half, pale vinaceous bay, blended with white; all the rest of the lower parts of the body, pure white; back, scapulars, and lesser wing-coverts, bright ferruginous, centred with black and interspersed with whitish; shoulders of the wing and quills, black; lower part of the back, the same, tinged with brown; tail, pale brown ash; inner vanes of all but the two middle feathers, white; legs and feet, dusky slate. The legs are placed far behind, which circumstance points out the species to be great divers. In some females, the upper parts are less ferruginous.

Some writers suppose the singular voice, or call, of this species to be occasioned by the remarkable construction of its windpipe; but the fact that the females are uniformly the most noisy, and yet are entirely destitute of the singularities of this conformation, overthrows the probability of this supposition.

SUMMER DUCK, OR WOOD DUCK. (*Anas sponsa.*)

PLATE LXX.—Fig. 3.

Le Canard d'Eté, *Briss.* vi. p. 351, 11, pl. 32, fig. 2.—Le Beau Canard Huppé,
　Buff. ix. p. 245.—*Pl. enl.* 980, 981.—Summer Duck, *Catesby,* i. pl. 97.—*Edw.*
　pl. 101.—*Arct. Zool.* No. 943.—*Lath. Syn.* iii. p. 546.—*Peale's Museum,*
　No. 2872.

DENDRONESSA SPONSA.—Richardson, Swainson.*

Anas sponsa, *Bonap. Synop.* p. 385.—Dendronessa sponsa, *North. Zool.* ii. 446.

This most beautiful of all our ducks has probably no superior
among its whole tribe for richness and variety of colours.　It
is called the wood duck from the circumstance of its breeding
in hollow trees, and the summer duck from remaining with us
chiefly during the summer.　It is familiarly known in every
quarter of the United States from Florida to Lake Ontario,
in the neighbourhood of which latter place I have myself met

* These lovely ducks may be said to represent an incessorial form
among the *Anatidæ ;* they build and perch on trees, and spend as much
time on land as upon the water.　Dr Richardson has given this group,
containing few members, the title of *Dendronessa* from their arboreal
habits.　Our present species is the only one belonging to America,
where it ranges rather to the south than north ; the others, I believe,
are all confined to India.　They are remarkable for the beauty and
splendour of their plumage, its glossy, silky, texture, and for the singular
form of the scapulars, which, instead of an extreme development in
length, receive it in the contrary proportion of breadth ; and instead
of lying flat, in some stand perpendicular to the back.　They are all
adorned with an ample crest, pendulous, and running down the back of
the neck.　They are easily domesticated, but I do not know that they
have been yet of much utility in this state, being more kept on account
of their beauty, and few have been introduced except to our menageries.
With a little trouble at first, they might form a much more common
ornament about our artificial pieces of water.　It is the only form of a
tree duck common to this continent ; in other countries there are, however, two or three others of very great importance in the natural system,
whose structure and habits have yet been almost entirely overlooked or
lost sight of.　These seem to range principally over India, and more
sparingly in Africa ; and the summer duck is the solitary instance, the
United States the nearly extreme limit, of its own peculiarities in this
division of the world.—Ed.

with it in October. It rarely visits the sea-shore or salt marshes, its favourite haunts being the solitary, deep, and muddy creeks, ponds, and mill-dams of the interior, making its nest frequently in old hollow trees that overhang the water.

The summer duck is equally well known in Mexico and many of the West India islands. During the whole of our winters, they are occasionally seen in the states South of the Potomac. On the 10th of January, I met with two on a creek near Petersburg, in Virginia. In the more northern districts, however, they are migratory. In Pennsylvania, the female usually begins to lay late in April or early in May. Instances have been known where the nest was constructed of a few sticks laid in a fork of the branches; usually, however, the inside of a hollow tree is selected for this purpose. On the 18th of May I visited a tree containing the nest of a summer duck on the banks of Tuckahoe river, New Jersey. It was an old grotesque white oak, whose top had been torn off by a storm. It stood on the declivity of the bank, about twenty yards from the water. In this hollow and broken top, and about six feet down, on the soft decayed wood, lay thirteen eggs, snugly covered with down, doubtless taken from the breast of the bird. These eggs were of an exact oval shape, less than those of a hen, the surface exceedingly fine grained, and of the highest polish, and slightly yellowish, greatly resembling old polished ivory. The egg measured two inches and an eighth by one inch and a half. On breaking one of them, the young bird was found to be nearly hatched, but dead, as neither of the parents had been observed about the tree during the three or four days preceding, and were conjectured to have been shot.

This tree had been occupied, probably by the same pair, for four successive years, in breeding-time. The person who gave me the information, and whose house was within twenty or thirty yards of the tree, said that he had seen the female, the spring preceding, carry down thirteen young, one by one, in less than ten minutes. She caught them in her bill by the wing or back of the neck, and landed them safely at the foot of the tree, whence she afterwards led them to the water. Under this same tree, at the time I visited it, a large sloop lay on the stocks, nearly finished; the deck was not more than twelve feet distant from the nest, yet, notwithstanding the presence and noise of the workmen, the ducks would not abandon their old breeding-place, but continued to pass out and in, as if no person had been near. The male usually perched on an adjoining limb, and kept watch while the female was laying, and also often while she was sitting. A tame goose had chosen a hollow space at the root of the same tree to lay and hatch her young in.

The summer duck seldom flies in flocks of more than three or four individuals together, and most commonly in pairs or singly. The common note of the drake is *peet, peet;* but when, standing sentinel, he sees danger, he makes a noise not unlike the crowing of a young cock, *oe eek! oe eek!* Their food consists principally of acorns, seeds of the wild oats, and insects. Their flesh is inferior to that of the blue-winged teal. They are frequent in the markets of Philadelphia.

Among other gaudy feathers with which the Indians ornament the calumet or pipe of peace, the skin of the head and neck of the summer duck is frequently seen covering the stem.

This beautiful bird has often been tamed, and soon becomes so familiar as to permit one to stroke its back with the hand. I have seen individuals so tamed in various parts of the Union. Captain Boyer, collector of the port of Havre-de-Grace, informs me that, about forty years ago, a Mr Nathan Nicols, who lived on the west side of Gunpowder Creek, had a whole yard swarming with summer ducks, which he had tamed and completely domesticated, so that they bred and were as familiar as any other tame fowls; that he (Captain Boyer) himself saw them in that state, but does not know what became of them. Latham says that they are often kept in European menageries, and will breed there.*

* General Synopsis, iii. 547.

The wood duck is nineteen inches in length, and two feet four inches in extent; bill, red, margined with black; a spot of black lies between the nostrils, reaching nearly to the tip, which is also of the same colour, and furnished with a large hooked nail; irides, orange red; front, crown, and pendant crest, rich glossy bronze green, ending in violet, elegantly marked with a line of pure white running from the upper mandible over the eye, and with another band of white proceeding from behind the eye, both mingling their long pendant plumes with the green and violet ones, producing a rich effect; cheeks and sides of the upper neck, violet; chin, throat, and collar round the neck, pure white, curving up in the form of a crescent, nearly to the posterior part of the eye; the white collar is bounded below with black; breast, dark violet brown, marked on the fore part with minute triangular spots of white, increasing in size until they spread into the white of the belly; each side of the breast is bounded by a large crescent of white, and that again by a broader one of deep black; sides under the wings, thickly and beautifully marked with fine undulating parallel lines of black, on a ground of yellowish drab; the flanks are ornamented with broad alternate semicircular bands of black and white; sides of the vent, rich light violet; tail-coverts, long, of a hair-like texture at the sides, over which they descend, and of a deep black, glossed with green; back, dusky bronze, reflecting green; scapulars, black; tail, tapering, dark glossy green above; below, dusky; primaries, dusky, silvery hoary without, tipt with violet blue; secondaries, greenish blue, tipt with white; wing-coverts, violet blue, tipt with black; vent, dusky; legs and feet, yellowish red; claws, strong and hooked.

The above is as accurate a description as I can give of a very perfect specimen now before me, from which the figure in the plate was faithfully copied.

The female has the head slightly crested; crown, dark purple; behind the eye, a bar of white; chin and throat, for two inches, also white; head and neck, dark drab; breast dusky brown, marked with large triangular spots of white; back, dark glossy bronze brown, with some gold and greenish reflections. Speculum of the wings nearly the same as in the male, but the fine pencilling of the sides, and the long hair-like tail coverts, are wanting; the tail is also shorter.

GREEN-WINGED TEAL. (*Anas crecca.*)

PLATE LXX.—Fig. 4.

Lath. Syn. iii. p. 554.—*Bewick's Br. Birds,* v. ii. p. 338.—*Peale's Museum,* No. 2832.

BOSCHAS CAROLINENSIS.—Jardine.*

Anas Carolinensis, *Lath. Ind. Ornith.* ii. p. 874.—Anas migratoria, Least Green-winged Teal, *Bart. Trav.* p. 293.—Anas crecca, varietas, *Forst. Phil. Trans.* lxii. p. 347.—American Teal, *Lath. Gen. Hist.* x. p. 371.—Anas crecca, *Bonap. Synop.* p. 386.—Anas (Boschas) crecca, var. *North. Zool.* ii. p. 443.

THE naturalists of Europe have designated this little duck by the name of the American teal, as being a species different from their own. On an examination, however, of the figure and description of the European teal by the ingenious and accurate Bewick, and comparing them with the present, no difference whatever appears in the length, extent, colour, or markings of either, but what commonly occurs among individuals of any other tribe; both undoubtedly belong to one and the same species.

* Most writers on the ornithology of America have considered this bird as a variety of the European teal. All, however, agree in their regarding the difference in the variety, and of its being constant in the northern specimens. Thus, Dr Latham mentions the white pectoral band. Foster says, "This is a variety of the teal, for it wants the two white streaks above and below the eyes; the lower one indeed is faintly expressed in the male, which has also a lunated bar of white over each shoulder; this is not to be found in the European teal." Pennant, "That it wants the white line which the European one has above each eye, having only one below; has over each shoulder a lunated bar." The authors of the "Northern Zoology" observe, "The only permanent difference that we have been able to detect, after comparing a number of specimens, is, that the English teal has a white longitudinal band on the scapulars, which the other wants. All the

This, like the preceding, is a fresh-water duck, common in our markets in autumn and winter, but rarely seen here in summer. It frequents ponds, marshes, and the reedy shores of creeks and rivers; is very abundant among the rice plantations of the southern States; flies in small parties, and feeds at night; associates often with the duck and mallard, feeding on the seeds of various kinds of grasses and water plants, and also on the tender leaves of vegetables. Its flesh is accounted excellent.

The green-winged teal is fifteen inches in length, and twenty-four inches in extent; bill, black; irides, pale brown; lower eyelid, whitish; head, glossy reddish chestnut; from the eye backwards to the nape, runs a broad band of rich silky green, edged above and below by a fine line of brownish white; the plumage of the nape ends in a kind of pendant crest; chin, blackish; below the chestnut, the neck, for three-quarters of an inch, is white, beautifully crossed with circular undulating lines of black; back, scapulars, and sides of the breast, white, thickly crossed in the same manner; breast, elegantly marked with roundish or heart-shaped spots of black, on a pale vinaceous ground, variegated with lighter tints; belly, white; sides waved with undulating lines; lower part of the vent-feathers, black; sides of the same, brownish white or pale reddish cream; lesser wing-coverts, brown ash; greater,

specimens brought home by the expedition have a broad transverse bar on the shoulder, which does not exist in the English one." And our author, in his plate, has most distinctly marked the differences. From the testimony of all its describers marking the variety as permanent and similar, I am certainly inclined to consider this bird, though nearly allied, to be distinct; and, as far as we yet know, peculiar to the northern parts of America. I have not been able to procure a specimen for immediate comparison, and only once had an opportunity of slightly examining a northern bird : in it the distinctions were at once perceptible. From their great similarity, no observers have yet particularly attended to the manners of the American bird, or to the marking of the females. If the above observations are the means of directing further attention to these points, they will have performed their intended end. I by no means consider the point decided.—Ed.

tipt with reddish cream; the first five secondaries deep velvety black, the next five resplendent green, forming the speculum or beauty spot, which is bounded above by pale buff, below by white, and on each side by deep black; primaries, ashy brown; tail, pointed, eighteen feathers, dark drab; legs and feet, flesh-coloured. In some a few circular touches of white appear on the breast near the shoulder of the wing. The windpipe has a small bony labyrinth where it separates into the lungs; the intestines measure three feet six inches, and are very small and tender.

The female wants the chestnut bay on the head, and the band of rich green through the eye, these parts being dusky white speckled with black; the breast is grey brown, thickly sprinkled with blackish or dark brown; the back, dark brown, waved with broad lines of brownish white; wing nearly the same as in the male.

This species is said to breed at Hudson's Bay, and to have from five to seven young at a time.* In France it remains throughout the year, and builds in April among the rushes on the edges of the ponds. It has been lately discovered to breed also in England, in the mosses about Carlisle.† It is not known to breed in any part of the United States. The teal is found in the north of Europe as far as Iceland, and also inhabits the Caspian Sea to the south. Extends likewise to China, having been recognised by Latham among some fine drawings of the birds of that country.

* Latham.   † Bewick.

CANVAS-BACK DUCK. (*Anas valisineria.*)

PLATE LXX.—Fig. 5.

Peale's Museum, No. 2816.

FULIGULA VALISNERIANA.—Stephens.*

Fuligula valisneriana, *Bonap. Synop.* p. 392.—*North. Zool.* iv. p. 450.—Anas
valisneriana, *Wilson.*

This celebrated American species, as far as can be judged
from the best figures and descriptions of foreign birds, is altogether unknown in Europe. It approaches nearest to the
pochard of England (*Anas ferina*), but differs from that bird
in being superior in size and weight, in the greater magnitude
of its bill, and the general whiteness of its plumage. A short
comparison of the two will elucidate this point. The canvas-
back measures two feet in length by three feet in extent,
and when in the best order weighs three pounds and upwards.
The pochard, according to Latham and Bewick, measures
nineteen inches in length, and thirty in extent, and weighs

---

* This species is now well established, and can never be mistaken.
I am not aware that anything can be added to Wilson's accurate
description, and the additional remarks of Mr Ord, which we have
printed, unless Wilson's own description, in poetry, of his first capture
of the *canvas-back.*—Ed.

> Slow round an opening point we softly steal,
> Where four large ducks in playful circles wheel,
> The far-famed *canvas-backs* at once we know,
> Their broad flat bodies wrapt in pencilled snow ;
> The burnished chestnut o'er their necks that shone
> Spread deepening round each breast a sable zone.
> Wary they gaze—our boat in silence glides,
> The slow-moved paddles steal along the sides ;
> Quick flashing thunders roar along the flood,
> And three lie prostrate, vomiting their blood !
> The fourth aloft on whistling pinions soared,
> One fatal glance the fiery thunders poured,
> Prone drops the bird amid the dashing waves,
> And the clear stream his glossy plumage laves.
> —*Foresters*, p. 39.

one pound twelve or thirteen ounces. The latter writer says
of the pochard, "The plumage, above and below, is wholly
covered with prettily-freckled slender dusky threads, disposed
transversely in close-set zigzag lines, on a pale ground, more
or less shaded off with ash;" a description much more
applicable to the bird figured beside it, the red-head, and
which very probably is the species meant. In the figure of
the pochard given by Mr Bewick, who is generally correct,
the bill agrees very well with that of our red-head; but is
scarcely half the size and thickness of that of the canvas-
back; and the figure in the *Planches Enluminées* corresponds
in that respect with Bewick's. In short, either these writers
are egregiously erroneous in their figures and descriptions,
or the present duck was altogether unknown to them.
Considering the latter supposition the more probable of the
two, I have designated this as a new species, and shall proceed
to detail some particulars of its history.

The canvas-back duck arrives in the United States from
the north about the middle of October; a few descend to the
Hudson and Delaware, but the great body of these birds
resort to the numerous rivers belonging to and in the
neighbourhood of the Chesapeake Bay, particularly the
Susquehannah, the Patapsco, Potomac, and James rivers,
which appear to be their general winter rendezvous. Beyond
this, to the south, I can find no certain accounts of them.
At the Susquehannah, they are called canvas-backs; on the
Potomac, white-backs; and on James river, sheldrakes. They
are seldom found at a great distance up any of these rivers,
or even in the salt-water bay; but in that particular part
of tide-water where a certain grass-like plant grows, on
the roots of which they feed. This plant, which is said to
be a species of *Valisineria*, grows on fresh-water shoals of
from seven to nine feet (but never where these are occasionally
dry), in long narrow grass-like blades of four or five feet in
length; the root is white, and has some resemblance to small
celery. This grass is in many places so thick, that a boat

can with difficulty be rowed through it, it so impedes the
oars. The shores are lined with large quantities of it, torn
up by the ducks, and drifted up by the winds, lying like hay
in wind-rows. Wherever this plant grows in abundance,
the canvas-backs may be expected either to pay occasional
visits or to make it their regular residence during the winter.
It occurs in some parts of the Hudson; in the Delaware,
near Gloucester, a few miles below Philadelphia; and in
most of the rivers that fall into the Chesapeake, to each of
which particular places these ducks resort; while in waters
unprovided with this nutritive plant they are altogether
unknown.

On the first arrival of these birds in the Susquehannah, near
Havre-de-Grace, they are generally lean; but such is the
abundance of their favourite food, that, towards the beginning
of November, they are in pretty good order. They are excellent divers, and swim with great speed and agility. They
sometimes assemble in such multitudes as to cover several
acres of the river, and when they rise suddenly, produce a noise
resembling thunder. They float about these shoals, diving and
tearing up the grass by the roots, which is the only part they
eat. They are extremely shy, and can rarely be approached,
unless by stratagem. When wounded in the wing, they dive
to such prodigious distances, and with such rapidity, continuing it so perseveringly, and with such cunning and active
vigour, as almost always to render the pursuit hopeless. From
the great demand for these ducks, and the high price they
uniformly bring in market, various modes are practised to get
within gunshot of them. The most successful way is said to
be decoying them to the shore by means of a dog, while the
gunner lies closely concealed in a proper situation. The dog,
if properly trained, plays backwards and forwards along the
margin of the water, and the ducks, observing his manœuvres,
enticed perhaps by curiosity, gradually approach the shore,
until they are sometimes within twenty or thirty yards of the
spot where the gunner lies concealed, and from which he rakes

them, first on the water, and then as they rise. This method is called *tolling them in.* If the ducks seem difficult to decoy, any glaring object, such as a red handkerchief, is fixed round the dog's middle, or to his tail, and this rarely fails to attract them. Sometimes, by moonlight, the sportsman directs his skiff towards a flock whose position he has previously ascertained, keeping within the projecting shadow of some wood, bank, or headland, and paddles along so silently and imperceptibly, as often to approach within fifteen or twenty yards of a flock of many thousands, among whom he generally makes great slaughter.

Many other stratagems are practised, and, indeed, every plan that the ingenuity of the experienced sportsman can suggest, to approach within gunshot of these birds; but of all the modes pursued, none intimidate them so much as shooting them by night; and they soon abandon the place where they have been thus repeatedly shot at. During the day they are dispersed about, but towards evening collect in large flocks, and come into the mouths of creeks, where they often ride, as at anchor, with their head under their wing asleep, there being always sentinels awake, ready to raise an alarm on the least appearance of danger. Even when feeding and diving in small parties, the whole never go down at one time, but some are still left above on the look-out.

When the winter sets in severely, and the river is frozen, the canvas-backs retreat to its confluence with the bay, occasionally frequenting air-holes in the ice, which are sometimes made for the purpose, immediately above their favourite grass, to entice them within gunshot of the hut or bush, which is usually fixed at a proper distance, and where the gunner lies concealed, ready to take advantage of their distress. A Mr Hill, who lives near James river, at a place called Herring Creek, informs me that, one severe winter, he and another person broke a hole in the ice about twenty by forty feet, immediately over a shoal of grass, and took their stand on the shore in a hut of brush, each having three guns well loaded with large shot. The ducks, which were flying up and down the river in great extremity, soon crowded to this place, so that the whole open space was not only covered with them, but vast numbers stood on the ice around it. They had three rounds, firing both at once, and picked up eighty-eight canvas-backs, and might have collected more, had they been able to get to the extremity of the ice after the wounded ones. In the severe winter of 1779–80, the grass on the roots of which these birds feed was almost wholly destroyed in James river. In the month of January, the wind continued to blow from W.N.W. for twenty-one days, which caused such low tides in the river, that the grass froze to the ice everywhere, and a thaw coming on suddenly, the whole was raised by the roots and carried off by the fresh. The next winter a few of these ducks were seen, but they soon went away again ; and, for many years after, they continued to be scarce, and even to the present day, in the opinion of my informant, have never been so plenty as before.

The canvas-back, in the rich juicy tenderness of its flesh, and its delicacy of flavour, stands unrivalled by the whole of its tribe in this or perhaps any other quarter of the world. Those killed in the waters of the Chesapeake are generally esteemed superior to all others, doubtless from the great abundance of their favourite food which these rivers produce. At our public dinners, hotels, and particular entertainments, the canvas-backs are universal favourites. They not only grace but dignify the table, and their very name conveys to the imagination of the eager epicure the most comfortable and exhilarating ideas. Hence, on such occasions, it has not been uncommon to pay from one to three dollars a pair for these ducks ; and, indeed, at such times, if they can, they must be had, whatever may be the price.

The canvas-back will feed readily on grain, especially wheat, and may be decoyed to particular places by baiting them with that grain for several successive days. Some few years since a vessel loaded with wheat was wrecked near the entrance of Great Egg Harbour in the autumn, and went to pieces. The wheat floated out in vast quantities, and the whole surface of the bay was in a few days covered with ducks of a kind altogether unknown to the people of that quarter. The gunners of the neighbourhood collected in boats in every direction, shooting them; and so successful were they, that, as Mr Beasley informs me, two hundred and forty were killed in one day, and sold among the neighbours at twelve and a half cents apiece, without the feathers. The wounded ones were generally abandoned, as being too difficult to be come up with. They continued about for three weeks, and during the greater part of that time a continual cannonading was heard from every quarter. The gunners called them sea-ducks. They were all canvas-backs, at that time on their way from the north, when this floating feast attracted their attention, and for a while arrested them in their course. A pair of these very ducks I myself bought in Philadelphia market at the time from an Egg Harbour gunner, and never met with their superior either in weight or excellence of flesh. When it was known among those people the loss they had sustained in selling for twenty-five cents what would have brought them from a dollar to a dollar and a half per pair, universal surprise and regret were naturally enough excited.

The canvas-back is two feet long, and three feet in extent, and, when in good order, weighs three pounds; the bill is large, rising high in the head, three inches in length, and one inch and three-eighths thick at the base, of a glossy black; eye, very small; irides, dark red; cheeks and forepart of the head, blackish brown; rest of the head and greater part of the neck, bright glossy reddish chestnut, ending in a broad space of black that covers the upper part of the breast, and spreads round to the back; back, scapulars, and tertials, white, faintly marked with an infinite number of transverse waving lines or points, as if done with a pencil; whole lower parts of the breast, also the belly, white, slightly pencilled in the same manner, scarcely perceptible on the breast, pretty thick towards the vent; wing-coverts, grey, with numerous specks of blackish; primaries and secondaries, pale slate, two or three of the latter of which nearest the body are finely edged with deep velvety black, the former dusky at the tips; tail, very short, pointed, consisting of fourteen feathers of a hoary brown; vent and tail-coverts, black; lining of the wing, white; legs and feet, very pale ash, the latter three inches in width, a circumstance which partly accounts for its great powers of swimming.

The female is somewhat less than the male, and weighs two pounds and three-quarters; the crown is blackish brown; cheeks and throat of a pale drab; neck, dull brown; breast, as far as the black extends on the male, dull brown, skirted in places with pale drab; back, dusky white, crossed with fine waving lines; belly, of the same dull white, pencilled like the back; wings, feet, and bill, as in the male; tail-coverts, dusky; vent, white, waved with brown.

The windpipe of the male has a large flattish concave labyrinth, the ridge of which is covered with a thin transparent membrane; where the trachea enters this, it is very narrow, but immediately above swells to three times that diameter. The intestines are wide, and measure five feet in length.

Mr Ord, in his reprint, has added the following interesting observations:—" It is a circumstance calculated to excite our surprise, that the canvas-back, one of the commonest species of our country, a duck which frequents the waters of the Chesapeake in flocks of countless thousands, should yet have been either overlooked by the naturalists of Europe, or confounded with the pochard, a species whose characters are so obviously different. But that this is the fact the editor feels well assured, since he has carefully examined every author of repute to which he has had access, and has not been enabled to find any description which will correspond to the subject before us. The species, then, we hope, will stand as Wilson's own; and it is no small addition to the fame of the " American Ornithology " that it contains the first scientific account of the finest duck that any country can boast of.

" The canvas-back frequents the Delaware in considerable numbers. The *Valisineria* grows pretty abundantly in various places from Burlington, New Jersey, to Eagle Point, a few miles below Philadelphia. Wherever this plant is found, there will the ducks be ; and they will frequently venture within reach of their enemies' weapons rather than abstain from the gratification of their appetite for this delicious food. The shooters in the neighbourhood of Philadelphia for many years were in the habit of supplying our markets with this species, which always bore the name of red-heads or red-necks ; and their ignorance of its being the true canvas-back was cunningly fostered by our neighbours of the Chesapeake, who boldly asserted that only their waters were favoured with this species, and that all other ducks which seemed to claim affinity were a spurious race, unworthy of consanguinity. Hence, at the same time, when a pair of legitimate canvas-backs, proudly exhibited from the mail-coach from Havre-de-Grace, readily sold for two dollars and fifty cents, a pair of the identical species, as fat, as heavy, as delicious, but which had been unfortunately killed in the Delaware, brought only one dollar ; and the lucky shooter thought himself sufficiently rewarded in obtaining twenty-five per cent. more for his *red-necks* than he could obtain for a pair of the finest mallards that our waters could afford. But the delusion is now past ; every shooter and huckster knows the distinctive characters of the canvas-back and the red-head, and prejudice no longer controverts the opinion that this species is a common inhabitant of the Delaware ; and epicures are compelled to confess that they can discern no difference between our canvas-back, when in season, and that from Spesutie or Carrol's Island, the notorious shooting ground of the *bon vivants* of Baltimore.

" The last-mentioned place, though commonly termed an island, is properly a peninsula, situated on the western side of the Chesapeake Bay, a few miles from Baltimore. It is a spot highly favourable for the shooting of waterfowl. It extends for a considerable distance into the bay ; and being connected to the main land by a narrow neck, the shooters are enabled to post themselves advantageously on the isthmus, and intercept the fowl, who, in roving from one feeding ground to another, commonly prefer crossing the land to taking a long flight around the peninsula. In calm weather the shooters have not much luck, the ducks keeping out in the coves, and when they do move, flying high ; but should a fresh breeze prevail, especially one from the eastward, rare sport may be anticipated ; and it is no unusual circumstance for a party of four or five gentlemen returning home, after a couple of days' excursion, with fifty or sixty canvas-backs, besides some other ducks of inferior note. The greatest flight of ducks commonly takes place between daybreak and sunrise ; and while it lasts,the roaring of the fowling-pieces, the bustle of the sportsmen, the fluttering of the fowl, and the plunging of the dogs, constitute a scene productive of intense interest. The dog in most esteem for this amusement is a large breed, partaking of the qualities of the Newfoundland variety. They trust altogether to their sight, and it is astonishing what sagacity they will manifest in watching a flock of ducks that had been shot at, and marking the birds that drop into the water, even at a considerable distance off. When at fault, the motion of their master's hand is readily obeyed by them ; and when unable to perceive the object of their search, they will raise themselves in the water for this purpose, and will not abandon the pursuit while a chance remains of succeeding. A generous, well-trained dog, has been known to follow a duck for more than half a mile ; and, after having been long beyond the reach of seeing or hearing his master, to return puffing and snorting under his load, which seemed sufficient to drag him beneath the waves. The editor having been an eye-witness of similar feats of these noble animals, can therefore speak with confidence as to the fact.

" On the Delaware, but few of this species, comparatively, are obtained, for the want of proper situations whence they may be shot on the wing. To attempt to approach them in open day with a boat is unproductive labour, except there be floating ice in the river, at which time, if the shooter clothe himself in white, and paint his skiff of the same colour, he may so deceive the ducks as to get within a few feet of them. At such times it is reasonable to suppose that these valuable birds get no quarter. But there is one caution to be observed, which experienced sportsmen never omit: it is to go always *with* the current; a duck being sagacious enough to know that a lump of ice seldom advances *against* the stream. They are often shot with us by moonlight in the mode related in the foregoing account; the first pair the editor ever killed was in this manner; he was then a boy, and was not a little gratified with his uncommon acquisition.

"As the *Valisineria* will grow in all our fresh-water rivers, in coves or places not affected by the current, it would be worth the experiment to transplant this vegetable in those waters where it at present is unknown. There is little doubt the canvas-backs would, by this means, be attracted; and thus would afford the lovers of good eating an opportunity of tasting a delicacy which, in the opinion of many, is unrivalled by the whole feathered race.

"In the spring, when the duck-grass becomes scarce, the canvas-backs are compelled to subsist upon other food, particularly shellfish; their flesh then loses its delicacy of flavour, and, although still fat, it is not esteemed by epicures; hence the ducks are not much sought after, and are permitted quietly to feed until their departure for the north.

"Our author states that he had no certain accounts of this species to the southward of James river, Virginia. In the month of January 1818, the editor saw many hundreds of these ducks feeding in the Savannah river, not far from Tybee lighthouse. They were known by the name of canvas-backs; but the inhabitants of that quarter considered them as fishing-ducks, not fit to be eaten: so said the pilot of the ship which bore the editor to Savannah. But a pair of these birds having been served up at table after his arrival, he was convinced by their delicate flavour that they had lost little by their change of residence, but still maintained their superiority over all the waterfowl of that region.   In the river St John, in East Florida, the editor also saw a few scattered individuals of this species; but they were too shy to be approached within gunshot.

"The canvas-backs swim very low, especially when fat; and when pursued by a boat, they stretch themselves out in lines, in the manner of the scaup ducks, so that some of the flock are always enabled to reconnoitre the paddler, and give information to the rest of his motions.   When the look-out ducks apprehend danger, the stretching up of their necks is the signal, and immediately the whole squadron, facing to the wind, rise with a noise which may be heard at the distance of half a mile."

RED-HEADED DUCK.   (*Anas ferina ?*)

PLATE LXX.—FIG. 6.

Peale's Museum, No. 2710.

FULIGULA FERINA.—STEPHENS.*

Steph. Cont. Sh. Zool. p. 193.—Fuligula ferina, *Bonap. Synop.* p. 392.—*North. Zool.* ii. p. 451.—Nyroca ferina, *Flem. Br. Anim.* p. 108.—Le Canard Miloun, *Temm. Man.* ii. 868.—Pochard, or Red-headed Widgeon, *Mont. Ornith. Dict.* ii. *and Supp.*—*Bew. Br. Birds,* ii. p. 356.—Red-headed Widgeon, *Selby's Illust. Br. Ornith.* pl. 63, fig. 1.

THIS is a common associate of the canvas-back, frequenting the same places, and feeding on the stems of the same grass, the latter eating only the roots; its flesh is very little inferior, and is often sold in our markets for the canvas-back to those unacquainted with the characteristic marks of each.   Anxious as I am to determine precisely whether this species be the

---

* A well-known duck common to both continents, keeping to the sea or large lakes, and only in very severe winters wandering to any extent inland.   Sometimes seen in the decoys, but very seldom taken, from their expertness in diving under the tunnel.—ED.

red-headed widgeon, pochard, or dun bird* of England, I have not been able to ascertain the point to my own satisfaction, though I think it very probably the same, the size, extent, and general description of the pochard agreeing pretty nearly with this.

The red-head is twenty inches in length, and two feet six inches in extent ; bill, dark slate, sometimes black, two inches long, and seven-eighths of an inch thick at the base, furnished with a large broad nail at the extremity ; irides, flame-coloured ; plumage of the head, long, velvety, and inflated, running high above the base of the bill ; head, and about two inches of the neck, deep glossy reddish chestnut ; rest of the neck and upper part of the breast, black, spreading round to the back ; belly, white, becoming dusky towards the vent by closely marked undulating lines of black ; back and scapulars, bluish white, rendered grey by numerous transverse waving lines of black ; lesser wing-coverts, brownish ash ; wing-quills, very pale slate, dusky at the tips ; lower part of the back and sides under the wings, brownish black, crossed with regular zigzag lines of whitish ; vent, rump, tail, and tail-coverts, black ; legs and feet, dark ash.

The female has the upper part of the head dusky brown, rest of the head and part of the neck, a light sooty brown ; upper part of the breast, ashy brown, broadly skirted with whitish ; back, dark ash, with little or no appearance of white pencilling ; wings, bill, and feet nearly alike in both sexes.

This duck is sometimes met with in the rivers of North and South Carolina, and also in those of Jersey and New York, but always in fresh water, and usually at no great distance from the sea ; is most numerous in the waters of the Chesapeake ; and, with the connoisseurs in good eating, ranks next in excellence to the canvas-back. Its usual weight is about a pound and three-quarters avoirdupois.

The red-head leaves the bay and its tributary streams in March, and is not seen till late in October.

* Local names given to one and the same duck. It is also called the poker.

The male of this species has a large flat bony labyrinth on the bottom of the windpipe, very much like that of the canvas-back, but smaller; over one of its concave sides is spread an exceeding thin transparent skin or membrane. The intestines are of great width, and measure six feet in length.

THE MALLARD. (*Anas boschas.*)

PLATE LXX.—FIG. 7.

Lath. Syn. iii. p. 489.—*Bewick,* ii. p. 291.—Le Canard Sauvage, *Briss.* vi. p. 318, 4.—*Buff.* ix. p. 415, pl. 7, 8.—*Peale's Museum,* No. 2864.

BOSCHAS MAJOR.—WILLOUGHBY.*

Anas boschas, *Linn. Syst.*—*Gmel.* i. p. 538.—*Bonap. Synop.* p. 382.—*Flem. Br. Anim.* p. 123.—Le Canard Sauvage, *Temm. Man.* p. 385.—Wild Duck, *Mont. Ornith. Dict.* ii. *and Supp.*—Common Wild Duck, *Selby, Illust. Br. Ornith.* pl. 5.—Anas (Boschas) domestica, *North. Zool.* ii. p. 442.

THE mallard, or common wild drake, is so universally known as scarcely to require a description. It measures twenty-four

---

* This well-known species becomes interesting when considered as the stock whence the most flourishing duckeries of the poultry-yard have sprung; it is most amply spread over Europe and America, and I have received it from India. Universally known, it is esteemed for the table, and will fetch a higher price in the markets than most of the others in this country, and in America seems only surpassed by the canvas-back. In structure and general economy, it presents a most interesting form, combining the peculiarities of the pelagic and more terrestrial. It will live and find a sustenance in the sea and its coasts, by lakes and rivers, and in the midst of extensive moors and fens; it possesses a powerful frame, and its wings are adapted to strong flight; it can derive its sustenance either from the waters or the more inland pastures and cultivated fields; it is an expert diver when necessity calls it; and its breeding places are chosen by the sides of lakes and marshes, on the stumps of aged trees, like the summer duck, and on precipitous cliffs. In the latter situation, I once took the nest of a wild duck within ten yards distant from that of a peregrine falcon. It was situated on a projecting knoll of heather, jutting from an ivied cliff, and the tenants must often have seen each other in their passage to and from their precious deposits In this species we have the type of the genus *Boschas*. The centre feathers

inches in length by three feet in extent, and weighs upwards
of two pounds and a half; the bill is greenish yellow; irides,
hazel; head, and part of the neck, deep glossy changeable
green, ending in a narrow collar of white; the rest of the
neck and breast are of a dark purplish chestnut; lesser wing-
coverts, brown ash; greater, crossed near the extremities with
a band of white, and tipt with another of deep velvety black;
below this lies the speculum, or beauty-spot, of a rich and
splendid light purple, with green and violet reflections,
bounded on every side with black; quills, pale brownish ash;
back, brown, skirted with paler; scapulars, whitish, crossed
with fine undulating lines of black; rump and tail-coverts,
black, glossed with green; tertials, very broad, and pointed at
the ends; tail, consisting of eighteen feathers, whitish, centred
with brown ash, the four middle ones excepted, which are
narrow, black, glossed with violet, remarkably concave, and
curled upwards to a complete circle; belly and sides, a fine
grey, crossed by an infinite number of fine waving lines,
stronger and more deeply marked as they approach the vent;
legs and feet, orange red.

The female has the plumage of the upper parts dark brown,
broadly bordered with brownish yellow; and the lower parts
yellow ochre, spotted and streaked with deep brown; the chin
and throat, for about two inches, plain yellowish white; wings,
bill, and legs, nearly as in the male.

The windpipe of the male has a bony labyrinth, or bladder-
like knob, puffing out from the left side. The intestines measure six feet, and are as wide as those of the canvas-back.
The windpipe is of uniform diameter until it enters the
labyrinth.

This is the original stock of the common domesticated duck,
reclaimed, time immemorial, from a state of nature, and now

of the tail are lengthened, but assume a different form, in being regularly rolled or curled up. Some specimens want the white ring round
the neck, and in some parts this variety is so common, as to be distinguished by the herds and country people.—ED.

become so serviceable to man. In many individuals, the general garb of the tame drake seems to have undergone little or no alteration ; but the stamp of slavery is strongly imprinted in his dull indifferent eye and grovelling gait, while the lofty look, long tapering neck, and sprightly action of the former, bespeak his native spirit and independence.

The common wild duck is found in every fresh-water lake and river of the United States in winter, but seldom frequents the sea-shores or salt marshes. Their summer residence is the north, the great nursery of this numerous genus. Instances have been known of some solitary pairs breeding here in autumn. In England these instances are more common. The nest is usually placed in the most solitary recesses of the marsh or bog, amidst coarse grass, reeds, and rushes, and generally contains from twelve to sixteen eggs, of a dull greenish white. The young are led about by the mother in the same manner as those of the tame duck, but with a superior caution, a cunning and watchful vigilance peculiar to her situation. The male attaches himself to one female, as among other birds in their native state, and is the guardian and protector of her and her feeble brood. The mallard is numerous in the rice-fields of the southern States during winter, many of the fields being covered with a few inches of water, and the scattered grains of the former harvest lying in abundance, the ducks swim about, and feed at pleasure.

The flesh of the common wild duck is in general and high estimation ; and the ingenuity of man, in every country where it frequents, has been employed in inventing stratagems to overreach these wary birds, and procure a delicacy for the table. To enumerate all these various contrivances would far exceed our limits ; a few, however, of the most simple and effective may be mentioned.

In some ponds frequented by these birds, five or six wooden figures, cut and painted so as to represent ducks, and sunk, by pieces of lead nailed on their bottoms, so as to float at the usual depth on the surface, are anchored in a favourable position for being raked from a concealment of brush, &c., on shore. The appearance of these usually attracts passing flocks, which alight, and are shot down. Sometimes eight or ten of these painted wood ducks are fixed on a frame in various swimming postures, and secured to the bow of the gunner's skiff, projecting before it in such a manner that the weight of the frame sinks the figures to their proper depth ; the skiff is then dressed with sedge or coarse grass in an artful manner, as low as the water's edge; and under cover of this, which appears like a party of ducks swimming by a small island, the gunner floats down sometimes to the very skirts of a whole congregated multitude, and pours in a destructive and repeated fire of shot among them. In winter, when detached pieces of ice are occasionally floating in the river, some of the gunners on the Delaware paint their whole skiff or canoe white, and, laying themselves flat at the bottom, with their hand over the side, silently managing a small paddle, direct it imperceptibly into or near a flock, before the ducks have distinguished it from a floating mass of ice, and generally do great execution among them. A whole flock has sometimes been thus surprised asleep, with their heads under their wings. On land, another stratagem is sometimes practised with great success. A large tight hogshead is sunk in the flat marsh or mud, near the place where ducks are accustomed to feed at low water, and where otherwise there is no shelter ; the edges and top are artfully concealed with tufts of long coarse grass and reeds or sedge. From within this the gunner, unseen and unsuspected, watches his collecting prey, and, when a sufficient number offers, sweeps them down with great effect. The mode of catching wild ducks, as practised in India,* China,† the island of Ceylon, and some parts of South America,‡ has been often described, and seems, if reliance may be placed on those accounts, only practicable in water of a certain depth. The sportsman,

* Naval Chronicle, vol. ii. p. 473.
† Du Halde, History of China, vol. ii. p. 142.
‡ Ulloa's Voyage, i. p. 53.

covering his head with a hollow wooden vessel or calabash
pierced with holes to see through, wades into the water, keeping his head only above, and, thus disguised, moves in among
the flock, which take the appearance to be a mere floating
calabash, while, suddenly pulling them under by the legs, he
fastens them to his girdle, and thus takes as many as he can
conveniently stow away, without in the least alarming the rest.
They are also taken with snares made of horse-hair, or with
hooks baited with small pieces of sheep's lights, which, floating on the surface, are swallowed by the ducks, and with them
the hooks.   They are also approached under cover of a stalk-
ing horse, or a figure formed of thin boards, or other proper
materials, and painted so as to represent a horse or ox.   But
all these methods require much watching, toil, and fatigue, and
their success is but trifling when compared with that of the
decoy now used both in France and England,* which, from
its superiority over every other mode, is well deserving the
attention of persons of this country residing in the neighbourhood of extensive marshes frequented by wild ducks, as by this
method mallard and other kinds may be taken by thousands
at a time.   The following circumstantial account of these
decoys, and the manner of taking wild ducks in them in England, is extracted from Bewick's " History of British Birds,"
vol. ii. p. 294 :—

"In the lakes where they resort," says the correspondent of
that ingenious author, " the most favourite haunts of the fowl
are observed ; then, in the most sequestered part of this haunt,
they cut a ditch about four yards across at the entrance, and
about fifty or sixty yards in length, decreasing gradually in
width from the entrance to the farther end, which is not more
than two feet wide.   It is of a circular form, but not bending
much for the first ten yards.   The banks of the lake, for about
ten yards on each side of this ditch (or pipe, as it is called),
are kept clear from reeds, coarse herbage, &c., in order that

* Particularly in Picardy in the former country, and Lincolnshire
in the latter.

the fowl may get on them to sit and dress themselves. Across
this ditch, poles on each side, close to the edge of the ditch,
are driven into the ground, and the tops bent to each other and
tied fast.   These poles at the entrance form an arch, from the
top of which to the water is about ten feet.   This arch is made
to decrease in height as the ditch decreases in width, till the
farther end is not more than eighteen inches in height.   The
poles are placed about six feet from each other, and connected
together by poles laid lengthwise across the arch and tied together.   Over them a net, with meshes sufficiently small to
prevent the fowl getting through, is thrown across, and made
fast to a reed fence at the entrance, and nine or ten yards up
the ditch, and afterwards strongly pegged to the ground.   At
the farther end of the pipe, a tunnel net, as it is called, is fixed,
about four yards in length, of a round form, and kept open by
a number of hoops about eighteen inches in diameter, placed
at a small distance from each other, to keep it distended.   Supposing the circular bend of the pipe to be to the right, when
you stand with your back to the lake, on the left-hand side a
number of reed fences are constructed, called shootings, for
the purpose of screening from sight the decoy-man, and in
such a manner that the fowl in the decoy may not be alarmed
while he is driving those in the pipe : these shootings are about
four yards in length, and about six feet high, and are ten in
number.   They are placed in the following manner :—

From the end of the last shooting, a person cannot see the
lake, owing to the bend of the pipe : there is then no further
occasion for shelter.   Were it not for those shootings, the
fowl that remain about the mouth of the pipe would be alarmed
if the person driving the fowl already under the net should be

exposed, and would become so shy as to forsake the place entirely. The first thing the decoy-man does when he approaches the pipe is to take a piece of lighted turf or peat, and hold it near his mouth, to prevent the fowl smelling him. He is attended by a dog taught for the purpose of assisting him ; he walks very silently about half-way up the shootings, where a small piece of wood is thrust through the reed fence, which makes an aperture just sufficient to see if any fowl are in ; if not, he walks forward to see if any are about the mouth of the pipe. If there are, he stops and makes a motion to his dog, and gives him a piece of cheese or something to eat ; upon receiving it, he goes directly to a hole through the reed fence, (No. 1), and the fowl immediately fly off the bank into the water ; the dog returns along the bank, between the reed fences and the pipe, and comes out to his master at the hole (No. 2). The man now gives him another reward, and he repeats his round again, till the fowl are attracted by the motions of the dog, and follow him into the mouth of the pipe. This operation is called working them. The man now retreats farther back, working the dog at different holes till the fowl are sufficiently under the net: he now commands his dog to lie down still behind the fence, and goes forward to the end of the pipe next the lake, where he takes off his hat and gives it a wave between the shooting ; all the fowl under the net can see him, but none that are in the lake can. The fowl that are in sight fly forward ; and the man runs forward to the next shooting and waves his hat, and so on, driving them along till they come to the tunnel net, where they creep in: when they are all in, he gives the net a twist, so as to prevent their getting back : he then takes the net off from the end of the pipe with what fowl he may have caught, and takes them out, one at a time, and dislocates their necks, and hangs the net on again ; and all is ready for working again.

" In this manner, five or six dozen have been taken at one drift. When the wind blows directly in or out of the pipe, the fowl seldom work well, especially when it blows in. If many pipes are made in a lake, they should be so constructed as to suit different winds.

"Duck and mallard are taken from August to June ; teal or widgeon from October to March; becks, smee, golden-eyes, arps, cricks, and pintails or sea-pheasants, in March and April.

"Poker ducks are seldom taken on account of their diving and getting back in the pipe.

"It may be proper to observe here, that the ducks feed during the night, and that all is ready prepared for this sport in the evening. The better to entice the ducks into the pipe, hempseed is strewn occasionally on the water. The season

REFERENCES TO THE CUT.

No.
1. Dog's hole, where he goes to unbank the fowl.
2. Reed fences on each side of the mouth of the pipe.
3. Where the decoy-man shows himself to the fowl first, and afterwards at the end of every shooting.
4. Small reed fence to prevent the fowl seeing the dog when he goes to unbank them.
5. The shootings.
6. Dog's holes between the shootings, used when working.
7. Tunnel net at the end of the pipe.
8. Mouth of the pipe.

Drawn from Nature by A. Wilson

1. Gadwal Duck. 2. Eider D. 3. Female. 4. Smew. 5. Ruddy D. 6. Female.

71.

Engraved by W.H. Lizars

allowed by Act of Parliament for catching these birds in this way is from the latter end of October till February.

" Particular spots or decoys in the fen countries are let to the fowlers at a rent of from five to thirty pounds per annum ; and Pennant instances a season in which thirty-one thousand two hundred ducks, including teals and widgeons, were sold in London only, from ten of these decoys near Wainfleet, in Lincolnshire. Formerly, according to Willoughby, the ducks while in moult, and unable to fly, were driven by men in boats, furnished with long poles, with which they splashed the water between long nets, stretched vertically across the pools in the shape of two sides of a triangle, into lesser nets placed at the point ; and in this way, he says, four thousand were taken at one driving in Deeping Fen ; and Latham has quoted an instance of two thousand six hundred and forty-six being taken in two days near Spalding, in Lincolnshire ; but this manner of catching them while in moult is now prohibited."

THE GADWALL. *(Anas strepera.)*

PLATE LXXI.—FIG. 1.

Le Chipeau, *Briss.* vi. p. 339, 8, pl. 33, fig. 1.—*Buff.* ix. 187.—*Pl. enl.* 958.—*Arct. Zool.* p. 575.—*Lath. Syn.* iii. p. 515.—*Peale's Museum*, No. 2750.

CHAULIODUS STREPERA.—Swainson?*

Anas strepera, *Linn. Syst.* i. p. 200.—*Lath. Ind. Ornith.* ii. p. 859.—*Bonap. Synop.* p. 383.—Canard Chipeau, ou Ridenne, *Temm. Man.* ii. p. 837.—Gadwall or Grey, *Mont. Ornith. Dict.* i. *and Supp.*—*Bew. Br. Birds*, ii. 350.—Gadwall, *Selby's Illust. Br. Ornith.* pl. 51.—Anas (Chauliodus) strepera, *North. Zool.* ii. p. 440.—Genus Chauliodus, *Swain. Journ. Royal Instit.* No. iv. p. 19.

THIS beautiful duck I have met with in very distant parts of the United States, viz., on the Seneca Lake, in New York,

---

* This beautiful duck is remarkable in presenting, next to the shovellers, the greatest developement of lateral laminæ of the bill ; it is also an expert diver.

In Britain they are rare, but appear more common in the lower

about the 20th of October, and at Louisville, on the Ohio, in February. I also shot it near Big Bone Lick, in Kentucky. With its particular manners or breeding-place, I am altogether unacquainted.

The length of this species is twenty inches ; extent, thirty-one inches ; bill, two inches long, formed very much like that of the mallard, and of a brownish black ; crown, dusky brown ; rest of the upper half of the neck, brownish white, both thickly speckled with black ; lower part of the neck and breast, dusky black, elegantly ornamented with large concentric semicircles of white ; scapulars, waved with lines of white on a dusky ground, but narrower than that of the breast ; primaries, ash ; greater wing-coverts, black, and several of the lesser coverts, immediately above, chestnut red ; speculum, white, bordered below with black, forming three broad bands on the wing, of chestnut, black, and white ; belly, dull white ; rump and tail-coverts, black, glossed with green ; tail, tapering, pointed, of a pale brown ash, edged with white ; flanks, dull white, elegantly waved ; tertials, long, and of a pale brown ; legs, orange red.

The female I have never seen. Latham describes it as follows :—" Differs in having the colours on the wings duller, though marked the same as the male ; the breast, reddish brown, spotted with black ; the feathers on the neck and back, edged with pale red ; rump, the same, instead of black ; and those elegant semicircular lines on the neck and breast wholly wanting."

The flesh of this duck is excellent, and the windpipe of the male is furnished with a large labyrinth.

The gadwall is very rare in the northern parts of the United States ; is said to inhabit England in winter, and various parts

countries of Europe, and towards the north. They seem very abundant in Holland ; in the months of September and October they were the most common duck in the market, and were often seen in abundance on the lakes. It will show Mr Swainson's genus *Chauliodus.*—ED.

of France and Italy; migrates to Sweden, and is found throughout Russia and Siberia.*

It is a very quick diver, so as to make it difficult to be shot;
flies also with great rapidity, and utters a note not unlike that
of the mallard, but louder. Is fond of salines and ponds overgrown with reeds and rushes. Feeds during the day, as well
as in the morning and evening.

EIDER DUCK. (*Anas mollissima.*)

PLATE LXXI.—Fig. 2, Male.

L'Oye à Duvet, ou l'Eider, *Briss.* vi. p. 294, pl. 29, 3.—*Buff.* ix. p. 103, pl. 6.—
Pl. enl. 209.—Great Black and White Duck, *Edw.* pl. 98.—*Bewick,* ii. p. 279.—
Arct. Zool. No. 480.—*Lath. Syn.* iii. p. 470.—*Peale's Museum,* No. 2706.

SOMATERIA MOLLISSIMA.—Leach.†

Anas mollissima, *Linn. Syst.* i. p. 198.—Canard Eider, *Temm. Man. d'Ornith.* ii.
p. 848.—Eider or Cuthbert's Duck, *Mont. Ornith. Dict.* i. *and Supp.*—Eider,
Selby, Illust. Br. Ornith. p. 70.—Fuligula (subgen. Somateria) mollissima,
Bonap. Synop. p. 388.—Somateria mollissima, *North. Zool.* ii. p. 448.

The eider duck has been long celebrated in Europe for the
abundance and excellence of its down, which, for softness,

---

* Latham.

† This other form among the *Anatidæ* was proposed by Dr Leach,
and will contain only two species, the eider and king ducks, both
common to Europe and America. It is very well marked, and possesses some peculiarities. The birds are truly sea-ducks, keep entirely to
that element, and breed on its shores or islands, and are never, as Dr
Richardson remarks, seen on fresh water. The form is thick, rather
flat and heavy; the plumage of the males possesses decided contrasting
colours of black and white; the females, reddish brown. The plumage
of the head projects far upon the base of the bill, and is of a thick silky
texture, which can be raised or swelled at pleasure, so as to increase the
apparent size of the head and neck, and in both species exhibits remarkable colours not often seen among birds, and very difficult for colourists
to represent—pistachia green, and a pleasing dull shade of blue verditer.
In the scapulars of the *Anatidæ*, we have already seen a variable
structure; they are here of considerable breadth, rigid texture, and
curve over the quills, as if curled with an iron. The feet are placed

warmth, lightness, and elasticity, surpasses that of all other
ducks. The quantity found in one nest more than filled the
crown of a hat, yet weighed no more than three-quarters of an
ounce ; * and it is asserted that three pounds of this down
may be compressed into a space scarce bigger than a man's
fist, yet is afterwards so dilatable as to fill a quilt five feet
square.†

The native regions of the eider duck extend from 45° N.
to the highest latitudes yet discovered, both in Europe and
America. Solitary rocky shores and islands are their favourite
haunts. Some wandering pairs have been known to breed on
the rocky islands beyond Portland, in the district of Maine,
which is perhaps the most southern extent of their breeding-
place. In England, the Fern Isles, on the coast of Northum-
berland, are annually visited by a few of these birds, being the
only place in South Britain where they are known to breed.
They occur again in some of the Western Isles of Scotland.
Greenland and Iceland abound with them, and here, in particular places, their nests are crowded so close together, that a
person can scarcely walk without treading on them. The
natives of these countries know the value of the down, and

far back, and show great powers for diving. The males undergo a
change of plumage, and leave the females as soon as they have commenced sitting, when they may be seen in large flocks by themselves ;
they commence their migrations much sooner than the females. It is
to this bird that we are principally indebted for the valuable eider
down, though many others of the northern aquatic fowl produce one
equally fine, which is often mixed with it. Lemmius remarks, that
the eiders are in immense profusion on the coasts of Norway and
Lapland ; when hatching, the eggs are often the prey of the crows and
of *Larus marinus,* who drag the female from her nest and destroy them
or the young. The male, however, if he perceives the assault, makes
furious attacks, and sometimes succeeds in beating them off, They are
very familiar, building close to the houses of the fishermen ; the female
will even allow herself to be lifted from the eggs and set down again ;
and sometimes a countryman will carry the young in his hat from the
nest to the sea, the duck running by his side, moaning gently with
anxiety.—Ed.

* Pennant.   † Salern. Ornith. p. 416.

carry on a regular system of plunder both of it and also of the eggs. The nest is generally formed outwardly of drift grass, dry seaweed, and such like materials; the inside composed of a large quantity of down plucked from the breast of the female. In this soft elastic bed she deposits five eggs, extremely smooth and glossy, of a pale olive colour; they are also warmly covered with the same kind of down. When the whole number is laid, they are taken away by the natives, and also the down with which the nest is lined, together with that which covers the eggs. The female once more strips her breast of the remaining down, and lays a second time; even this, with the eggs, is generally taken away, and it is said that the male, in this extremity, furnishes the third quantity of down from his own breast; but if the cruel robbery be a third time repeated, they abandon the place altogether. One female, during the whole time of laying, generally gives half a pound of down; and we are told, that in the year 1750, the Iceland Company sold as much of this article as amounted to three thousand seven hundred and forty-five banco dollars, besides what was directly sent to Gluckstadt.* The down from dead birds is little esteemed, having lost its elasticity.

These birds associate together in flocks, generally in deep water, diving for shellfish, which constitute their principal food. They frequently retire to the rocky shores to rest, particularly on the appearance of an approaching storm. They are numerous on the coast of Labrador, and are occasionally seen in winter as far south as the Capes of Delaware. Their flesh is esteemed by the inhabitants of Greenland, but tastes strongly of fish.

The length of this species is two feet three inches, extent, three feet; weight, between six and seven pounds; the head is large, and the bill of singular structure, being three inches in length, forked in a remarkable manner, running high up in the forehead, between which the plumage descends nearly to the nostril; the whole of the bill is of a dull yellowish horn

* Letters on Iceland, by Uno Van Troil, p. 146.

colour, somewhat dusky in the middle ; upper part of the
head, deep velvet black, divided laterally on the hind head by
a whitish band ; cheeks, white ; sides of the head, pale pea-
green, marked with a narrow line of white dropped from the ear-
feathers ; the plumage of this part of the head, to the throat,
is tumid, and looks as if cut off at the end, for immediately
below the neck it suddenly narrows, somewhat in the manner
of the buffel-head, enlarging again greatly as it descends, and
has a singular hollow between the shoulders behind ; the
upper part of the neck, the back scapulars, lesser wing-
coverts, and sides of the rump, are pure white ; lower part of
the breast, belly, and vent, black ; tail, primaries, and secondaries, brownish black ; the tertials curiously curved, falling
over the wing ; legs, short, yellow ; webs of the feet, dusky.

Latham has given us the following sketch of the gradual
progress of the young males to their perfect colours :—" In
the first year the back is white, and the usual parts, except
the crown, black ; but the rest of the body is variegated with
black and white.   In the second year, the neck and breast are
spotted black and white, and the crown black.   In the third,
the colours are nearly as when in full plumage, but less vivid,
and a few spots of black still remaining on the neck ; the
crown, black, and bifid at the back part.

" The young of both sexes are the same, being covered
with a kind of hairy down ; throat and breast, whitish ; and
a cinereous line from the bill through the eyes to the hind
head." *

FEMALE EIDER DUCK.

PLATE LXXI.— Fig. 3.

Peale's Museum, No. 2707.

SOMATERIA MOLLISSIMA.—Leach.

The difference of colour in these two birds is singularly great.
The female is considerably less than the male, and the bill

* Synopsis, iii. 471.

does not rise so high in the forehead; the general colour is a dark reddish drab, mingled with lighter touches, and everywhere spotted with black ; wings, dusky, edged with reddish ; the greater coverts, and some of the secondaries, are tipt with white; tail, brownish black, lighter than in the male; the plumage in general is centred with bars of black, and broadly bordered with rufous drab ; cheeks and space over the eye, light drab ; belly, dusky, obscurely mottled with black ; legs and feet, as in the male.

Van Troil, in his " Letters on Iceland," observes respecting this duck, that " the young ones quit the nest soon after they are hatched, and follow the female, who leads them to the water, where, having taken them on her back, she swims with them a few yards, and then dives, and leaves them floating on the water ! In this situation they soon learn to take care of themselves, and are seldom afterwards seen on the land, but live among the rocks, and feed on insects and seaweed."

Some attempts have been made to domesticate these birds, but hitherto without success.

THE SMEW, OR WHITE NUN. (*Mergus albellus.*)

PLATE LXXI.—Fig. 4.

Le Petit Harle Huppé, ou la Piette, *Briss.* vi. p. 243, 3, pl. 24, fig. 1.—*Buff.* viii. p. 275, pl. 24.—*Pl. enl.* 449.—*Bewick,* ii. p. 238.—*Lath. Syn.* iii. p. 428.— *Arct Zool.* No. 468.

MERGUS ALBELLUS.—Linnæus.*

Mergus albellus, *Linn. Syst.* i. p. 209.—*Bonap. Synop.* p. 397.—Harle Piette, *Temm. Man. d'Ornith.* ii. p. 887.—Minute Merganser, *Mont. Ornith. Dict.* i. *and Supp.*—Lough Diver, and Red-headed Smew, *Penn.*, for young and female.—Smew, *Selby, Illust. Br. Ornith.* pl. 69.

This is another of those mergansers commonly known in this country by the appellation of fishermen, fisher ducks, or divers.

* The male of this merganser is one of the cleanest and most delicate-looking of the genus, the colours being entirely of the purest black and white. The bill presents a shorter and more dilated form than its con-

The present species is much more common on the coast of New England than farther to the south. On the shores of New Jersey it is very seldom met with. It is an admirable diver, and can continue for a long time under water. Its food is small fry, shellfish, shrimps, &c. In England, as with us, the smew is seen only during winter; it is also found in France, in some parts of which it is called *la piette*, as in parts of England it is named the magpie driver. Its breeding-place is doubtless in the arctic regions, as it frequents Iceland, and has been observed to migrate with other mergansers and several kinds of ducks up the river Wolga in February.*

The smew, or white nun, is nineteen inches in length, and two feet three inches in extent; bill, black, formed very much like that of the red-breasted merganser, but not so strongly toothed; irides, dark; head, crested; crown, white; hind head, black; round the area of the eye, a large oval space of black; whole neck, breast, and belly, white, marked on the upper and lower part of the breast with a curving line of black; back, black; scapulars, white, crossed with several faint, dusky bars; shoulder of the wing and primaries, black; secondaries and greater coverts, black, broadly tipt with white; across the lesser coverts, a large band of white; sides and flanks, crossed with waving lines; tail, dark ash; legs and feet, pale bluish slate.

The female is considerably less than the male; the bill, a dark lead colour; crest of the same peculiar form as that of the male, but less, and of a reddish brown; marked round the area of the eyes with dusky; cheeks, fore part of the neck, and belly, white; round the middle of the neck, a collar of pale

geners, approaching almost to some of the more aberrant ducks. It is very rare in this country, and appears only in winter. The propagation and extent of the breeding migrations are only surmised, and we possess no very authentic authority upon the subject; they are said, however, to resemble the others.—ED.

* Dec. Russ. ii. p. 145.

brown; breast and shoulders, dull brown and whitish inter-mixed: wings and back, marked like those of the male, but of a deep brownish ash in those parts which in him are black; legs and feet, pale blue. The young birds, as in the other three species, strongly resemble the female during the first and part of the second year. As these changes of colour, from the garb of the female to that of the male, take place in the re-mote regions of the north, we have not the opportunity of detecting them in their gradual progress to full plumage. Hence, as both males and females have been found in the same dress, some writers have considered them as a separate species from the smew, and have given to them the title of the red-headed smew.

In the ponds of New England, and some of the lakes in the State of New York, where the smew is frequently observed, these red-headed kind are often found in company, and more numerous than the other, for very obvious reasons, and bear, in the markings, though not in the colours, of their plumage, evident proof of their being the same species, but younger birds or females. The male, like the Muscovy drake and many others, when arrived at his full size, is nearly one-third heavier than the female; and this disproportion of weight and difference of colour, in the full-grown males and females, are characteristic of the whole genus.

RUDDY DUCK. (*Anas rubidus.*)

PLATE LXXI.—Fig. 5, Male.

Peale's Museum, No. 2808.

FULIGULA RUBIDA.—Bonaparte.*

Fuligula (Oxyura) rubida, *Bonap. Synop.* p. 391.—Fuligula rubida, *North. Zool.* ii. p. 455.—Anas Jamaicensis, *Ord's edit.* p. 133.

This very rare duck was shot some years ago on the river Delaware, and appears to be an entirely new species. The

* Bonaparte has proposed this form as the type of a subgenus, under the name of *Oxyura*, from the form of the tail; and Mr Swainson ob-

specimen here figured, with the female that accompanies it, and which was killed in the same river, are the only individuals of their kind I have met with. They are both preserved in the superb museum of my much respected friend, Mr Peale, of this city.

On comparing this duck with the description given by Latham of the Jamaica shoveller, I was at first inclined to believe I had found out the species; but a more careful examination of both satisfied me that they cannot be the same, as the present differs considerably in colour; and, besides, has some peculiarities which the eye of that acute ornithologist could not possibly have overlooked in his examination of the species said to have been received by him from Jamaica. Wherever the general residence of this species may be, in this part of the world, at least, it is extremely rare, since among the many thousands of ducks brought to our markets during winter, I have never heard of a single individual of the present kind having been found among them.

The ruddy duck is fifteen inches and a half in length, and twenty-two inches in extent; the bill is broad at the tip, the under mandible much narrower, and both of a rich light blue; nostrils small, placed in the middle of the bill; cheeks and chin, white; front, crown, and back part of the neck, down nearly to the back, black; rest of the neck, whole back, scapulars, flanks, and tail-coverts, deep reddish brown, the colour of bright mahogany; wings, plain pale drab, darkest at the points; tail, black, greatly tapering, containing eighteen narrow-pointed feathers; the plumage of the breast and

serves, "We suspect that this bird, and one or two others of similar form, found by us in tropical Brazil, will constitute a subgenus." There are many modifications from the *Fuligulæ* in this bird, which would, with additional species, entitle a subgenus, and, in that case, *Oxyura* may be adopted. They seem very rare, and Wilson has the merit of first distinguishing them; the bill becomes much broader at the tip, and the lamellæ are more prominent than in the *Fuligula;* the feet are placed very far back, and the hind toe is furnished with a much narrower membrane.—ED.

upper part of the neck is of a remarkable kind, being dusky olive at bottom, ending in hard bristly points of a silvery grey, very much resembling the hair of some kinds of seal-skins; all these are thickly marked with transverse curving lines of deep brown; belly and vent, silver grey, thickly crossed with dusky olive; under tail-coverts, white; legs and feet, ash-coloured.

FEMALE RUDDY DUCK.

PLATE LXXI.— Fig. 6.

Peale's Museum, No. 2809.

FULIGULA RUBIDA.—Bonaparte.—Young.

This is nearly of the same size as the male; the front, lores, and crown, deep blackish brown; bill, as in the male, very broad at the extremity, and largely toothed on the sides, of the same rich blue; cheeks, a dull cream; neck, plain dull drab, sprinkled about the auriculars with blackish; lower part of the neck and breast, variegated with grey, ash, and reddish brown; the reddish dies off towards the belly, leaving this last of a dull white, shaded with dusky ash; wings, as in the male; tail, brown; scapulars, dusky brown, thickly sprinkled with whitish, giving them a grey appearance; legs, ash.

A particular character of this species is its tapering, sharp-pointed tail, the feathers of which are very narrow; the body is short; the bill, very nearly as broad as some of those called shovellers; the lower mandible much narrower than the upper.

Mr Ord has added a very elaborate description in his edition of this work, completing the history of this bird, which we have thought best to print, as showing many points of discussion; we, however, consider the species established as above named.

"In the first edition of this work, the author states that the two ducks of this species figured in the plate as male and female were the only individuals that he had ever met with. They had been shot on the river Delaware, and were deposited in Peale's Museum. 'On comparing this duck,' he observes, 'with the description given by Latham of the *Jamaica shoveller*, I was at first inclined to believe I had found out the species ; but a more careful examination of both satisfied me that they cannot be the same, as the present differs considerably in colour ; and, besides, has some peculiarities which the eye of that acute ornithologist could not possibly have overlooked in his examination of the species said to have been received by him from Jamaica. Wherever the general residence of this species may be, in this part of the world, at least, it is extremely rare, since among the many thousands of ducks brought to our markets during winter, I have never heard of a single individual of the present kind having been found among them.'

" It is a circumstance in ornithology well worthy of note, that migratory birds frequently change their route, and, consequently, become common in those districts where they had been either unknown, or considered very rare. Of the *Sylvia magnolia*, Wilson declares that he had seen but two individuals, and these in the western country; the *Muscicapa cucullata* he says is seldom observed in Pennsylvania and the northern States ; the *Muscicapa pusilla*, and the *Muscicapa Canadensis*, he considered rare birds with us ; notwithstanding, in the month of May 1815, all of these were seen in our gardens ; and the editor noted the last-mentioned as among the most numerous of the passenger birds of that season.

" The subject of this chapter affords a case in point. The year subsequent to the death of our author, this duck began to make its appearance in our waters. In October 1814, the editor procured a female, which had been killed from a flock, consisting of five, at Windmill Island, opposite to Philadelphia. In October 1818, he shot three individuals, two females and a male ; and in April last another male, all of which, except one, were young birds. He has also, at various times, since 1814, seen several other male specimens of this species, not one of which was an adult. In effect, the only old males which he has ever seen are that in Peale's Museum, and another in the Cabinet of the Academy of Natural Sciences of Philadelphia.

" The duck figured in the plate as the female was a young male, as the records of the Museum show; the great difference between its colours and markings and those of the full-plum-aged male having induced the author to conclude it was a female, although he was perfectly familiar with the fact that the young males of several species of this genus so nearly re-semble the other sex, that it requires a very accurate eye, aided by much experience, to distinguish them by their external char-acters. This is precisely the case with the present species; the yearlings of both sexes are alike; and it is not until the succeeding spring that those characters appear in the males which enable one to indicate them, independent of dissection.

" The opinion of our author that this species is not the Jamaica shoveller of Latham, the editor cannot subscribe to, it appearing to him that the specimen from which Latham took his description was a young male of the duck now before us. The latter informs us that the species appears in Jamaica in October or November, remains till March, and then retires to the north. This account coincides with ours: we see the bird on its way to the south in October; it reaches Jamaica in November: it departs thence in March, and revisits us, in regular progression, in April. Where its summer residence is we are not informed; and we are equally ignorant whether the species is numerous in any part of our continent or not.

" Judging from the descriptions of the Ural duck of Euro-pean writers, there should seem to be a great affinity between that and the present. Through the polite attention of Mr Charles Bonaparte, the editor was enabled to examine a female specimen of the former; and as he perceived some differences, he will here note them. The bill of the Ural duck, from the angle of the mouth, is two inches long; that of our duck is

one inch and three-quarters; it is also less gibbous at the base
than in the former, and it is less depressed above: the tail-
feathers of the Ural duck are guttered their whole length;
those of the ruddy duck are slightly canaliculated at their tips;
the lateral membrane of the inner toe of the latter is not half
the breadth of that of the former. In other respects the females
of the two species much resemble each other. In order to
draw a just parallel, it would be necessary to examine a male
specimen of the European bird, which our cabinets do not
possess.

"The adult male, figured in the plate, is thus described
by our author:—'Length, fifteen inches and a half; extent,
twenty-two inches; bill, broad at the tip, the under mandible
much narrower than the upper, and *both* of a rich light blue; *
nostrils small, placed in the middle of the bill; cheeks and
chin white; front, crown, and back part of the neck, down
nearly to the back, black; rest of the neck, whole back,
scapulars, flanks, and tail-coverts, deep reddish brown, the
colour of bright mahogany; wings, pale plain drab, darkest
at the points; tail, black, greatly tapering, containing eighteen
narrow-pointed feathers; the plumage of the breast and
upper part of the neck is of a remarkable kind, being dusky
olive at bottom, ending in hard bristly points of a silvery
grey, very much resembling the hair of some kinds of seal-
skins: this plumage is thickly marked with transverse curving
lines of deep brown; belly and vent, silver grey, thickly
crossed with dusky olive; under tail-coverts, white; legs and
feet, ash-coloured.'

"The female is fifteen inches in length; bill, to the angle
of the mouth, one inch and three-quarters long, its lower half
very broad, of a deep dusky olive, the nail resembling a narrow
clasp of iron; nostrils, oval, with a curved furrow below them;
eyes, small and dark; the upper part of the head, from the
bill to the hind head, variegated with shining bronze and

* So coloured in Peale's specimen, but there is reason to conjecture
that the colour of the upper mandible alone was a blue ash.

blackish brown, the latter crossing the head in lines; cheeks, white, mixed with dusky, and some touches of bronze; lores, drab and dusky, mixed with a small portion of white; neck, short and thick, its lower half above, extending between the shoulders, drab, mixed with dusky; throat, and whole lower parts, dusky ash, the plumage tipt with dull white, having a silver grey appearance; the upper parts are dusky, marked or pencilled with pale ferruginous and dull white; breast, slightly tinged with reddish brown; the wings are small, greatly concave, and, when closed, are short of the extremities of the tail-coverts about three-quarters of an inch; they are dusky, their coverts finely dotted or powdered with white; tail, dusky, marked at its extremity with a few very fine dots of reddish white; it extends beyond its upper coverts two inches and a half; under tail-coverts, white; legs and feet, dusky slate; weight, sixteen ounces and a half. The gizzard of the above contained sand and some small seeds. Her eggs were numerous, and tolerably large; hence, as she was shot in the month of October, it was conjectured that she was a bird of the preceding year.

" The young male, shot in April last, measured fifteen inches in length; its irides were dark brown; bill, elevated at the base, slightly gibbous, and blue ash, from the nostrils to the tip, mixed with dusky; lower mandible, yellowish flesh colour, marbled with dusky; crown, brown black; throat and cheeks, as far as the upper angle of the bill, white, stained with bright yellow ochre; auriculars, almost pure white; the black from the crown surrounded the eyes, and passed round the white of the auriculars; hind head, black, mixed with ferruginous; breast and shoulders, bright ferruginous; belly, ash and silver white; back and scapulars, liver brown, finely pencilled with grey and reddish white; rump and upper tail-coverts, the same ground colour, but the markings not so distinct; wings, light liver brown, the lesser coverts finely powdered with grey; on the back and scapulars, the flanks, and around the base of the neck, the brownish red, or bright mahogany-coloured plumage, which distinguishes the adult male, was coming out; inner webs of the tail, partly dusky, outer webs, for two-thirds of their length, and the tip, dirty ferruginous; legs, blue ash in front, behind, the toes and webs, dusky. When the tail is not spread, it is somewhat conical, and its narrow-pointed feathers are slightly guttered at their tips; when spread, it is wedge-shaped. The trachea is of nearly equal diameter throughout; and has no labyrinth or enlargement at its lower part.

" Another young male, shot in October, measured fifteen and a quarter inches in length, and twenty-three inches in breadth; bill, greenish black, lower mandible, yellowish flesh colour, mixed with dusky; from the bill to the hind head a deep liver brown, the tips of the plumage bronzed; whole upper parts, dark umber brown, pencilled with pale ferruginous, buff, and white; from the corner of the mouth, a brown marking extended towards the eye; tail, dusky, ash-coloured at its extremity; legs and feet, dusky ash; toes paler, having a yellowish tinge; webs, dusky; claws sharp.

"The shafts of the tail-feathers of all these specimens, except that shot in April, projected beyond the webs; in one specimen, the shaft of one of the middle feathers projected an inch, and was ramified into rigid bristles, resembling those of the tail of Buffon's *Sarcelle à queue épineuse de Cayenne,* Pl. Enl. 967; in all the specimens, there was the appearance of the tail-feathers having been furnished with the like process, but which had been rubbed off. Can it be that this duck makes use of its tail in climbing up the fissures of rocks, or the hollows of trees? Its stiff narrow feathers, not unlike those of the tail of a woodpecker, would favour this supposition. It is worthy of note, that the tail of Mr Bonaparte's female specimen alluded to above is thus rubbed.

"The plumage of the neck and breast, which Wilson says is of a remarkable kind, that is, stiff and bristly at the tips, is common to several ducks, and therefore is no peculiarity.

"The body of this species is broad, flat, and compact; its

Drawn from Nature by A.Wilson

1.Brant. 2.Scoter Duck. 3.Velvet D. 4.Harlequin.D. 5.Dusky D. 6.Marsh Tern. 7.Sooty T.

72.

Engraved by W.H.Lizars.

wings, short and concave ; its legs placed far behind, and its feet uncommonly large ; it consequently is an expert diver. It flies with the swiftness and in the manner of the buffel-head ; and it swims precisely as Latham reports the Ural duck to swim, with the tail immersed in the water as far as the rump ; but whether it swims thus low with the view of employing its tail as a rudder, as Latham asserts of the Ural, or merely to conceal itself from observation, as the scaup duck is wont to do when wounded, and as all the divers do when pursued, I cannot determine.

"This is a solitary bird, and with us we never see more than five or six together, and then always apart from other ducks. It is uncommonly tame ; so much so, that, by means of my skiff, I have never experienced any difficulty in approaching within a few yards of it. Its flesh I do not consider superior to that of the buffel-head, which, with us, is a duck not highly esteemed.

"I should not be surprised if Buffon's *Sarcelle à queue épineuse de Cayenne* should turn out to be this species. The characters of the two certainly approximate ; but as I have not been enabled to settle the question of their identity in my own mind, I shall, for the present, let the affair rest."

THE BRANT. (*Anas bernicla.*)

PLATE LXXII.—Fig. 1.

Le Cravant, *Briss.* vi. p. 304, 16, pl. 31.—*Buff.* ix. p. 87.—*Bew.* ii. p. 277.—*Lath. Syn.* iii. p. 467.—*Arct. Zool.* No. 478.—*Peale's Museum*, No. 2704.

BERNICLA BRENTA.—Stephens.*

Bernicla brenta, *Steph. Cont. Sh. Zool.* xii. p. 46.—Oie Cravant, *Temm. Man.* ii. p. 824.—Anser brenta, *Flem. Br. Anim.* p. 127.—Anser bernicla, *North. Zool.* ii. p. 469.—Brent, or Boord Goose, *Mont. Orn. Dict. and Supp.*—*Bew. Br. Birds*, ii. p. 311.—Brent Bernicle, *Selby, Illust. Br. Orn.* pl. 65.

THE brant, or, as it is usually written, *brent*, is a bird well known on both continents, and celebrated in former times

* Stephens first applied this title, as a generic one, to a considerable number of birds, and gives, as their characters, "distinguished from the

throughout Europe for the singularity of its origin, and the strange transformations it was supposed to undergo previous to its complete organisation. Its first appearance was said to be in the form of a barnacle shell adhering to old water-soaked logs, trees, or other pieces of wood taken from the sea. Of this goose-bearing tree, Gerard, in his "Herbal," published in 1597, has given a formal account, and seems to have reserved it for the conclusion of his work, as being the most wonderful of all he had to describe. The honest naturalist, however, though his belief was fixed, acknowledges that his own personal information was derived from certain shells which adhered to a rotten tree that he dragged out of the sea between Dover and Romney, in England; in some of which he found "living things without forme or shape; in others, which were nearer come to ripeness, living things that were very naked, in shape like a birde; in others, the birds covered with soft downe, the shell half open, and the birde readie to fall out, which no doubt were the foules called Barnakles." * Ridiculous and chimerical as this notion was, it had many advocates, and was at that time as generally believed, and with about as much reason too, as the present opinion of the annual submersion of swallows, so tenaciously insisted on by some of our philosophers, and which, like the former absurdity, will in its turn disappear before the penetrating radiance and calm investigation of truth.

The brant and barnacle goose, though generally reckoned two different species, I consider to be the same. Among those large flocks that arrive on our coasts about the beginning of October, individuals frequently occur corresponding in their markings with that called the bernacle of Europe; that is, in

geese by their shorter and slenderer beak, the edges of which are reflected over the lamellæ, and obstruct the view of them." We shall consider the form to which that title should be restricted to be that of the present—the *B. erythropus*, and *B. ruficollis*. Many of those admitted by Stephen show very different characters, and will range elsewhere.— ED.

* See Gerard's Herbal, Art. Goose-bearing Tree.

having the upper parts lighter, and the front, cheeks, and chin whitish. These appear evidently a variety of the brant, probably young birds: what strengthens this last opinion is the fact, that none of them are found so marked on their return northward in the spring.

The brant is expected at Egg Harbour, on the coast of New Jersey, about the 1st of October, and has been sometimes seen as early as the 20th of September. The first flocks generally remain in the bay a few days, and then pass on to the south. On recommencing their journey, they collect in one large body, and, making an extensive spiral course, some miles in diameter, rise to a great height in the air, and then steer for the sea, over which they uniformly travel; often making wide circuits to avoid passing over a projecting point of land. In these aerial routes, they have been met with many leagues from shore, travelling the whole night. Their line of march very much resembles that of the Canada goose, with this exception, that frequently three or four are crowded together in the front, as if striving for precedency. Flocks continue to arrive from the north, and many remain in the bay till December, or until the weather becomes very severe, when these also move off southwardly. During their stay, they feed on the bars at low water, seldom or never in the marshes; their principal food being a remarkably long and broad-leaved marine plant, of a bright green colour, which adheres to stones, and is called by the country people sea-cabbage; the leaves of this are sometimes eight or ten inches broad, by two or three feet in length: they also eat small shellfish. They never dive, but wade about, feeding at low water. During the time of high water, they float in the bay in long lines, particularly in calm weather. Their voice is hoarse and honking, and, when some hundreds are screaming together, reminds one of a pack of hounds in full cry. They often quarrel among themselves, and with the ducks, driving the latter off their feeding ground. Though it never dives in search of food, yet, when wing-broken, the brant will go one hundred yards at a stretch under water; and is considered, in such circumstances, one of the most difficult birds to kill. About the 15th or 20th of May, they reappear on their way north; but seldom stop long, unless driven in by tempestuous weather.

The breeding-place of the brant is supposed to be very far to the north. They are common at Hudson's Bay, very nume-rous in winter on the coasts of Holland and Ireland; are called in Shetland, Harra geese, from their frequenting the sand of that name; they also visit the coast of England. Buffon relates, that in the severe winters of 1740 and 1765, during the prevalence of a strong north wind, the brant visited the coast of Picardy, in France, in prodigious multitudes, and committed great depredations on the corn, tearing it up by the roots, trampling, and devouring it; and, notwithstanding the exertions of the inhabitants, who were constantly employed in destroying them, they continued in great force until a change of weather carried them off.

The brant generally weighs about four pounds avoirdupois, and measures two feet in length, and three feet six inches in extent; the bill is about an inch and a half long, and black; the nostril large, placed nearly in its middle; head, neck, and breast, black; the neck marked with a spot of white, about two inches below the eye; belly, pale ash, edged with white; from the thighs backwards, white; back and wing-coverts, dusky brownish black, the plumage lightest at the tips; rump and middle of the tail-coverts, black; the rest of the tail-coverts, pure white, reaching nearly to the tip of the tail, the whole of which is black, but usually concealed by the white coverts; primaries and secondaries, deep black; legs, also black; irides, dark hazel.

The only material difference observable between the plumage of the male and female is, that in the latter the white spot on the neck is less, and more mottled with dusky. In young birds it is sometimes wanting, or occurs on the front, cheeks, and chin; and sometimes the upper part of the neck only is black ; * but in full-plumaged birds of both sexes, the markings are very much alike.

The brant is often seen in our markets for sale. Its flesh, though esteemed by many, tastes somewhat sedgy, or fishy.†

SCOTER DUCK. (*Anas nigra.*)

PLATE LXXII.—Fig. 2.

Le Macreuse, *Briss.* vi. p. 420, pl. 38, fig. 2.—*Buff.* ix. p. 234, pl. 16.—*Pl. enl.* 978.—*Bewick,* ii. p. 288.—*Arct. Zool.* No. 484.—*Lath. Syn.* iii. p. 480.—*Peale's Museum,* No. 2658.

OIDEMIA NIGRA.—Fleming.‡

Oidemia nigra, *Flem. Br. Anim.* p. 119.—*North. Zool.* ii. p. 450.—*Bonap. Synop.* p. 390.—Canard Macreuse, *Temm. Man.* ii. p. 856.—Scoter, or Black Diver, *Mont. Ornith. Dict.* ii. *and Supp.*—*Bew. Br. Birds,* ii. p. 325.—Black Scoter, *Selby, Illust. Br. Orn.* pl. 68.

THIS duck is but little known along our sea-coast, being more usually met with in the northern than southern districts, and only during the winter. Its food is shellfish, for which it is almost perpetually diving. That small bivalve so often mentioned, small mussels, spout-fish, called on the coast, razor handles, young clams, &c., furnish it with abundant fare; and wherever these are plenty, the scoter is an occasional visitor. They swim, seemingly at ease, amidst the very roughest of the surf, but fly heavily along the surface, and to no great distance. They rarely penetrate far up our rivers

* The figure of this bird given by Bewick is in that state.

† Mr Ord adds :—" The individual figured in the plate was shot by the editor, at Great Egg Harbour, on the 19th of May 1813. It had been compelled to land by a storm, which surprised it while travelling to the north. The procuring of this bird was a fortunate occurrence for Mr Wilson, he having omitted to make a drawing while specimens were to be obtained during their sojourn on our coast. The following day, the author and his friend returned home from their last shooting expedition to the sea-shore ; and on the 23d of August, the ingenious and excellent Wilson bade adieu to this world for ever."—ED.

‡ The plumage on the head and neck of this bird is remarkable for its rigid texture and the narrow hackled shape of the feathers.—ED.

but seem to prefer the neighbourhood of the ocean, differing in this respect from the cormorant, which often makes extensive visits to the interior.

The scoters are said to appear on the coasts of France in great numbers, to which they are attracted by a certain kind of small bivalve shellfish, called *vaimeaux*, probably differing little from those already mentioned. Over the beds of these shellfish the fishermen spread their nets, supporting them, horizontally, at the height of two or three feet from the bottom. At the flowing of the tide the scoters approach in great numbers, diving after their favourite food, and soon get entangled in the nets. Twenty or thirty dozen have sometimes been taken in a single tide. These are sold to the Roman Catholics, who eat them on those days on which they are forbidden by their religion the use of animal food, fish excepted; these birds and a few others of the same fishy flavour, having been exempted from the interdict, on the supposition of their being cold-blooded, and partaking of the nature of fish.*

The scoter abounds in Lapland, Norway, Sweden, Russia, and Siberia. It was also found by Osbeck, between the islands of Java and St Paul, lat. 30 and 34, in the month of June.†

This species is twenty-one inches in length, and thirty-four in extent, and is easily distinguished from all other ducks by the peculiar form of its bill, which has at the base a large elevated knob of a red colour, divided by a narrow line of yellow, which spreads over the middle of the upper mandible, reaching nearly to its extremity, the edges and lower mandible are black; the eyelid is yellow; irides, dark hazel; the whole plumage is black, inclining to purple on the head and neck; legs and feet, reddish.

The female has little or nothing of the knob on the bill; her plumage above a sooty brown, and below of a greyish white.

    * Bewick.   † Voyage, i. p. 120.

VELVET DUCK. *(Anas fusca.)*

PLATE LXXII.—FIG. 3.

Le Grande Macreuse, *Briss.* vi. p. 423, 29.—*Buff.* ix. p. 242.—*Pl. enl.* 956.—
 Arct. Zool. No. 482.—*Bewick*, ii. p. 286.—*Lath. Syn.* iii. p. 482.—*Peale's
 Museum*, No. 2658, female.

OIDEMIA FUSCA.—FLEMING.*

Oidemia fusca, *Flem. Br. Anim.* p. 119.—*Bonap. Synop.* p. 390.—*North. Zool.* ii.
 p. 450.—Canard Double Macreuse, *Temm. Man.* ii. p. 854.—Velvet Duck,
 Mont. Ornith. Dict.—*Bew. Br. Birds*, ii. 322.—Velvet Scoter, *Selby, Illust.
 Br. Ornith.* pl. 67.

THIS and the preceding are frequently confounded together
as one and the same species by our gunners on the sea-coast.
The former, however, differs in being of greater size; in
having a broad band of white across the wing; a spot of the
same under the eye ; and in the structure of its bill.   The
habits of both are very much alike ; they visit us only during
the winter ; feed entirely on shellfish, which they procure by
diving ; and return to the northern regions early in spring to
breed.   They often associate with the scoters, and are taken
frequently in the same nets with them.   Owing to the rank
fishy flavour of its flesh, it is seldom sought after by our sportsmen or gunners, and is very little esteemed.

The velvet duck measures twenty-three inches in length,
and two feet nine inches in extent, and weighs about three

---

* This, with the preceding, and the *O. perspicillata*, constitute the
American species of Fleming's genus *Oidemia*.   They are all visitants
also of the European continent during winter, and, with the exception
of the last, are of rather common occurrence.   They are truly sea-ducks,
and never almost leave that element except during the season of incubation.   They are expert divers, and feed on fish and marine moluscæ ;
we find, therefore, the foot expanded, the hallux furnished with a broad
membrane, and the legs placed far back.   The bill is expanded, and
generally swollen at the base ; the plumage thick and compact, and of
glossy smoothness ; the wings short, but firm, and sharp-pointed,
capable, apparently, of a strong flight for a short while, but unfitted for
any prolonged exertion.—ED.

pounds; the bill is broad, a little elevated at the base, where it is black, the rest red, except the lower mandible, which is of a pale yellowish white; both are edged with black, and deeply toothed; irides, pale cream; under the eye is a small spot of white; general colour of the plumage brownish black, the secondaries excepted, which are white, forming a broad band across the wing; there are a few reflections of purple on the upper plumage; the legs are red on the outside, and deep yellow, sprinkled with blackish, on the inner sides; tail, short and pointed.

The female is very little less than the male, but differs considerably in its markings. The bill is dusky; forehead and cheeks, white; under the eye, dull brownish; behind that, a large oval spot of white; whole upper parts and neck, dark brownish drab; tips of the plumage, lighter; secondaries, white; wing-quills, deep brown; belly, brownish white; tail, hoary brown; the throat is white, marked with dusky specks; legs and feet, yellow.

Latham informs us, that this species is sometimes seen on the coast of England, but is not common there; that it inhabits Denmark and Russia, and in some parts of Siberia is very common. It is also found at Kamtschatka, where it is said to breed, going far inland to lay; the eggs are eight or ten, and white; the males depart, and leave the females to remain with the young until they are able to fly. In the river Ochotska they are so numerous that a party of natives, consisting of fifty or more, go off in boats, and drive these ducks up the river before them, and, when the tide ebbs, fall on them at once, and knock them on the head with clubs, killing such numbers that each man has twenty or thirty for his share.*

* History of Kamtschatka, p. 160.

HARLEQUIN DUCK. (*Anas histrionica.*)

PLATE LXXII.—Fig. 4.

Le Canard à collier de Terre Neuve, *Briss.* vi. p. 362, 14.—*Buff.* ix. p. 250.—
Pl. enl. 798.—*Arct. Zool.* No. 490.—*Lath. Syn.* iii. p. 484.

CLANGULA HISTRIONICA.—Leach.*

Clangula histrionica, *Bonap. Synop.* p. 394.—*North. Zool.* ii. p. 459.—Canard à
collier, ou Histrion, *Temm. Man.* ii. p. 878.

THIS species is very rare on the coasts of the middle and
southern States, though not unfrequently found off those of
New England, where it is known by the dignified title of *the
lord,* probably from the elegant crescents and circles of white
which ornament its neck and breast. Though an inhabitant
of both continents, little else is known of its particular manners
than that it swims and dives well, flies swift, and to a great
height, and has a whistling note. Is said to frequent the
small rivulets inland from Hudson's Bay, where it breeds.
The female lays ten white eggs on the grass ; the young are
prettily speckled. It is found on the eastern continent as far
south as Lake Baikal, and thence to Kamtschatka, particularly
up the river Ochotska, and was also met with at Aoonalashka
and Iceland.† At Hudson's Bay, it is called *the painted
duck ;* at Newfoundland, and along the coast of New England,
the lord. It is an active vigorous diver, and often seen in deep
water, considerably out at sea.

The harlequin duck, so called from the singularity of its
markings, is seventeen inches in length, and twenty-eight
inches in extent; the bill is of moderate length, of a lead
colour, tipt with red ; irides, dark ; upper part of the head,
black ; between the eye and bill, a broad space of white, ex-

* Dr Richardson observes of this duck—" *C. histrionica* haunts eddies
under cascades, and rapid streams. It takes wing at once when disturbed, and is very vigilant. We never saw it associating with any
other duck, and it is a rare bird."—ED.
  † Latham.

tending over the eye, and ending in reddish ; behind the ear, a similar spot ; neck, black ; ending below in a circle of white ; breast, deep slate ; shoulders or sides of the breast, marked with a semicircle of white ; belly, black ; sides, chestnut ; body above, black or deep slate, some of the scapulars, white ; greater wing-coverts, tipt with the same ; legs and feet, deep ash ; vent and pointed tail, black.

The female is described as being less ; " the forehead, and between the bill and eye, white, with a spot of the same behind the ear ; head, neck, and back, brown, palest on the fore part of the neck ; upper part of the breast and rump, red brown ; lower breast and belly, barred pale rufous and white ; behind the thighs, rufous and brown ; scapulars and wing-coverts, rufous brown ; outer greater ones, blackish ; quills and tail, dusky, the last inclining to rufous ; legs, dusky." *

The few specimens of this duck which I have met with were all males ; and from the variation in their colours it appears evident that the young birds undergo a considerable change of plumage before they arrive at their full colours. In some the white spot behind the eye was large, extending irregularly half way down the neck ; in others confined to a roundish spot.

The flesh of this species is said to be excellent.

DUSKY DUCK.   (*Anas obscura.*)

PLATE LXXII.—Fig. 5.

Arct. Zool. No. 469.—*Lath. Syn.* iii. p. 545.—*Peale's Museum*, No. 2880.

BOSCHAS? OBSCURA.—Jardine.†

Anas obscura, *Bonap. Synop.* p. 384.

THIS species is generally known along the sea-coast of New Jersey and the neighbouring country by the name of the *black*

* Latham.

† Having now arrived at the conclusion of a group which holds a very prominent rank in the ornithology of Northern America, a few

duck, being the most common and most numerous of all those of its tribe that frequent the salt marshes. It is only partially

general observations regarding their economy, with an enumeration of those species omitted by Wilson which have been since discovered, may not be deemed improper.

The *Anatidæ,* or those birds generally known under the denomination of ducks, geese, and swans, taken as a family, will range with groups of great extent and varied form, as the falcons, the parrots, or pigeons, and will present similar modifications. The characters of the greater part of the groups which inhabit the northern and temperate regions of the world have been already drawn by Dr Leach and Dr Fleming, and one sub-family has been more lately analysed by Mr Swainson, as far as our knowledge of them extends, apparently with tolerable accuracy. They, however, want comparison with the tropical forms, which depart so much in their manners from those we are accustomed to see, and by which our opinions have hitherto been led. The wood ducks, constituting Mr Swainson's genus *Dendronessa,* the long-legged whistling ducks of India, those birds allied to the little Gambia goose, and those approaching in their form to the *Grallatores,* all want our close examination.

In distribution, the *Anatidæ* extend over the world, from the warmest tropics to the extreme arctic cold, but exist in greatest abundance near the confines of temperate regions and in northern latitudes. Their habits may be called truly aquatic, as the presence of water is necessary, even in the most aberrant forms, for their healthy support. Some groups are exclusively aquatic, and never quit the sea or large inland lakes, except during the season when the duties of incubation for a while call them to the shore. These may be termed pelagic or sea-ducks, and feed on fish and moluscæ; others delight in lakes and rivers as well as the sea, resort more frequently to the land, seek the same nourishment, and both are expert divers. Some hold a middle way, are as much on land as on water, and, in addition to the food of the truly sea species, live on the spawn of fresh-water fish, insects peculiar to muddy banks and slimy pools, with vegetables, such as the tender shoots of the grasses or newly-sown grains, or, while on the shores, upon the *Zostera marina;* while one or two forms resemble the grallatorial birds, and are more independent of water and aquatic nourishment. In their breeding-places, they show a like variety, choosing the reedy banks of lakes and rivers, the treacherous morass, the cliffs and desert sands of the sea-shore, the burrows of various animals, the hollows of decaying trees in the stupendous forests of America, or in India the welcome shade of the sacred banyan.

Their uses are various and extensive, either as food, or their skins, feathers, and down, for commerce and articles of wearing apparel, or

migratory. Numbers of them remain during the summer, and breed in sequestered places in the marsh, or on the sea islands

household comfort. Many species are also domesticated, and in a way less precarious lend their aid to the wants and luxuries of their owners. In the northern parts of America, this extensive family is most bountifully supplied, and her sealike lakes and majestic rivers are suitable nurseries for the innumerable multitudes that annually resort to, and reassemble to perform the duties of incubation. In the warmer parts, many remain at all seasons ; but it is in what is called Arctic America and the Fur Countries that the prodigious concourse annually arrive, and are so much hunted, both for food and a profitable emolument. Several of the spring months have received appellations from the birds which are most plentiful during them. The expected visitants arrive with remarkable precision, nearly at the same period of the month. They extend over a large space in breadth, and continue flying, without intermission, for many days. The native tribes are prepared by experience and the signals of their watches for their appearance ; and the first bird, for there are generally a scattered few before, gives notice that the havoc should commence.

"They are," says Dr Richardson, "of great importance in the Fur Countries, as they furnish, at certain seasons in the year, in many extensive districts, almost the only article of food that can be procured. The arrival of the waterfowl marks the commencement of spring, and diffuses as much joy among the wandering hunters of the arctic regions as the harvest or vintage excites in more genial climes. The period of their migration southwards again, in large flocks, at the close of summer, is another season of plenty, bountifully granted to the natives, and fitting them for encountering the rigours and privations of a northern winter."

To the species of *Anatidæ* which were known to Wilson as inhabitants of Northern America, with which his eighth volume has been almost wholly occupied, the researches of later ornithologists and travellers have added considerably, and the following enumeration of them will serve to fill up the list to last discoveries.

Somateria, Leach.

1. *S. spectabilis,* Leach. —King, Eider.—Common to both continents, and has much of the habits of the common eider. One or two specimens have been killed on the northern shores of Great Britain.

Clangula, Leach.

2. *C. Barrovii,* Swain. and Richards.—Rocky Mountain Garrot.—See note to p. 476 of the second volume.

of the beach.   The eggs are eight or ten in number, very nearly resembling those of the domestic duck.   Vast numbers, however, regularly migrate farther north on the approach of spring.

Cygnus, Steph.

Wilson, in his List of Birds, mentions the "swan;" but from three species at least being natives of the arctic countries, it is impossible to say whether or not he was aware of any distinctions.

3. *C. musicus*, Bechst., or Wild Swan.—Inhabits the arctic circle, whence it migrates to both continents.

4. *C. buccinator*, Richardson.—Trumpeter Swan.—Discovered to be undescribed by Dr Richardson during the last Overland Expedition; distinguished by the bill being entirely black, longer and more depressed than in the common wild swan, the tail containing twenty-four feathers, and by a difference in the folding of the windpipe. The Doctor remarks, it is the most common swan in the interior of the Fur Countries. It breeds as far south as lat. 61 deg., but principally within the arctic circle, and in its migrations generally precedes the geese a few days. It is to the trumpeter the bulk of the swan-skins imported by the Hudson's Bay Company belong.

5. *C. Bewickii*, Yarrel.—Bewick's Swan.—This bird has lately been discovered as a migratory visitant to Britain. Dr Richardson met with it during the last expedition, and remarks :—"This swan breeds on the sea-coast with in the arctic circle, and is seen in the interior of the Fur Countries in its passage only. It makes its appearance among the latest of the migratory birds in the spring, while the trumpeter swans are, with the exception of the eagles, the earliest."

Lewis and Clarke, Lawson, and Hearne, were all aware of the difference among the American swans, but they have never, till lately, been really distinguished and characterised.

Anser, Bechst.

6. *A. albifrons*, Bechst.—White-fronted Goose.—Is mentioned by Bonaparte, and is introduced in the "Northern Zoology." Its breeding-places are the woody districts skirting the Mackenzie, to the north of the sixty-seventh parallel, and also the islands of the Arctic Sea.

7. *A. segetum*, Meyer.—Common Bean Goose.—Inhabiting the more arctic regions. Bonaparte mentions also four additional species as probably accidental inhabitants of the United States and the arctic countries—*A. cinercus*, Meyer ; *A. rufescens*, Brehm ; *A. medius*, Temm. ; and *A. ceneraceus*, Brehm.

Bernicla, Steph.

8. *B. leucopsis* (*Anas erythropus*, Linn.—*A. leucopsis*, Temm.)—Inhabiting the arctic circle, migrating during winter to more temperate regions, and very rare and accidental in the United States.

9. *B. Hutchinsii* (*Anser Hutchinsii*, Richards.—*Hutchin's Bernacle*, *North. Zool.* ii. p. 470).—Described by Dr Richardson as a variety of the brant in the appendix to Captain Parry's second volume, and distinguished from it during the last Arctic Expedition.—ED.

During their residence here in winter they frequent the marshes, and the various creeks and inlets with which those extensive flats are intersected. Their principal food consists of those minute snail-shells so abundant in the marshes. They occasionally visit the sandy beach in search of small bivalves, and on these occasions sometimes cover whole acres with their numbers. They roost at night in the shallow ponds, in the middle of the salt marsh, particularly on islands, where many are caught by the foxes. They are extremely shy during the day; and, on the most distant report of a musket, rise from every quarter of the marsh in prodigious numbers, dispersing in every direction. In calm weather they fly high, beyond the reach of shot; but when the wind blows hard, and the gunner conceals himself among the salt grass, in a place over which they usually fly, they are shot down in great numbers; their flight being then low. Geese, brant, and black ducks are the common game of all our gunners along this part of the coast during winter; but there are at least ten black ducks for one goose or brant, and probably many more. Their voice resembles that of the duck and mallard; but their flesh is greatly inferior, owing to the nature of their food. They are, however, large, heavy-bodied ducks, and generally esteemed.

I cannot discover that this species is found in any of the remote northern parts of our continent; and this is probably the cause why it is altogether unknown in Europe. It is abundant from Florida to New England; but it is not enumerated among the birds of Hudson's Bay or Greenland. Its chief residence is on the sea-coast, though it also makes extensive excursions up the tide waters of our rivers. Like the mallard, they rarely dive for food, but swim and fly with great velocity.

The dusky or black duck is two feet in length, and three feet two inches in extent; the bill is of a dark greenish ash formed very much like the mallard, and nearly of the same length; irides, dark; upper part of the head, deep dusky brown, intermixed on the fore part with some small streaks of drab; rest of the head and greater part of the neck, pale yellow ochre, thickly marked with small streaks of blackish brown; lower part of the neck, and whole lower parts, deep dusky, each feather edged with brownish white, and with fine seams of rusty white; upper parts of the same, but rather deeper; the outer vanes of nine of the secondaries, bright violet blue, forming the beauty-spot, which is bounded on all sides by black; wings and tails, sooty brown; tail-feathers, sharp-pointed; legs and feet, dusky yellow; lining of the wings, pure white.

The female has more brown on her plumage; but in other respects differs little from the male, both having the beauty-spot on the wing.

MARSH TERN. (*Sterna aranea.*)

PLATE LXXII.—Fig. 6.

Peale's Museum, No. 3521.

STERNA ARANEA.—Wilson.*

Sterna aranea, *Bonap. Synop.* p. 354.

This new species I first met with on the shores of Cape May, particularly over the salt marshes, and darting down after a

* The Prince of Musignano writes the following observations in his " Nomenclature : "—

" A new species of Wilson, referred by Temminck to a bird which he calls *Sterna Anglica,* thinking that it is no other than *S. Anglica* of Montagu. But, as Brehm proves in his late work, the *S. Anglica* of Temminck is not the *S. Anglica* of Montagu. To the latter he gives the name of *S. risoria* (which cannot be adopted), and he calls the former *S. meridionalis.* He does not decide to which of the two species the American *S. aranea* belongs, and expresses the possibility of its being an independent species, but seems inclined to believe it identical with his *S. meridionalis.* Whether this bird is the *S. Anglica,* Mont., the *S. meridionalis,* Brehm, *Anglica,* Temm., or a distinct species peculiar to the north and south of this continent, it shall be the object of

kind of large black spider, plenty in such places. This spider can travel under water, as well as above, and, during summer at least, seems to constitute the principal food of the present tern. In several which I opened, the stomach was crammed with a mass of these spiders alone; these they frequently pick up from the pools, as well as from the grass, dashing down on them in the manner of their tribe. Their voice is sharper and stronger than that of the common tern; the bill is dif-

those observations to determine. The specimen deposited by Wilson in the Philadelphia Museum (a single glance at which would have enabled us to decide the question) being unfortunately destroyed, and Wilson's figure and description being too unessential to justify any conclusion, we should have been obliged to have left the matter unsettled, had it not been for the successful zeal of Mr Titian Peale, whose practical knowledge (the most important) of North American birds is equalled by none. Their favourite haunts, their note, their flight, are perfectly familiar to him. He succeeded in procuring a fine specimen at Long Beach, N.J., just as we were in want of one, and thus enabled us to give with more security the following opinion, which we had previously formed :—

"*S. aranea*, Wils., was a nondescript, different from *S. Anglica*, Mont., but the same with *S. Anglica*, Temm., and *S. meridionalis*, Brehm, and therefore common to both continents. Wilson's name having the priority, must be exclusively retained, and Brehm's name of *meridionalis* must be rejected. Thus has our author here also first named and described a European bird.

"Mr Ord was therefore right in not finding himself authorised to change the name. He was right in believing Montagu's bird distinct, but wrong in thinking Temminck's bird different, though Temminck had positively stated the specimens he had received from the United States and Brazil differed in nothing from his south Europeans. Even as respects the discrepance of *S. Anglica*, Mont., his reasons resting upon the slight difference of an unpublished drawing of Wilson respecting measurements of parts, to which Wilson did not attach great importance, were by no means conclusive. In fact, these measurements are incorrect, with the exception of the tarsus, which corresponds within a trifle of the bird. The bill is two and one-eighth inches to the corners of the mouth, and about one and one-half inches to the feathers of the forehead ; thus bearing more in favour of Mr Ord's argument that it is not the *Anglica*, Mont., than he himself supposed ; but proving that it is no other than *S. Anglica*, Temm. (*meridionalis*, Brehm), to which, as above stated, Wilson's name of *aranea* must be exclusively applied.

ferently formed, being shorter, more rounded above, and thicker ; the tail is also much shorter, and less forked. They do not associate with others, but keep in small parties by themselves.

The marsh tern is fourteen inches in length, and thirty-four in extent ; bill, thick, much rounded above, and of a glossy blackness ; whole upper part of the head and hind neck, black ; whole upper part of the body, hoary white; shafts of the quill and tail-feathers, pure white ; line from the nostril under the

" The principal character we should assign for a ready distinction between these two closely related species (in addition to the shorter, thicker, less compressed, and straighter bill, with its edges turned inwards in *Anglica*), consists in the tarsus, which in *aranea* (owing to its shortness and the extraordinary length of the nail) is of the same length as the middle toe, including the nail, whilst in *Anglica* it is nearly twice the length (owing to its superior length, and the shortness of the nail). The membranes of our bird are also much more scalloped. The habits of the two species are very different. The *S. Anglica*, confined to the seashores, feeds almost exclusively on strand birds,* and their eggs, sometimes on fishes ; whilst the *S. aranea*, generally found on marshes, feeds exclusively on insects."

Bonaparte, and the authors of the " Northern Zoology," have mentioned the following species as also found in North America :—

1. *S. cyana*, Lath.—Inhabiting the tropical seas of America; common on the coasts of the southern States.
2. *S. Arctica*, Temm.—*North. Zool.*, p. 114.—Bonaparte expresses a doubt that this is the true *Arctica* of Temm.; and the description in the " Northern Zoology " points out some discrepancies.
3. *S. stolida.*—Migrates to the North American coasts.

Phæton, Linn.

These birds, from general appearance, approach near to the terns (*S. Caspia*) ; but from the want of specimens, I am unable to enter into the proper situation of the form, except from the authority of others. Bonaparte places it between *sula* and *plotus*. The only American species is—

1. *P. thereus*, Linn., tropic bird of Wilson's List.—Common during summer on the coasts of the southern States.—ED.

---

* Is this correct? Does this tern kill other seafowl, and plunder their nests ? —ED.

eye, and whole lower parts, pure white ; tail, forked, the outer feathers about an inch and three-quarters longer than 'the middle ones ; the wings extend upwards of two inches beyond the tail ; legs 'and feet, black ; hind toe, small, straight, and pointed.

The female, as to plumage, differs in nothing from the male. The yearling birds, several of which I met with, have the plumage of the crown white at the surface, but dusky below ; so that the boundaries of the black, as it will be in the perfect bird, are clearly defined ; through the eye a line of black passes down the neck for about an inch, reaching about a quarter of an inch before it ; the bill is not so black as in the others ; the legs and feet, dull orange, smutted with brown or dusky ; tips and edges of the primaries, blackish ; shafts, white.

This species breeds in the salt marshes ; the female drops her eggs, generally three or four in number, on the dry drift grass, without the slightest appearance of a nest ; they are of a greenish olive, spotted with brown.

A specimen of this tern has been deposited in the Museum of this city (Philadelphia).

SOOTY TERN. (*Sterna fuliginosa.*)

PLATE LXXII.—Fig. 7.

La Hirondelle de Mer à grande enverguer, *Buff.* viii. p. 345.—Egg-bird, *Forst. Voy.* p. 113.—Noddy, *Damp. Voy.* iii. p. 142.—*Arct. Zool.* No. 447.—*Lath. Syn.* iii. p. 352.—*Peale's Museum*, No. 3459.

STERNA FULIGINOSA.—Latham.

S. fuliginosa, *Bonap. Synop.* p. 355.

This bird has been long known to navigators, as its appearance at sea usually indicates the vicinity of land ; instances, however, have occurred in which they have been met with one hundred leagues from shore.[*] The species is widely dispersed over the

[*] Cook's Voyages, i. 275.

various shores of the ocean. They were seen by Dampier in New Holland ; are in prodigious numbers in the Island of Ascension and in Christmas Island; are said to lay in December one egg on the ground ; the egg is yellowish, with brown and violet spots.* In passing along the northern shores of Cuba and the coast of Florida and Georgia in the month of July, I observed this species very numerous and noisy, dashing down headlong after small fish. I shot and dissected several, and found their stomachs uniformly filled with fish. I could per-ceive little or no difference between the colours of the male and female.

Length of the sooty tern, seventeen inches ; extent, three feet six inches ; bill, an inch and a half long, sharp pointed and rounded above, the upper mandible serrated slightly near the point ; nostril, an oblong slit ; colour of the bill, glossy black ; irides, dusky ; forehead, as far as the eyes, white ; whole lower parts and sides of the neck, pure white ; rest of the plumage, black ; wings, very long and pointed, extending, when shut, nearly to the extremity of the tail, which is greatly forked, and consists of twelve feathers, the two exterior ones four inches longer than those of the middle, the whole of a deep black, except the two outer feathers, which are white, but towards the extremities a little blackish on the inner vanes ; legs and webbed feet, black ; hind toe, short.

The secondary wing-feathers are eight inches shorter than the longest primary.

This bird frequently settles on the rigging of ships at sea, and, in common with another species, *S. stolida*, is called by sailors the *noddy*.

* Turton.

CINEREOUS COOT.   (*Fulica Americana.*)

PLATE LXXIII.—Fig. 1.

Fulica Americana, *Gmel. Syst.* i. p. 704, 23.—*Lath. Ind. Orn.* p. 779, 5.—Cinereous Coot, *Gen. Syn.* iii. p. 279.—*Peale's Museum*, No. 4322.

FULICA AMERICANA.—Gmelin.*

Fulica Americana, *Sab. Append. to Capt. Frank. Exp.* p. 690.—*Bonap. Synop.* p. 338.—Fulica atra, *Wilson's List.*

THIS species makes its appearance in Pennsylvania about the 1st of October. Among the muddy flats and islands of the river Delaware, which are periodically overflowed, and which are overgrown with the reeds or wild oats or rushes, the coots are found. They are not numerous, and are seldom

* This description commences the ninth and supplementary volume of the original printed by Mr Ord, after the decease of Wilson, from his notes. The volume was published in 1814, and a second edition appeared in 1825, correcting several mistakes which had occurred in the first. Our present bird was there described as identical with that of Europe, and a detail of the habits of our native species given as belonging to it ; these Mr Ord has corrected. The distinctions, I believe, were first pointed out by Mr Sabine, in the Appendix to Captain Franklin's Narrative, and I now add them in that gentleman's words :—

" They are of the same length, though there is a general inferiority in the size of the body, as well as of the legs, head, and bill of the American ; the bill is smaller, less thick and strong, and shorter by a quarter of an inch ; the callus, independent of the difference in colour in the American bird, extends only half an inch over the head, but in the European, above an inch ; the whole head is smaller ; the plumage, generally, is similar in colour and character ; the outer margin of the first primary feathers of the wing is more conspicuously marked with white, and there are a few white feathers on the upper edge of the wing ; the secondaries in both are tipped with white ; the principal difference in the plumage is, that in the American the feathers at the vent are quite black, and the under tail-coverts white ; in the European coot, these correspond with the rest of the plumage ; the legs are much more slender in the American bird ; the tarse of the European measures near two inches and a half, that of the American not quite two inches ; the toes are smaller in like proportion ; the middle toe, including the claw, of the European coot is three inches and three-quarters long ; of the American, three inches and one quarter only."—ED.

1.Common Coot. 2.Purple Gallinule. 3.Gray Phalarope. 4.Red P. 5.Wilsons Plover.

73.

seen, except their places of resort be covered with water; in that case they are generally found sitting on the fallen reed, waiting for the ebbing of the tide, which will enable them to feed. Their food consists of various aquatic plants, seeds, insects, and, it is said, small fish. The coot has an aversion to take wing, and can seldom be sprung in its retreat at low water; for although it walks rather awkwardly, yet it contrives to skulk through the grass and reeds with great speed, the compressed form of its body, like that of the rail genus, being well adapted to the purpose. It swims remarkably well, and, when wounded, will dive like a duck. When closely pursued in the water, it generally takes to the shore, rising with apparent reluctance, like a wounded duck, and fluttering along the surface, with its feet pattering on the water.* It is known in Pennsylvania by the name of the *mud-hen*.

I have never yet discovered that this species breeds with us; though it is highly probable that some few may occupy the marshes of the interior, in the vicinity of the ponds and lakes, for this purpose, those retired situations being well adapted to the hatching and rearing of their young. In the southern States, particularly South Carolina, they are well known; but the Floridas appear to be their principal rendezvous for the business of incubation. "The coot," says William Bartram, "is a native of North America, from Pennsylvania to Florida. They inhabit large rivers, fresh-water inlets or bays, lagoons, &c., where they swim and feed amongst the reeds and grass of the shores, particularly in the river St Juan, in East Florida; where they are found in immense flocks. They are loquacious and noisy, talking to one another night and day; are constantly on the water, the broad lobated membranes on their toes enabling them to swim and dive like ducks." †

I observed this species to be numerous during the winter

* In Carolina, they are called *flusterers*, from the noise they make in flying over the surface of the water.—*A Voyage to Carolina*, by John Lawson, p. 149.

† Letter from Mr Bartram to the author.

in the fresh-water ponds situated in the vicinity of the river St Juan or St John, in East Florida; but I did not see them in the river. The food which they obtain in these places must be very abundant and nutritious, as the individuals which I shot were excessively fat. One male specimen weighed twenty-four ounces avoirdupois. They associate with the common gallinule (*Gallinula chloropus*), but there is not, perhaps, one of the latter for twenty of the former.

The cinereous coot is sixteen inches in length, and twenty-eight in extent; bill, one and a half inch long, white, the upper mandible slightly notched near the tip, and marked across with a band of chestnut, the lower mandible marked on each side with a squarish spot of the like colour, edged on the lower part with bright yellow or gamboge, thence to the tip, pale horn colour; membrane of the forehead, dark chestnut brown; irides, cornelian red; beneath the eyes, in most specimens, a whitish spot; the head and neck are of a deep shining black, resembling satin; back and scapulars, dirty greenish olive; shoulders, breast, and wing-coverts, slate blue; the under parts are hoary; vent, black; beneath the tail, pure white; primaries and secondaries, slate, the former tipt with black, the latter with white, which does not appear when the wing is closed; outer edges of the wings, white; legs and toes, yellowish green, the scalloped membrane of the latter, lead colour; middle toe, including the claw, three inches and three-quarters long.

The bird from which the foregoing description was taken was shot in the Delaware, below Philadelphia, the 29th of October 1813. It was an old male, an uncommonly fine specimen, and weighed twenty-three ounces avoirdupois. It is deposited in Peale's Museum.

The young birds differ somewhat in their plumage, that of the head and neck being of a brownish black; that of the breast and shoulders, pale ash; the throat, grey or mottled; the bill, bluish white; and the membrane on the forehead considerably smaller.

The young females very much resemble the young males ; all the difference which I have been enabled to perceive is as follows :—Breast and shoulders, cinereous ; markings on the bill, less ; upper parts of the head, in some specimens, mottled ; and being less in size.

The lower parts of these birds are clothed with a thick down, and, particularly between the thighs, covered with close fine feathers. The thighs are placed far behind, are fleshy, strong, and bare above the knees.

The gizzard resembles a hen's, and is remarkably large and muscular. That of the bird which has been described was filled with sand, gravel, shells, and the remains of aquatic plants.

Buffon describes the mode of shooting coots in France, particularly in Lorraine, on the great pools of Tiaucourt and of Indre ; hence we are led to suppose that they are esteemed as an article of food. But with us, who are enabled, by the abundance and variety of game, to indulge in greater luxuries in that season when our coots visit us, they are considered as of no account, and are seldom eaten.

The European ornithologists represent the membrane on the forehead of the *Fulica atra* as white, except in the breeding season, when it is said to change its colour to pale red. In every specimen of the cinereous coot which I have seen, except one, the membrane of the forehead was of a dark chestnut brown colour. The one alluded to was a fine adult male, shot in the Delaware, at Philadelphia; on the 11th of May : the membrane was of a pure white ; no white marking beneath the eye ; legs and feet of a bright grass green.

In Wilson's figure of the coot accompanying this volume there are some slight errors ; the auriculars are designated, which should not have been done, as they are not distinguishable from the rest of the plumage of the head and neck, which is all of a fine satiny texture ; and the outline of the bill is not correct.

Latham states that the common European coot (*F. atra*) is

"met with in Jamaica, Carolina, and other parts of North America." This, I presume, is a mistake, as I have never seen but one species of coot in the United States. Brown, in speaking of the birds of Jamaica, mentions a coot which, in all probability, is the same as ours. The coot mentioned by Sloan is the common gallinule. So is also that spoken of in the "Natural History of Barbadoes," by Hughes, p. 71.

In Lewis and Clark's history of their expedition, mention is made of a bird which is common on the Columbia ; is said to be very noisy, to have a sharp, shrill whistle, and to associate in large flocks ; it is called the black duck.* This is doubtless a species of coot, but whether or not different from ours cannot be ascertained. How much is it to be regretted that, in an expedition of discovery, planned and fitted out by an enlightened Government, furnished with every means for safety, subsistence, and research, not one naturalist, not one draughtsman, should have been sent to observe and perpetuate the infinite variety of natural productions, many of which are entirely unknown to the community of science, which that extensive tour must have revealed !

The coot leaves us in November for the southward.

The foregoing was prepared for the press, when the author, in one of his shooting excursions on the Delaware, had the good fortune to kill a full-plumaged female coot. This was on the 20th of April. It was swimming at the edge of a *cripple*, or thicket of alder bushes, busily engaged in picking something from the surface of the water, and while thus employed it turned frequently. The membrane on its forehead was very small, and edged on the fore part with gamboge. Its eggs were of the size of partridge-shot. And on the 13th of May, another fine female specimen was presented to him,

* History of the Expedition, vol. ii. p. 194. Under date of November 30th, 1805, they say—"The hunters brought in a few black ducks of a species common in the United States, living in large flocks, and feeding on grass ; they are distinguished by a sharp white beak, toes separated, and by having no craw."

which agreed with the above, with the exception of the membrane on the forehead being nearly as large and prominent as that of the male. From the circumstance of the eggs of all these birds being very small, it is probable that the coots do not breed until July.

MARTINICO GALLINULE. (*Gallinula Martinica.*)

PLATE LXXIII.—Fig. 2.

Gallinula Martinica, *Lath. Ind. Orn.* p. 769, 9; *Gen. Syn.* iii. p. 255, 7, pl. 88.—
  Fulica Martinica, *Linn. Syst.* ed. 12, i. p. 259, 7.—Fulica Martinicensis,
  Gmel. Syst. p. 700, 7.—La Petite Poule-sultane, *Briss. Orn.* v. p. 526, pl. 42,
  fig. 2.—*Buff. Ois.* viii. p. 206.—La Favourite de Cayenne, *Pl. enl.* No. 897,
  young?—*Peale's Museum*, No. 4294.

GALLINULA ? MARTINICA.—Latham.*

Gallinula Martinica, *Bonap. Synop.* p. 336.

This splendid bird is a native of the southern parts of the continent of North America. I have never learnt that it migrates as far north as Virginia, though it is probable that it may be occasionally seen in that State. It makes its appearance in the Sea Islands of Georgia in the latter part of April, and after spending the summer, it departs, with its young, in the autumn. The marshes of Mexico appear to be its winter residence. It frequents the rice-fields and fresh-water ponds, in company with the common gallinule; but the latter, being of a more hardy nature, remains all winter both in Georgia and Florida.

During its migration, this bird is frequently driven to sea; and I have known two or three instances of its having sought refuge on board of vessels. On the 24th May 1824, a brig

---

* This species in form runs very much into the *Porphyrio* of Brisson; but without specimens, I cannot decide whether it should rank there, or on the confines of *Gallinula*. The characters of the former group are the much greater strength of the bill, being almost as high as long, the greater proportional length of legs, and the splendid and metallic lustre of the plumage. In their manners they are partly granivorous, and live more upon land than the water-hens.—Ed.

arrived at Philadelphia from New Orleans, bringing a fine living specimen, which had flown on board of her in the Gulf Stream. This bird is now (1825) alive in the Philadelphia Museum. In the month of August 1818, a storm drove another individual on board of a vessel in her passage from Savannah to Philadelphia. This also lived for some time in Peale's Museum.

The Martinico gallinule is a vigorous and active bird. It bites hard, and is quite expert in the use of its feet. When it seizes upon any substance with its toes, it requires a considerable effort to disengage it. Its toes are long, and spread greatly. It runs with swiftness ; and, when walking, it jerks its tail in the manner of the common rail. Its manners and food are somewhat similar to those of the far-famed purple gallinule, whose history is so beautifully detailed in the works of Buffon.

In its native haunts, it is vigilant and shy ; and it is not easy to spring it without the assistance of a dog.

Length, from the tip of the bill to the end of the tail, fourteen inches ; bill, an inch and a quarter long, vermilion, greenish yellow at the tip ; irides, pale cornelian ; naked crown, dull azure ; head, part of the neck, throat, and breast, of a rich violet purple ; back and scapulars, olive green ; rump, tail, and its coverts, brownish green ; sides of the neck and wings, ultra-marine, the latter tinged with green ; shoulders of wing, rich azure ; inner webs of the quills and tail-feathers, dusky brown ; belly and thighs, dull purplish black ; vent, pure white ; tail, rounded ; legs and feet, greenish yellow ; claws, long, sharp, and of a pale flesh colour ; span of the foot, five inches.

BROWN PHALAROPE. (*Phalaropus lobatus.*)

PLATE LXXIII.—Fig. 3.

Tringa lobata, *Linn. Syst.* ed. 10, tom. i. p. 148, 5.—T. hyperborea, *Id.* ed. 12, tom. i. p. 249, 9.—Tringa lobata, *Gmel. Syst.* i. p. 764, 6.—T. fusca, *Id.* p. 675, 33.—T. hyperborea, *Id.* No. 9.—Phalaropus cinereus, *Briss. Orn.* vi. p. 15.—P. fuscus, *Id.* p. 18.—Le Phalarope Cendré, *Buff. Ois.* viii. p. 224.—*Pl. enl.* 766.—Coot-footed Tringa, *Edwards,* pl. 46.—Cock Coot-footed Tringa, *Id.* pl. 143.—Red Phalarope, *Penn. Brit. Zool.* No. 219.—Brown Phalarope, *Arct. Zool.* No. 414.—Phalaropus hyperboreus, *Lath. Ind. Orn.* p. 775, 1.— P. fuscus, *Id.* p. 776, 4.—Red Phalarope, *Gen. Syn.* iii. p. 270, 1.—*Id.* p. 272, var. A.—Brown Phalarope, *Id.* p. 274, 4.—Red Phalarope, *Montagu, Orn. Dict. Id. Supp. and Appendix.*—Phalaropus hyperboreus, *Temm. Man. d'Orn.* p. 709.—Le Lobipède à hausse-col, *Cuv. Reg. Anim.* i. p. 495.

LOBIPES WILSONII.—Jardine.*

Phalaropus (subgen. Holopodius), *Bonap. Synop.* p. 342.—Phalaropus Wilsonii, *Sab. App. to Frank. Narrat.* p. 691.—Lobipes incanus, *Jard. and Selby, Illust. Ornith.* pl. 25.—Phalaropus Wilsonii, *North. Zool.* ii. pl. 69.

Of this species, only one specimen was ever seen by Wilson, and that was preserved in Trowbridge's Museum, at Albany,

---

* In the small group known as *Phalaropus* we have two forms, distinguished by the stouter make, the flat-formed bill, and the development of the webs to the toes in the one, and by the slender bill and greater alliance of the other to the *Totani.* The Prince of Musignano has instituted another sub-group from what appears to me to be only the greater development of the latter form. Following the arrangement of Cuvier, I have retained *Lobipes* for those of slender make, and *Phalaropus* for that of this country, and only one yet discovered.

I have little hesitation in considering the *L. incanus* of ornithological illustrations to be this bird in imperfect plumage. Bonaparte is of opinion that the American bird was a new species ; Mr Ord, that it is some undescribed state of *P. hyperboreus :* with the former of these opinions I agree, and have accordingly adopted the specific name which Sabine had previously chosen for it, but have referred it to the genus *Lobipes* of Cuvier. This plate of our author is one of the very few exceptions where an imperfect representation of the bird is given, the figure being much too stout and thick, and not of that more elegant form, one of the *characteristics* of the *Totani.*

Under this division will also range the little red-necked phalarope, *Lobipes hyperboreus* of Temminck, and the present type of the genus. According to Bonaparte, this species is exceedingly rare and accidental in the United States. It will appear in the forthcoming volumes of that gentleman's illustrations.—Ed.

in the State of New York.   On referring to Wilson's Journal, I found an account of the bird, there called a *tringa*, written with a lead pencil, but so scrawled and obscured, that parts of the writing were not legible.   I wrote to Mr Trowbridge soliciting a particular description, but no answer was returned. However, having had the good fortune, since publishing the first edition, of examining a fine recent specimen of this rare bird, I hope I shall be enabled to fix the species by such characters as will prevent any ornithologist in future from confounding it with the species which follows ;—two birds which, owing to a want of precision, were involved in almost inextricable confusion, until Temminck applied himself to the task of disembroiling them ; and this ingenious naturalist has fully proved that the seven species of authors constituted in effect only two species.

Temminck's distinctive characters are drawn from the bill ; and he has divided the genus into two sections—an arrangement of which the utility is not evident, seeing that each section contains but one species, unless we may consider that the barred phalarope of Latham constitutes a third, a point not yet ascertained, and not easy to be settled, for the want of characters.

In my examination of these birds, I have paid particular attention to the feet, which possess characters equally striking with those of the bill ; hence, a union of all these will afford a facility to the student, of which he will be fully sensible when he makes them the subject of his investigation.

Our figure of this species betrays all the marks of haste ; it is inaccurately drawn, and imperfectly coloured ; notwithstanding, by a diligent study of it, I have been enabled to ascertain that it is the coot-footed tringa of Edwards, plate 46 and 143, to which bird Linnæus gave the specific denomination of *lobata*, as will be seen in the synonyms at the head of this article.   In the twelfth edition of the "Systema Naturæ," the Swedish naturalist, conceiving that he might have been in error, omitted, in his description of the *lobata*, the synonym of Edward's, cock coot-footed tringa, No. 143, and recorded the latter bird under the name of *hyperborea*—a specific appellation, which Temminck and other ornithologists have sanctioned, but which the laws of methodical nomenclature prohibit us from adopting, as, beyond all question, *hyperborea* is only a synonym of *lobata*, which has the priority, and must stand.

M. Temminck differs from us in the opinion that the *T. lobata* of Gmelin, vol. i. p. 674, is the present species, and refers it to that which follows. But, if this respectable ornithologist will take the trouble to look into the twelfth edition of Linnæus, vol. i. p. 249, No. 8, he will there find two false references, Edward's No. 308, and Brisson's No. 1, which gave rise to Gmelin's confusion of synonyms, and a consequent confusion in his description, as the essential character in both authors being nearly in the same words (*rostro subulato apice inflexo, &c.*), we are at no loss to infer that both descriptions have reference to the same bird ; and we are certain that the *lobata* of the twelfth edition of the former is precisely the same as that of the tenth edition, which cites for authority Edward's 46 and 143, as before mentioned.

I shall now give the short description of the bird figured in the plate, as I find it in Wilson's notebook :—

Bill, black, slender, and one inch and three-eighths in length. In the original, the bill is said to be one inch and three-quarters long. Lores, front, crown, hind head, and thence to the back, very pale ash, nearly white ; from the anterior angle of the eye, a curving stripe of black descends along the neck for an inch or more, thence to the shoulders, dark reddish brown, which also tinges the white on the side of the neck next to it ; under parts, white ; above, dark olive ; wings and legs, black. Size of the turnstone.

The specimen from which the following description was taken was kindly communicated to me by my friend Mr Titian R. Peale, while it was yet in a recent state, and before it was prepared for the museum. It was this individual which enabled me to ascertain the species figured by Wilson. It was shot in the neighbourhood of Philadelphia, on the 7th of May 1818.

Bill, narrow, slender, flexible, subulate, of equal width; nostrils, basal, and linear; lobes of the toes, thick, narrow, and but slightly scalloped. Outer toe connected to the middle one as far as the first joint; inner toe divided nearly to its base; hind toe resting on the ground. Bill, black, one inch and three-eighths in length; head above, of an ash grey; hind head, whitish, which colour extends a short distance down the neck; over the eyes, a white stripe, below them, a white spot; throat and lower parts, white; a line of black passes through the eyes, spreads out towards the hind head, and descends along the neck; lower part of the neck, pale ferruginous; back part of the neck, deep ferruginous, which descends on each side, and mingles with the plumage of the back and scapulars, which are of a clove brown, the feathers tipt with whitish; wings and tail, dark clove brown, some of the lesser coverts having a reddish tinge; the upper tail-feathers, tinged with red at their tips, the under feathers marked with white on their inner webs; irides, dark brown; legs and feet, dark plumbeous; claws, long, of a dark horn colour; hind toe, independent of the claw, five-sixteenths of an inch long; the tertials, when the wing is closed, extend to within three-eighths of an inch of the tip of the primaries; weight, an ounce and three-quarters; length, nine inches and a half; breadth, sixteen inches. This was a female; her eggs very small.

In the grand chain of animated nature, the phalaropes constitute one of the links between the waders and the web-footed tribes, having the form of the sandpipers, with some of the habits of the gulls; the scalloped membranes on their toes enabling them to swim with facility. They are clothed with a thick coat of feathers, beneath which, as in the ducks, lies a mass of down, to protect them from the rigours of the northern climates, of which they are natives. They do not appear to be fond of the neighbourhood of the ocean, and are generally found in the interior, about the lakes, ponds, and streams of fresh water, where they delight to linger, swimming near the margin in search of seeds and insects. They are nowhere numerous, are commonly seen in pairs, and are so extremely tame and unsuspicious, that one may approach to within a few feet of them.

The genus *Lobipes* of the Baron Cuvier is founded upon this species ; and it must be confessed that its characters are sufficiently distinct from those of the bird which follows to authorise such a separation ; but unless some new species should be discovered, we see no impropriety in associating the two birds already known, taking care, however, to preserve a consistency in the generic characters, which Temminck, in his "Manuel," has not sufficiently observed.

In the Appendix to Montagu's "Supplement to the Ornithological Dictionary," we find the following remarks on this species, there named *fulicaria :*—"We have before mentioned that this bird had been observed in the Orkneys in considerable abundance in the summer, and that no doubts were entertained of its breeding there, although the nest had not been found. To Mr Bullock, therefore, we are indebted for the farther elucidation of the natural history of this elegant little bird. In a letter to the author, this gentleman says, 'I found the red phalarope common in the marshes of Sanda and Westra, in the breeding season, but which it leaves in the autumn. This bird is so extremely tame, that I killed nine without moving out of the same spot, being not in the least alarmed at the report of a gun. It lays four eggs, of the shape of that of a snipe, but much less, of an olive colour, blotched with dusky. It swims with the greatest ease, and when on the water, looks like a beautiful miniature of a duck, carrying its head close to the back, in the manner of a teal.' Mr Bullock further observes, 'That the plumage of the female is much lighter, and has less of the rufous than the other sex.'" *

* All observations referring to European birds in this description cannot apply. *L. Wilsonii* is yet known only as American. The last may be referred with propriety, however, to *Lobipes hyperboreus.*—ED.

GREY PHALAROPE.* (*Phalaropus fulicarius.*)

PLATE LXXIII.—Fig. 4.

Tringa fulicaria, *Linn. Syst.* ed. 10, tom. i. p. 148, 6.—Tringa glacialis, *Gmel. Syst.* i. p. 675, 2.—T. hyperborea, var. B, *Id.* p. 676.—Le Phalarope, *Briss. Orn.* vi. p. 12, No. 1.—Phalaropus rufescens, *Id.* p. 20.—Phalaropus lobatus, *Lath. Ind. Orn.* p. 776, 2.—P. glacialis, *Id.* No. 3.—Red Phalarope, fem. *Gen. Syn.* iii. p. 271.—Grey Phalarope, *Id.* p. 272, 2.—Plain Phalarope, *Id.* p. 273, 3.—Grey Phalarope, *Penn. British. Zool.* No. 218.—*Arct. Zool.* No. 412.— Red Phalarope, *Id.* No. 413.—Plain Phalarope, *Id.* 415.—Red Coot-footed Tringa, *Edw.* pl. 142.—Grey Coot-footed Tringa, *Id. Gleanings,* pl. 308.—Le Phalarope Rouge, *Buffon, Ois.* viii. p. 225.—Le Phalarope à festons dentelés, *Id.* p. 226.—Grey Phalarope, *Montagu, Orn. Dict. and Appendix to Supp.*— *Bewick,* ii. p. 132.—Le Phalarope Gris, *Cuv. Reg. Anim.* i. p. 492.—Le Phala- rope Rouge, *Id. ibid.*—Phalaropus platyrhinchus, *Temm. Man. d'Orn.* p. 712. —*Peale's Museum,* No. 4088.

PHALAROPUS FULICARIUS.—Bonaparte.†

Phalaropus hyperborea, *Ord,* 1st edit. of *Supp.* p. 75.—Phalaropus fulicarius, *Bonap. Synop.* p. 341.—*Nomenclature,* No. 232.—Phalaropus lobatus, Grey Phalarope, *Flem. Br. Anim.* p. 100.—Grey Phalarope, *Selby, Illust. Br. Ornith.* 2d ed. pl. 28.

Bill pretty stout and wide, slightly compressed at the tip, depressed on the lower half; upper mandible, carinate; nos- trils, subovate, a short distance from the base; feet, semipal- mate; lobes of the toes, broad, and greatly scalloped; hind toe, barely touching the ground; bill, reddish orange at the

---

* Named in the plate, red phalarope.

† This bird is here represented in the change from the summer or breeding state to that of the grey plumage of its winter dress, in which alternations it bears a strong resemblance to the knots and godwits, &c. It will show an example of the genus *Phalaropus;* indeed, it is the only one hitherto discovered.* The form appears more stout, from the short- ness of the legs, and it is also distinguished from *Lobipes* by the flattened or depressed bill, and more than usually fleshy tongue. They are expert swimmers, are often found out at sea, and their whole manners on the water resemble more those of a truly aquatic bird than of a form allied to the *Tringæ.* Bonaparte mentions, that this bird is rare and accidental, and during winter only found in the United States, extend- ing its migrations to Florida.—Ed.

* Dr Richardson thinks that another species will be found in the plain phala- rope of Pennant, and proposes the name of *P. glacialis* for it.—Ed.

base, the remainder black, an inch long; front and crown, black, barred transversely with lines of white; throat, sides of the neck, and lower parts, white, thickly and irregularly barred with curving dashes of reddish chocolate; upper parts, of a deep cinereous blue, streaked with brownish yellow and black; the back scapulars, broadly edged with brownish yellow; wings and rump, dark cinereous; greater wing-coverts, broadly tipt with white, forming a large band; primaries, nearly black, and crossed with white below their coverts; tail, plain olive, middle of its coverts, black, their sides bright brownish yellow; vent, white, those feathers immediately next to the tail, reddish chocolate; legs, black on the outside, yellowish within. Length, nine inches; breadth, fifteen inches and a half; length of hind toe, independent of the claw, one-eighth of an inch. Male. The inner toe is connected to the middle one by a membrane as far as the first joint, the outer toe much farther; hence the feet may be properly termed semipalmate; webs and lobes, finely pectinated. This conformation of the feet is pretty accurately exhibited in Edwards's plate, No. 308. The grey phalarope is a rare bird in Pennsylvania, and is not often met with in any part of the United States. The individual from which our description was taken was shot in a pond, in the vicinity of Philadelphia, in the latter part of May 1812. There were three in company. The person who shot it had never seen one of the species before, and was struck with their singular manners. He described them as swimming actively near the margin of the pond, dipping in their bill very often, as if feeding, and turning frequently. In consequence of our specimen being in a state of putridity when received, it was preserved with considerable difficulty, and the sex could not be ascertained.

In the spring of the year 1816, my friend Mr Le Sueur shot, in Boston Bay, a young individual of this species. Crown, dark slate, tinged with yellowish brown; front, throat, line over the eye, belly, and vent, white; shoulders, breast, and sides, tawny or fawn colour; back, dark slate, paler near the rump, the feathers, edged with bright yellow ochre; wings, pale cinereous, some of the lesser coverts edged with white, the greater coverts largely so, forming the bar; primaries and tail, black; the latter edged with yellowish brown, the shafts of the former white; bill and feet, as in the first described.

On the 20th of March 1818, I shot, in the river St John, in East Florida, an immature female specimen; irides, dark brown; around the base of the bill, a slight marking of dark slate; front and crown, white, mottled with pale ash; at the interior part of each eye, a black spot; beneath the eyes, dark slate, which extends over the auriculars, the hind head, and upper part of the neck; upper parts, cinereous grey, with a few faint streaks of slate; throat, breast, whole lower parts, and under tail-coverts, pure white; flanks, with a few faint ferruginous stains; wings, slate brown, the coverts of the secondaries, and a few of the primary coverts, largely tipt with white, forming the bar as usual; tail, brown, edged with cinereous; legs and feet, pale plumbeous; the webs, and part of the scalloped membranes, yellowish; bill and size as in the first specimen.

The tongue of this species is large, fleshy, and obtuse.

This bird has been described under a variety of names. What could induce that respectable naturalist, M. Temminck, to give it a new appellation, we are totally at a loss to conceive. That his name (*Phalaropus platyrhinchus*) is good,—that it is even better than all the rest, we are willing to admit,—but that he had no right to give it a new name we shall boldly maintain, not only on the score of expediency, but of justice. If the right to change be once conceded, there is no calculating the extent of the confusion in which the whole system of nomenclature will be involved. The study of methodical natural history is sufficiently laborious, and whatever will have a tendency to diminish this labour, ought to meet the cordial support of all those who are interested in the advancement of the natural sciences.

"The study of natural history," says the present learned

president of the Linnæan Society, "is, from the multitude of objects with which it is conversant, necessarily so encumbered with names, that students require every possible assistance to facilitate the attainment of those names, and have a just right to complain of every needless impediment. Nor is it allowable to alter such names, even for the better. In our science, the names established throughout the works of Linnæus are become current coin, nor can they be altered without great inconvenience." *

That there is a property in names as well as in things will not be disputed; and there are few naturalists who would not feel as sensibly a fraud committed on their nomenclature as on their purse. The ardour with which the student pursues his researches, and the solicitude which he manifests in promulgating his discoveries under appropriate appellations, are proofs that at least part of his gratification is derived from the supposed distinction which a name will confer upon him; deprive him of this distinction, and you inflict a wound upon his self-love which will not readily be healed.

To enter into a train of reasoning to prove that he who first describes and names a subject of natural history agreeable to the laws of systematic classification is for ever entitled to his name, and that it cannot be superseded without injustice, would be useless, because they are propositions which all naturalists deem self-evident. Then how comes it, whilst we are so tenacious of our own rights, we so often disregard those of others?

I would now come to the point. It will be perceived that I have ventured to restore the long-neglected name of *fulicaria*. That I shall be supported in this restoration I have little doubt, when it shall have been made manifest that it was Linnæus himself who first named this species. A reference to the tenth edition of the "Systema Naturæ" † will show that

* An Introduction to Physiological and Systematical Botany, chap. 12.

† Of all the editions of the "Systema Naturæ," the tenth and the twelfth are the most valuable; the former being the first which contains the

the authority for *Tringa fulicaria* is Edwards's red coot-footed tringa, pl. 142, and that alone, for it does not appear that Linnæus had seen the bird.   The circumstance of the change of the generic appellation can in nowise affect the specific name ; the present improved state of the science requires the former, justice demands that the latter should be preserved.

In this work I have preserved it ; and I flatter myself that this humble attempt to vindicate the rights of Linnæus will be approved by all those who love the sciences of which he was so illustrious a promoter.

WILSON'S PLOVER.   (*Charadrius Wilsonius.*)

PLATE LXXIII.—Fig. 5.

Peale's Museum, No. 4159, male ; 4160, female.

CHARADRIUS WILSONIUS.—Ord.

Charadrius Wilsonius, *Bonap. Synop.* p. 296.—*Nomenclature*, No. 221.

Of this neat and prettily marked species I can find no account, and have concluded that it has hitherto escaped the eye of the naturalist.   The bird from which this description was taken was shot the 13th of May 1813, on the shore of Cape Island, New Jersey, by my ever-regretted friend ; and I have honoured it with his name.*   It was a male, and was accompanied by

Synonyma, and the latter being that which received the finishing hand of its author.   In the United States, Linnæus is principally known through two editors—Gmelin, whose thirteenth edition of the " Systema Naturæ" has involved the whole science in almost inextricable confusion ; and Turton, whose English translation of Gmelin is a disgrace to science and letters.   All writers on zoology and botany should possess Linnæus's tenth and twelfth editions ; they will be found to be of indispensable use in tracing synonyms and fixing nomenclature.

* Bonaparte thus observes in his "Nomenclature : "—" A very rare species, established by the editor (Mr Ord), and dedicated to Wilson.   It is the first homage of the kind paid to the memory of this great and lamented self-taught naturalist.   The descriptions of several species in the works of former authors come more or less near to it, but after a careful investigation we are satisfied that it is new."—Ed.

another of the same sex and a female, all of which were fortunately obtained.

This bird very much resembles the ring plover, except in the length and colour of the bill, its size, and in wanting the yellow eyelids. The males and females of this species differ in their markings, but the ring plovers nearly agree. We conversed with some sportsmen of Cape May, who asserted that they were acquainted with these birds, and that they sometimes made their appearance in flocks of considerable numbers; others had no knowledge of them. That the species is rare we were well convinced, as we had diligently explored the shore of a considerable part of Cape May, in the vicinity of Great Egg Harbour, many times at different seasons, and had never seen them before. How long they remain on our coast, and where they winter, we are unable to say. From the circumstance of the oviduct of the female being greatly enlarged, and containing an egg half grown, apparently within a week of being ready for exclusion, we concluded that they breed there. Their favourite places of resort appear to be the dry sand flats on the sea-shore. They utter an agreeable piping note.

This species is seven inches and three-quarters in length, and fifteen and a half in extent; the bill is black, stout, and an inch long, the upper mandible projecting considerably over the lower; front, white, passing on each side to the middle of the eye above, and bounded by a band of black of equal breadth; lores, black; eyelids, white; eye, large and dark; from the middle of the eye backwards the stripe of white becomes duller, and extends for half an inch; the crown, hind head, and auriculars, are drab olive; the chin, throat, and sides of the neck, for an inch, pure white, passing quite round the neck, and narrowing to a point behind; the upper breast below is marked with a broad band of jet black; the rest of the lower parts, pure white; upper parts, pale olive drab; along the edges of the auriculars and hind head, the plumage, where it joins the white, is stained with raw terra sienna; all the plumage is

darkest in the centre; the tertials are fully longer than the
primaries, the latter brownish black, the shafts and edges of
some of the middle ones, white; secondaries and greater coverts,
slightly tipt with white; the legs are of a pale flesh colour;
toes bordered with a narrow edge; claws and ends of the toes,
black; the tail is even, a very little longer than the wings, and
of a blackish olive colour, with the exception of the two exterior feathers, which are whitish; but generally the two middle
ones only are seen.

The female differs in having no black on the forehead, lores,
or breast, those parts being pale olive.

DARTER, OR SNAKE-BIRD.* (*Plotus anhinga.*)

PLATE LXXIV.—FIG. 1, MALE.

Plotus anhinga, *Linn. Syst.* ed. 12, tom. i. p. 218.—*Gmel. Syst.* i. p. 580, 1.—
   Ind. Orn. p. 895, 1.—Plotus melanogaster, *Id.* p. 896, var. B, var. C.—An-
   hinga Brasiliensis Tupinamb. *Marcgrav. Hist. Nat. Bras.* p. 218.—L'Anhinga,
   Briss. vi. p. 476.—*Salerne,* p. 375.—*Buff. Ois.* viii. p. 448.—Anhinga Noir de
   Cayenne, *Pl. enl.* 960.—White-bellied Darter, *Lath. Gen. Syn.* iii. p. 622, 1.
   —Black-bellied Darter, *Id.* p. 624, var. A, pl. 106.—*Id.* p. 625, var. B.—
   Colymbus colubrinus, Snake-bird, *Bartram,* p. 132, 295.—*Peale's Museum,*
   No. 3188, male.

PLOTUS ANHINGA.—LINNÆUS.†

Plotus anhinga, *Bonap. Synop.* p. 411.—Plotus melanogaster, *Ord,* 1st edit. of
   Supp. p. 79.

HEAD, neck, whole body, above and below, of a deep shining
black, with a green gloss, the plumage extremely soft and

---

* Named in the plate, black-bellied darter.

† This very curious genus contains only two known species—that of
our author, common to both continents of America, and the *Plotus
Vaillantii* of Temminck, a native of India, Africa, and the South Seas.
It has been placed among the *Pelicanidæ* by most ornithologists; but
how far all the forms which are at present included in that family have
a right to be there, I am not at present prepared to determine: if they
are, that of *Plotus* will hold a very intermediate rank, particularly in
habits which may lead to some discoveries in the relations to each other.
The economy is in a considerable measure arboreal, and in their own family,
as now constituted, they show the greatest development of the power

Drawn from Nature by A. Wilson.

Engraved by W. H. Lizars.

1. Black-bellied Darter.   2. Female D.   3. Great Northern Diver.   4. Black-headed Gull.   5. Little Auk.

74.

agreeable to the touch ; the commencement of the back is ornamented with small, oblong, ashy white spots, which pass

of diving, and activity in the water. They show also the extreme structure in the power of darting and suddenly again withdrawing their head. The cormorants and herons possess this power to a great extent, and they all possess a peculiar bend of the neck, observed in certain circumstances of the bird's economy, and into which that part at once puts itself when the bird is dead. This is produced chiefly by the action of two muscles ; the one inserted within the cavity of the breast, and running up with a long tendon to the vertebra beneath the bend ; the other inserted in the joint above the bend, and running far down with another slender tendon. The action of these two powers, resisted by the muscles on the back part, produce the peculiar angular bend, and enable the head to be thrown forward with great force. The effect may be easily seen, and produced, by a jointed stick having cords affixed, and acted on in this way. We may here introduce the genera *Pelicanus, Phalacracorax, Tachypetes, Sula,* and *Heliornis,* with a short notice of the species of America, as pointed out by the ornithologists who have described the productions of that country.

Pelicanus, Linn.

1. *P. onocrotalus.*—White Pelican.—According to Bonaparte, rare and accidental on the coasts of the middle States, and said by Dr Richardson to be numerous in the interior of the Fur Countries, up to the 61st deg. parallel.
2. *P. fuscus,* Linnæus.—Brown Pelican.—Common in the southern States, where it breeds.

Phalacracorax, Briss.

The species of this genus amount to a considerable number, and are distributed over the known world, but there yet exists confusion among them, from the near alliance of many to each other. The Prince of Musignano seems to have taken the authority of Dumont for the species he enumerates. They are as follow :—

1. *P. carbo.*—Cormorant of Wilson's List.—Tail of fourteen feathers; rare and migratory in the United States.
2. *P. graculus.*—Tail, twelve feathers; not uncommon in spring and autumn in the middle States; very common in Florida, where it breeds ; though very abundant in the arctic and antarctic circles.
3. *P. cristatus.*—Rather rare, and found during winter only in the United States.
4. *P. pygmœus.*—Inhabiting the north of both continents.
5. *P. Africanus.*—Inhabiting Africa and America; not found in Europe.

The Prince of Musignano is doubtful whether the two last are entitled to any place in the ornithology of America, the specimens which he has seen of both being only *reported* to have been killed in that country.

down the shoulders, increasing in size according to the size of the feathers, and running down the scapulars; wings and tail of a shining black, the latter broadly tipt with dirty white; the lesser coverts are glossed with green, and are spotted with ashy white; the last row of the lesser coverts, and the coverts of the secondaries, are chiefly ashy white, which forms a large bar across the wing; the outer web of the large scapulars is crimped; tail, rounded, the two under feathers the shortest; the two upper feathers, for the greater part of their length, beautifully crimped on their outer webs, the two next feathers in a slight degree so; bill, dusky at the base and above; the upper mandible brownish yellow at the sides, the lower man-

He mentions also another inhabiting the United States, which he has not examined, but thinks may turn out *P. Brazilianus.*

The first four species are common to Europe and America; the three first are also British. In addition to these, Mr Swainson has described another in the "Northern Zoology," under the title *dilophus*, or double-crested cormorant, which he cannot reconcile to any of these already described. His characters are :—"Tail of twelve feathers; bill, three inches and a half long; a crested tuft of feathers behind each eye."

Tachypetes, Vieill.

1. *T. aquilus*, Vieill.—Not uncommon during summer on the coasts of the United States, as far south as Carolina.

Sula, Briss.

1. *Sula Bassana*, Briss.—Common during summer over the coasts of the United States, especially the southern.
2. *L. fusca*, Briss.—Booby.—Common in summer on the coasts of the southern States.

Heliornis, Bonat.

1. *H. Surinamensis*, Surinam Heliornis.—An accidental visitant in summer in the middle States.

I have introduced *Heliornis* here, but without at all placing it in this station from my own opinion of its real place; the form of the birds contained in it (amounting yet to only two species) is very curious, and though showing the form of the body, and, according to Bonaparte, of the skeleton of *Plotus*, yet the habits are much more that of the *Grebes*. This agrees with the arrangement by the Prince of Musignano in one range, but I do not so easily see its connection in the opposite direction with *Phaeton* and *Sula*, the immediately preceding genera.—ED.

dible yellow ochre ; inside of the mouth, dusky ; irides, dark crimson ; the orbit of the eye, next to the plumage of the head, is of a greenish blue colour, this passes round, in the form of a zigzag band, across the front,—the next colour is black, which entirely surrounds the eye ; eyelids, of a bright azure, running into violet next the eyeball ; lores, greenish blue ; naked skin in front, black ; jugular pouch, jet black ; hind head, subcrested ; along the sides of the neck there runs a line of loose unwebbed feathers of a dingy ash colour, resembling the plumage of callow young ; here and there, on the upper part of the neck, one perceives a feather of the same ; on the forehead there is a small knob or protuberance ; the neck, near its centre, takes a singular bend, in order to enable the bird to dart forward its bill with velocity when it takes its prey ; legs and feet of a yellowish clay colour, the toes and hind part of the legs with a dash of dusky ; claws greatly falcated ; when the wings are closed they extend to the centre of the tail.

Length, from the tip of the bill to the end of the tail, two feet ten inches,* breadth three feet ten inches ; bill to the angle of the mouth, full four inches ; tail, ten inches and a half, composed of twelve broad and stiff feathers ; weight, three pounds and a half.

The serratures of the bill are extremely sharp, so much so, that when one applies tow, or such like substance, to the bird's mouth, it is with difficulty disengaged.

The lower mandible and throat, as in the divers, are capable of great expansion to facilitate the swallowing of fish, which constitute the food of this species. The position of these birds, when standing, is like that of the gannets.

The above description was taken from a fine adult male

---

* The admeasurement of the specimen described in the first edition of the ninth volume was made by Wilson himself from the stuffed bird in Peale's Museum. It differs considerably from that described above ; but as our specimen was a very fine one, there is room to conjecture that there was some error in the admeasurement of the former, ours being described immediately after death.

specimen, which was shot by my fellow-traveller Mr T. Peale, on the 1st of March 1818, in a creek below the Cow Ford, situated on the river St John, in East Florida. We saw some others in the vicinity, but owing to their extreme vigilance and shyness, we could not procure them.

From the description of the white-bellied darter of Latham and others, which is unquestionably this species, one would be inclined to conjecture that the bird figured as the female is the young male. But this point it is not in my power to ascertain. All the darters which I saw while in Florida were males.

The snake-bird is an inhabitant of the Carolinas, Georgia, the Floridas, and Louisiana, and is common in Cayenne and Brazil. It seems to have derived its name from the singular form of its head and neck, which at a distance might be mistaken for a serpent. In those countries where noxious animals abound, we may readily conceive that the appearance of this bird, extending its slender neck through the foliage of a tree, would tend to startle the wary traveller, whose imagination had portrayed objects of danger lurking in every thicket. Its habits, too, while in the water, have not a little contributed to its name. It generally swims with its body immerged, especially when apprehensive of danger, its long neck extended above the surface, and vibrating in a peculiar manner. The first individual that I saw in Florida was sneaking away, to avoid me, along the shore of a reedy marsh which was lined with alligators, and the first impression on my mind was that I beheld a snake, but the recollection of the habits of the bird soon undeceived me. On approaching it, it gradually sank, and my next view of it was at many fathoms' distance, its head merely out of the water. To pursue these birds at such times is useless, as they cannot be induced to rise, or even expose their bodies.

Wherever the limbs of a tree project over, and dip into the water, there the darters are sure to be found, these situations being convenient resting-places for the purpose of sunning and preening themselves, and probably giving them a better opportunity than when swimming of observing their finny prey. They crawl from the water upon the limbs, and fix themselves in an upright position, which they maintain in the utmost silence. If there be foliage, or the long moss, they secrete themselves in it in such a manner that they cannot be perceived, unless one be close to them. When approached, they drop into the water with such surprising skill, that one is astonished how so large a body can plunge with so little noise, the agitation of the water being apparently not greater than that occasioned by the gliding of an eel.

Formerly the darter was considered by voyagers as an anomalous production, a monster partaking of the nature of the snake and the duck ; and, in some ancient charts which I have seen, it is delineated in all the extravagance of fiction.

From Mr William Bartram we have received the following account of the subject of our history :—

" Here is in this river,* and in the waters all over Florida, a very curious and handsome bird,—the people call them snake-birds ; I think I have seen paintings of them on the Chinese screens and other Indian pictures ; they seem to be a species of *Colymbus*, but far more beautiful and delicately formed than any other that I have ever seen. They delight to sit in little peaceable communities, on the dry limbs of trees, hanging over the still waters, with their wings and tails expanded, I suppose to cool and air themselves, when at the same time they behold their images in the watery mirror. At such times, when we approach them, they drop off the limbs into the water as if dead, and for a minute or two are not to be seen ; when on a sudden, at a great distance, their long slender head and neck appear, like a snake rising erect out of the water ; and no other part of them is to be seen when swimming, except sometimes the tip end of their tail. In the heat of the day they are seen in great numbers, sailing very high in the air over lakes and rivers.

" I doubt not but if this bird had been an inhabitant of the Tiber in Ovid's days, it would have furnished him with a sub-

* The river St Juan, East Florida.

ject for some beautiful and entertaining metamorphoses. I believe they feed entirely on fish, for their flesh smells and tastes intolerably strong of it : it is scarcely to be eaten, unless one is constrained by insufferable hunger. They inhabit the waters of Cape Fear river, and, southerly, East and West Florida."*

FEMALE DARTER, OR SNAKE-BIRD.†

PLATE LXXIV.—Fig. 2.

Anhinga de Cayenne, *Pl. enl.* 959.—*Peale's Museum*, No. 3189, female.

PLOTUS ANHINGA.—Linnæus.

The female darter measures three feet five inches in length, and differs in having the neck before of a roan colour, or iron grey ; the breast the same, but lighter, and tinged with pale chestnut ; the belly as in the male ; where the iron grey joins the black on the belly there is a narrow band of chestnut ; upper head and back of the neck, dark sooty brown, streaked with blackish ; cheeks and chin, pale yellow ochre ; in every other respect the same as the male, except in having only a few slight tufts of hair along the side of the neck ; the tail is twelve inches long to its insertion, generally spread out like a fan, and crimped like the other on the outer vanes of the middle feathers only.

The above is a description of the supposed female darter which is preserved in Peale's Museum.

The author having written to Mr John Abbott of Georgia relative to this species, and some others, received from this distinguished naturalist a valuable communication, from which the following extract is made :—" Both the darters I esteem as but one species. I have now by me a drawing of the male or black-bellied only, but have had specimens of both at the same time. I remember that the upper parts of the female were similar to those of the male, except that the colour and

* Bartram's Travels, p. 132.—MS. in the possession of the author [Mr Ord].

† This article was written by Mr Ord.

markings were not so pure and distinct ; length, thirty-six inches, extent, forty-six. These birds frequent the ponds, rivers, and creeks during the summer ; build in the trees of the swamps, and those of the islands in the ponds ; they construct their nests of sticks ; eggs of a sky-blue colour. I inspected a nest, which was not very large ; it contained two eggs and six young ones, the latter varying much in size ; they will occupy the same tree for a series of years. They commonly sit on a stump which rises out of the water in the mornings of the spring, and spread their wings to the sun, from which circumstance they have obtained the appellation of sun-birds. They are difficult to be shot when swimming, in consequence of only their heads being above the water."

Never having seen a specimen of the black-bellied darter of Senegal and Java, I cannot give an opinion touching its identity with ours.

GREAT NORTHERN DIVER, OR LOON.
(*Colymbus glacialis.*)

PLATE LXXIV.—Fig. 3.*

Colymbus glacialis, *Linn. Syst.* ed. 12, tom. i. p. 221, 5.—C. immer, *Id.* p. 222, No. 6.—*Ind. Orn.* p. 799, 1.—C. immer, *Id.* p. 800, 2.—Le Grand Plongeon, *Briss.* vi. p. 105, pl. 10, fig. 1.—Le Grand Plongeon Tacheté, *Id.* p. 120, pl. 11, fig. 2.—Le Grand Plongeon, *Buff. Ois.* viii. p. 251.—L'Imbrim, ou Grand Plongeon de la Mer du Nord, *Id.* p. 258, tab. 22.—*Pl. enl.* 952.—Northern Diver, *Lath. Gen. Syn.* iii. p. 337.—Imber Diver, *Id.* p. 340.—*Penn. Brit. Zool.* No. 237, 238.—*Arct. Zool.* No. 439, 440.—*Bewick*, ii. p. 168, 170.—*Montagu, Orn. Dict. Supp. App.*—*Low, Fauna Orcadensis*, p. 108, 110.—Plongeon Imbrim, *Temm. Man. d'Orn.* p. 910.—*Peale's Museum*, No. 3262, male and young ; 3263, female.

COLYMBUS GLACIALIS.—Linnæus.†

Colymbus glacialis, *Bonap. Synop.* p. 420.—*Flem. Brit. Anim.* p. 132.—*North. Zool.* ii. p. 474.

THIS bird in Pennsylvania is migratory. In the autumn it makes its appearance with the various feathered tribes that

* This article is by Mr Ord.
† The genus *Colymbus*, or the loons, have been restricted to those

frequent our waters ; and, when the streams are obstructed with ice, it departs for the southern States.*   In the months

large divers, of which our present species will point out a good example. They are all birds of a large size, truly aquatic, are seldom on land except during incubation, and though endowed with a considerable power, seldom fly, unless very much pressed by necessity.   The great northern diver is very frequent in the Firth of Forth, and there I have never been able either to make up with, or cause one to fly from the sea.   I have pursued this bird in a Newhaven fishing-boat with four sturdy rowers, and, notwithstanding it was kept almost constantly under water, by firing as soon as it appeared, the boat could not succeed in making one yard upon it.   They are sometimes caught in the herring-nets, and at set lines, when diving.

The loons and guillemots approach very near in their characters, except in lesser size, and a particular modification of habit in the one preferring the sea-shores or the reedy banks of inland lakes for breeding-places, while the others are gregarious, and choose the most precipitous cliffs on the sea, and deposit their eggs, without the least preparation, on the bare rock.   The construction of the feet and tarse at once points out in the large birds their great facility of diving, and rapid progression under water, the proportional expanse of web is much greater, and the form of it runs into that of *Phalacracorax* and *Sula ;* the legs are placed very far back, and the muscles possess very great power ; the tarsus is flat-tened laterally, and thus presents a small surface of resistance, and the whole plumage of the bird is close and rigid, presenting a smooth and almost solid resistance in passing through the water.   The adults require at least the first season to attain maturity.   Dr Richardson mentions the following method of shooting them during the winter :—" They arrive in that season when the ice of the lakes continues entire, except, perhaps, a small basin of open water where a rivulet happens to flow in, or where the discharge of the lake takes place.   When the birds are observed to alight in these places, the hunter runs to the margin of the ice, they instantly dive, but are obliged, after a time, to come to the surface to breathe, when he has an opportunity of shooting them.   In this way upwards of twenty were killed at Fort Enterprise in the spring of 1821, in a piece of water only a few yards square."

The present species is the only one described in Wilson's volumes as a native of America.   Bonaparte mentions two others, which are also described in the " Northern Zoology,"—the black-throated diver, *Colymbus Arcticus,* common in Arctic America, but rare, and only found

---

* The loon is said to winter in the Chesapeake Bay.

of March and April it is again seen, and, after lingering awhile, it leaves us for the purpose of breeding. The loons are found along the coast as well as in the interior ; but in the summer they retire to the fresh-water lakes and ponds. We have never heard that they breed in Pennsylvania, but it is said they do in Missibisci pond, near Boston, Massachusetts. The female lays two large brownish eggs. They are commonly seen in pairs, and procure their food, which is fish, in the deepest water of our rivers, diving after it, and continuing under for a length of time. Being a wary bird, it is seldom they are killed, eluding their pursuers by their astonishing faculty of diving. They seem averse from flying, and are but seldom seen on the wing. They are never eaten.

The loon is restless before a storm ; and an experienced master of a coasting vessel informed me that he always knew when a tempest was approaching by the cry of this bird, which is very shrill, and may be heard at the distance of a mile or more. The correctness of this observation I have myself since experienced in a winter voyage on the southern coasts of the United States.

during winter, in the middle States ; and *Colymbus septentrionalis*, red-throated diver : all are common also to Europe and Great Britain.

The vast lakes and rivers of America, and her interminable swamps, would seem proper nurseries for another family, the grebes ; and their recluse yet active aquatic manners must either have yet prevented the discovery of more species, or this form is comparatively wanting to that division of the world. Two species only are mentioned in Wilson's "History," and Bonaparte adds other two. They are as follows, from that gentleman's "Synopsis :"—

Podiceps.

1. *P. cristatus*, Lath.—Crested grebe of Wilson's List ; rare in the middle States, and only during winter common in the interior and on the lakes.
2. *P. rubricollis*, Lath.—Rare, and during winter only, in the middle States ; very common in Arctic America,
3. *P. cornutus*, Lath.—Common during winter, the young especially, in the middle States.
4. *P. Carolinensis*, Lath.—Little grebe, of Wilson's List ; inhabits the whole continent of America, not extending far to the north. Common from Canada to Louisiana, migrating in the middle States.—Ed.

This species seldom visits the shores of Britain, except in very severe winters ; but it is met with in the north of Europe, and spreads along the arctic coast as far as the mouth of the river Ob, in the dominions of Russia.   It is found about Spitz-bergen, Iceland, and Hudson's Bay.   Makes its nest, in the more northern regions, on the little isles of fresh-water lakes : every pair keep a lake to themselves.   It sees well, flies very high, and, darting obliquely, falls secure into its nest.   Appears in Greenland in April or the beginning of May, and goes away in September or October, on the first fall of snow.*   It is also found at Nootka Sound † and Kamtschatka.

The Barabinzians, a nation situated between the river ¡Ob and the Irtisch, in the Russian dominions, tan the breasts of this and other waterfowl, whose skins they prepare in such a manner as to preserve the down upon them, and, sewing a number of these together, they sell them to make pelisses, caps, &c.   Garments made of these are very warm, never imbibing the least moisture, and are more lasting than could be imagined. ‡

The natives of Greenland use the skins for clothing, and the Indians about Hudson's Bay adorn their heads with cir-clets of their feathers. §

Lewis and Clark's party, at the mouth of the Columbia, saw robes made of the skins of loons,‖ and abundance of these birds during the time that they wintered at Fort Clatsop on that river.¶

The Laplanders, according to Regnard, cover their heads with a cap made of the skin of a loom (loon), which word signifies, in their language, lame, because the bird cannot walk well.   They place it on their head in such a manner that the bird's head falls over their brow, and its wings cover their ears.

"Northern divers," says Hearne, "though common in

* Pennant.   † Cook's Last Voyage, vol. ii. p. 237, Am. ed.
‡ Latham.   § Arctic Zoology.   ‖ Gass's Journal.
¶ History of the Expedition, vol. ii. p. 189.

Hudson's Bay, are by no means plentiful; they are seldom found near the coast, but more frequently in fresh-water lakes, and usually in pairs. They build their nests at the edge of small islands, or the margins of lakes or ponds; they lay only two eggs, and it is very common to find only one pair and their young in one sheet of water; a great proof of their aversion to society. They are known in Hudson's Bay by the name of loons." *

The great northern diver measures two feet ten inches from the tip of the bill to the end of the tail, and four feet six inches in breadth; the bill is strong, of a glossy black, and four inches and three-quarters long to the corner of the mouth; the edges of the bill do not fit exactly into each other, and are ragged, the lower mandible separates into two branches, which are united by a thin elastic membrane, and are easily movable horizontally, or receding from each other, so as to form a wider gap to facilitate the swallowing of large fish; tongue, bifid; irides, dark blood red; the head and half of the length of the neck are of a deep black with a green gloss, and purple reflections; this is succeeded by a band consisting of interrupted white and black lateral stripes, which encompasses the neck, and tapers to a point on its fore part, without joining,—this band measures about an inch and a half in its widest part, and, to appearance, is not continuous on the back part of the neck, being concealed by some thick, overhanging, black feathers, but, on separating the latter, the band becomes visible: the feathers which form these narrow stripes are white, streaked down their centre with black, and, what is a remarkable peculiarity, their webs project above the common surface; below this, a broad band of dark glossy green and violet, which is blended behind with the plumage of the back; the lower part of the neck, and the sides of the breast, are ribbed in the same manner as the band above; below the chin, a few stripes of the same; the whole of the upper parts are of a deep black, slightly glossed with green, and thickly spotted with white, in

* Hearne's Journey, p. 429, 4to.

regular transverse or semicircular rows, two spots on the end
of each feather—those on the upper part of the back, shoulders,
rump, and tail-coverts, small and roundish, those on the
centre of the back, square and larger ; those on the scapulars
are the largest, and of an oblong square shape ; the wing-
feathers and tail are plain brown black, the latter composed
of twenty feathers ; the lower parts are pure white, a slight
dusky line across the vent ; the scapulars descend over the
wing when closed, and the belly feathers ascend so as to meet
them, by which means every part of the wing is concealed,
except towards the tip ; the outside of the legs and feet is
black, inside, lead colour ; the leg is four inches in length, and
the foot measures, along the exterior toe to the tip of its claw,
four inches and three-quarters ; both legs and feet are marked
with five-sided polygons ; weight of the specimen described,
eight pounds and a half.

The female diver is somewhat less than the male ; the bill
is yellowish ; crown, back part of the neck, and whole upper
parts, pale brown ; the plumage of part of the back and scapulars is tipt with pale ash ; the throat, lower side of the neck,
and whole under parts, are white, but not so pure as that of
the male, having a yellowish tinge ; the quill-feathers, dark
brown. She has no appearance of bands on her neck, or of
spots on her body.

The young males do not obtain their perfect plumage until
the second or third year. One which we saw, and which was
conjectured to be a yearling, had some resemblance to the
female, with the exception of its upper parts being of a darker
and purer brown, or mouse colour, and its under parts of a
more delicate white ; it had likewise a few spots on the back
and scapulars ; but none of those markings on the neck which
distinguish the full-grown male.

The conformation of the ribs and bones of this species is
remarkable, and merits particular examination.

In the account which some of the European ornithologists
give of their northern diver, we presume there is an inaccu-

racy. They say it measures three feet six inches in length, and four feet eight in breadth, and weighs sixteen pounds. If this be a correct statement, it would lead to the surmise that our diver is a different species; for, of several specimens which we examined, the best and largest has been described for this work, the admeasurement of which bird comes considerably short of that of the European mentioned above. The weight, as has been stated, was eight pounds and a half.

According to Temminck, the adult male and female are alike in plumage. All the females which have passed under my examination differed from the old males; and it is the universal opinion among our sportsmen who reside on the coast, where the loons are common, that the adults of both sexes may always be distinguished by their garb. However, in confirmation of Temminck's opinion, I can adduce the authority of the Prince of Musignano, Charles Lucian Bonaparte, who has informed me that he has in his collection a female which was shot in the Delaware, and which differs in no respect from the adult male.

On a re-examination of the " Supplement to the Ornithological Dictionary " of Montagu, I find upon this subject the following remarks, which should seem to put the question at rest respecting the identity of the European and American species :—" It should appear that the size of this species has been commonly exaggerated, or they must vary very materially, since those which have come under our examination did not exceed ten pounds; and an old or matured male measured only two feet eight inches. A young female, before the plumage was perfected, weighed eight pounds six ounces, and measured two feet seven inches in length.

" A northern diver, taken alive, was kept in a pond for some months, which gave us an opportunity of attending to its manners. In a few days it became extremely docile, would come at the call from one side of the pond to the other, and would take food from the hand. The bird had received an injury in the head, which had deprived one eye of its sight, and the other was a little impaired; but, notwithstanding, it could, by incessantly diving, discover all the fish that was thrown into the pond. In defect of fish, it would eat flesh.

"It is observable that the legs of this bird are so constructed and situated as to render it incapable of walking upon them. This is probably the case with all the divers, as well as the grebes.

"When this bird quitted the water, it shoved its body along upon the ground, like a seal, by jerks, rubbing the breast against the ground, and it returned again to the water in a similar manner. In swimming and diving, only the legs are used, and not the wings, as in the guillemot and auk tribes; and by their situation so far behind, and their little deviation from the line of the body, the bird is enabled to propel itself in the water with great velocity, in a straight line, as well as turn with astonishing quickness."

LAUGHING GULL.* (*Larus atricilla.*)

PLATE LXXIV.—FIG. 4.

Larus atricilla, *Linn. Syst.* ed. 10, tom. i. p. 136, 5.—*Gmel. Syst.* i. p. 600, 8.— *Ind. Orn.* p. 813, 4.—Laughing Gull, *Catesby,* i. pl. 89.—*Lath. Gen. Syn.* iii. p. 383, 12.—*Arct. Zool.* No. 454.—La Mouette Rieuse, *Briss.* vi. p. 192, 13, pl. 18, fig. 1.—Mouette à capuchon plombé, *Temm. Man. d'Orn.* p. 779.— *Peale's Museum,* No. 3881.

LARUS ATRICILLA.—Linnæus.†

Larus ridibundus, *Ord.* 1st edit. of *Supp.* p. 89.—Larus atricilla, *Bonap. Synop.* p. 359.

LENGTH, seventeen inches; extent, three feet six inches; bill, thighs, legs, feet, sides of the mouth, and eyelids, dark blood

* Named in the plate, black-headed gull.

† This gull is the only one figured by Wilson, though several are mentioned in his list, and, no doubt, had he survived to complete his great undertaking, many others would have been both added and figured. I have introduced a short description of those which have

red; inside of the mouth, vermilion; bill, nearly two inches and a half long; the nostril is placed rather low; the eyes are

been since noticed by writers on arctic and northern zoology, but any observations will be confined, for the present, to the form now before us, perhaps more familiar in the black-headed gull of Britain.

The gulls are distributed over the whole world, and present various forms. They are mostly, however, of graceful appearance, and perform their motions with ease and lightness; their plumage is often of snowy whiteness, or tinged with a pale blush, adding to its delicacy. By the poets they are employed as emblems of purity, when riding buoyantly on the waves, and weaving a sportive dance, or as accessories to the horrors of a storm, by their shrieks and wild piercing cries. In their manners they are the vultures of the ocean, feed indiscriminately on fish or on carrion, and frequently attack birds of inferior power. A dead horse, newly cast upon the beach, will present a picture little inferior to that drawn by Audubon of the American vultures on the discovery of some putrid carcass.

Our present bird will rank under the genus *Xema* of Bojè, which will contain those of swallow-like form, apparently both a natural and well-defined group. They are not so truly pelagic as many of the other forms —ascend the course of rivers in search of food, and breed by the inland lochs or marshes—are extremely clamorous and intrepid in defence of their young, but during winter are one of the most shy and wary. They undergo an annual change of plumage during the breeding season, obtaining the whole or part of the head of a dark and decided colour from the rest of the body, generally shades of deep and rich brown or grey; in winter this entirely disappears, and is succeeded by pure white, except on the auriculars, which retain a trace of the darker shade. They feed on fish and insects, and some follow the plough in search of what it may turn up. In fishing, they exhibit occasionally the same manner of seizing their prey as the terns, hovering above, and striking it under water with the wings closed.

The species which are noticed by the Prince of Musignano and the authors of the "Northern Zoology," as inhabiting North America, are—

1. *L. Sabinii* (*Xema Sabinii*, Leach).—Discovered by Captain Edward Sabine, breeding in company with the arctic tern, on the west coast of Greenland; they seem confined to high latitudes.
2. *Larus minutus*, Pall.—Inhabiting the north, but seldom seen in the United States.
3. *Larus capistratus*, Temm.—Inhabiting the north, and not very rare during autumn on the Delaware and Chesapeake, and found as far inland as Trenton. These will all rank in *Xema*, and Swainson and Richardson have described two under the titles of *L. Franklinii* and *L. Bonapartii*. These gentlemen seem to think that the American *L. atricilla* is confounded with Temminck's

black; above and below each eye there is a spot of white; the
head and part of the neck are black; remainder of the neck,

atricilla, and that they embrace two species. I have added the descriptions
from Dr Richardson and Mr Swainson's notes in their own words. I have
no means at present of deciding this point.

4. *L. Franklinii*, Swain. and Richards.—Franklin's rosy gull, with vermilion
bill and feet; mantle, pearl grey; five exterior quills broadly barred with
black, the first one tipped with white for an inch; tarsus, twenty lines
long; hood, black in summer.

"This is a very common gull in the interior of the Fur Countries, where it frequents the shores of the larger lakes. It is generally seen in flocks, and is very
noisy. It breeds in marshy places. Ord's description of his black-headed gull
(Wilson, vol. ix. p. 89—present edition, vol. iii. p. 114) corresponds with our
specimens, except that the conspicuous white end of the first quill is not noticed:
the figure (pl. 74, fig. 4) differs in the primaries being entirely black.* The
Prince of Musignano gives the totally black primaries, and a tarsus nearly two
inches long, as part of the specific character of his *L. atricilla*, to which he refers
Wilson's bird; though, in his "Observations," he states that the adult specimens
have the primaries, with the exception of the first and second, tipped with white.
L. Franklinii cannot be referred either to the *L. atricilla* or *L. melanocephalus*
of M. Temminck: the first has a lead-coloured hood, and deep black quill-feathers,
untipped by white; and the black hood of the second does not descend lower on
the throat than on the nape; its quill-feathers are also differently marked, and
its tarsus is longer. His *L. ridibundus* and *capistratus* have brown heads, and
the interior of the wings grey; the latter has also a much smaller bill than our
L. Franklinii."

5. *L. Bonapartii*, Swain. and Richards.—Bonapartian Gull.—*North. Zool.* ii.
p. 425.—"With a black bill; the mouth and feet, carmine red; wings,
bordered with white anteriorly; posteriorly, together with the back, pearl
grey; six exterior quills, black at the end, slightly tipped with white; the
first quill entirely black exteriorly; tarsus, scarcely an inch and a half
long; head, greyish black in summer.

"This handsome small gull is common in all parts of the Fur Countries, where
it associates with the terns, and is distinguished by its peculiar shrill and
plaintive cry. The *L. capistratus* of the Prince of Musignano differs, according to his description, in the first quill being white exteriorly, pale ash
interiorly; in the light brown colour of its head, and in its tail being slightly
emarginated; while the tail of *L. Bonapartii* is more inclined to be rounded
laterally than notched in the middle."

6. *L. roseus*, Macgilliv.—A rare species confined to high latitudes, discovered
during Sir Edward Parry's second voyage, when two specimens were obtained; the one is now in the Edinburgh Museum, the other was presented
to Mr Sabine, whose collection has been lately sold to the Andersonian
Museum in Glasgow.

---

* "Four American specimens of *L. atricilla* are now before me. It is a larger
and a totally different species. The three outer quills are wholly black; the
fourth tipped for about one inch, and the fifth for half an inch, with black; the
extreme white spot at the point of the five first quills is very small in some, and
not seen in adult specimens, having these feathers worn."—Sw.

breast, whole lower parts, tail-coverts and tail, pure white ;
the scapulars, wing-coverts, and whole upper parts are of a

7. *L. tridactylus*, Linn.—Kittiewake, Wilson's List.—Inhabiting both continents.

8. *L. canus*, Linn.—Common Gull, Wilson's List.—Inhabiting both continents, and numerous during winter in the middle States of America.

9. *L. eburneus*, Gmel.—Inhabits the arctic circle ; migrating occasionally
to the temperate regions. A few specimens have been killed in Britain.

10. *L. fuscus*, Linn.—Very common during winter near Philadelphia and
New York.

11. *L. argentatoides*, Brehm.—This bird is separated from *Larus argentatus*
by Bonaparte, who mentions having shot it on the southern coasts of
England. At the same time that he separates it from the herring-gull,
he expresses a doubt of its being the *L. argentatoides* of Brehm. This I
cannot at present decide, but have appended, without any abridgment, the
observations and description of a bird referred to this, from the "Northern
Zoology ;" it is very closely allied, at all events, to the *L. argentatus ;* and
it is of importance that the characters of a species said to be killed on our
coasts should be properly investigated.

Larus argentatoides.—Arctic Silvery Gull.

"*Larus argentatus*, Richards. *Append. Parry's Second Voy.*, p. 358, No. 22.—
Larus argentatoides, Bonap. Syn., No. 299.—*Novya*, Esquimaux.

"The Prince of Musignano has distinguished this gull from *Larus argentatus*,
with which it had been confounded by most other writers. It is impossible,
therefore, to separate its history, or to cite the descriptions of other authors
correctly. It was found breeding on Melville Peninsula ; and the eggs that were
brought home have an oil-green colour, marked with spots and blotches of blackish brown and subdued purplish grey. It preys much on fish, and is noted at
Hudson's Bay for robbing the nets set in the fresh-water lakes. I have seen no
specimens from Arctic America which I can unequivocally refer to the *Larus
argentatus*, as characterised by the Prince of Musignano."

Description of a Male in the Edinburgh Museum, killed on Melville Peninsula, June 29, 1822.

"Colour, mantle pearl grey. Six outer quills crossed by a brownish-black band,
which takes in nearly the whole of the first one, but becoming rapidly narrower
on the others, terminates in a spot near the tip of the sixth. The first quill has
a white tip an inch and a half long, marked interiorly with a brown spot; the
second has a round white spot on its inner web, and, together with the rest of the
quill-feathers, is tipped with white. Head, neck, rump, tail, and all the under
plumage, pure white. Bill, wine-yellow, with an orange-coloured spot near the
tip of the under mandible. Irides, primrose-yellow. Legs, flesh-coloured.

"FORM.—Bill moderately strong, compressed ; upper mandible, arched from
the nostrils ; nostrils, oblong oval ; wings, about an inch longer than the tail ;
thighs, naked for three-quarters of an inch ; hind toe, articulated rather high.

"The young have the upper plumage hair-brown, with reddish brown borders ;
the head and under plumage, grey, thickly spotted with pale brown ; the tail
mostly brown, tipped with white.

fine blue ash colour ; the first five primaries are black towards
their extremities ; the secondaries are tipt largely with white,

"DIMENSIONS.—Length, total, 23 inches ; of tail, 7 inches, 3 lin. ; of wing, 16
inches, 6 lin. ; of bill above, 2 inches ; of bill to rictus, 3 inches ; from nostrils to
tip, 11 lin. ; of nostrils, 4½ lin. ; of tarsus, 2 inches, 4½ lin. ; of middle toe, 2
inches, 1 lin. ; of middle nail, 5 lin. ; of inner toe, 1 inch, 6 lin. ; of inner nail, 4
lin. ; of hind toe, 3 lin. ; length of hind nail, 2½ lin.

"Six individuals, killed on Melville Peninsula in June, July, and September,
varied in total length from 23 to 25 inches, and in the length of their tarsi from
27 to 31 lines.

"Bonaparte thus gives the distinctive characters of the two species :—

"*L. argentatoides.*—Back and wings, bluish grey ; quills, black at the point,
tipped with white, reaching but little beyond the tail ; shafts, black ; first
primary, broadly white at tip ; second, with a round white spot besides ; tarsus,
less than two and a half inches ; nostrils, oval ; length, twenty inches.

"*L. argentatus.*—Mantle, bluish grey ; quills, black at the point, tipped with
white, reaching much beyond the tail ; shafts, black ; first primary only with a
white spot besides the narrow tip ; tarsus, nearly three inches ; nostrils, linear ;
length two feet. They are closely allied, and may at once be distinguished by
the size."

12. *L. argentatus*, Brunn.—Herring Gull, Wilson's List.—Common to both
continents, and not uncommon near New York and Philadelphia.

13. *L. leucopterus*, Faber.—Inhabiting the arctic circle, whence it migrates
in winter to the boreal regions of both continents, advancing farther south
in America : not rare in the northern and middle States.

14. *L. glaucus*, Brunn.—Inhabiting the arctic regions, and exceedingly rare
in the United States.

15. *L. marinus*, Linn.—Black-backed Gull, Wilson's List.—Not uncommon
during winter in the middle States.

16. *L. zonorhynchus*, Richards.—Ring-billed Mew Gull, a new species, described in "Northern Zoology."—Bill, ringed, rather longer than the tarsus,
which measures two and a half inches ; mantle, pearl grey ; ends of the
quills and their shafts, blackish ; a short white space on the two exterior
ones.

17. *L. bachyrhynchus*, Richards.—Short-billed Mew Gull.—Another species,
described as new in the "Northern Zoology."—From the description of the
present bird copied from that work, it will be seen that the authors themselves are not decided in their opinions as to the absolute distinction of this
and the preceding from *L. canus*, and I have placed them here for the
same reason that they are admitted into that valuable work. It is not
unlikely that they, or at least the same varieties, may be discovered on
our own coasts.

"Short-billed mew gull, with a short, thickish bill ; a tarsus scarcely two inches
long ; quills, not tipped with white ; a short white space on the two exterior ones,
and blackish shafts.

"Our specimen of this gull is a female, killed on the 23d of May 1826, at Great
Bear Lake. Some brown markings on the tertiaries, primary coverts, and bastard
wing, with an imperfect sub-terminal bar on the tail, point it out as a young bird,
most probably commencing its second spring. The rest of its plumage corresponds
with that of *L. zonorhynchus*, except that it wants the extreme white tips of the

and almost all the primaries slightly ; the bend of the wing
is white, and nearly three inches long ; the tail is almost even ;

quill-feathers, which, on the third and following ones, are very conspicuous in
L. zonorhynchus. It differs, however, remarkably, in its bill being shorter, though
considerably stouter, than that of our *L. canus ;* and, like it, it is wax-yellow,
with a bright yellow rictus and point. Its tarsus is nearly one-third shorter than
that of *L. zonorhynchus.* Many may be disposed to consider this and the preceding as merely local varieties of *L. canus ;* and it might be urged, in support of
this opinion, that there are considerable differences in the length and thickness
of the bills of individuals of the common and winter gulls killed on the English
coasts, which are all usually referred to *L. canus.* We have judged it advisable,
however, to call the attention of ornithologists to these American birds, by giving
them specific names, leaving it to future observation to determine whether they
ought to retain the rank of species, or be considered as mere varieties."—
RICHARDS.

In this place must be introduced the genus *Lestris* or *Skua,* of which
only one species was enumerated by Wilson in his List—the *L. cataractes,*
Illiger—the common *Skua gull* of British ornithologists. The Prince of
Musignano mentions, in addition, the now well-known European and
British species, *L. parasiticus* and *pomarinus,* another somewhat allied,
but not yet well distinguished, *L. Buffonii,* Bojè ; and a fifth species is
described as new in the "Northern Zoology," and is dedicated to Dr
Richardson—*L. Richardsonii.* It seems closely allied to *L. Buffonii,*
but the distinctions yet want clearness and confirmation. It was found
breeding in considerable numbers in the barren grounds at a distance
from the coast. The following are Bonaparte's characters of *L. Buffonii,*
by which it is alone known :—

Lestris Buffonii, Bojè.—Bill, one inch and a quarter from the front, straight,
notched ; middle tail-feathers, gradually tapering, narrow for several inches,
ending in a point ; tarsus, one inch and a half long, almost smooth. Adult,
brown ; neck and beneath, white, the former tinged with yellow. Young,
wholly brownish.

"Arctic bird, Edw. pl. 148 ; Buff. *Pl. enl.* 762 ; *Lestris crepidata,* Brehm."—
BONAP. *Syn.* No. 306.

And I add the observations of Mr Swainson regarding *L. Richardsonii :*—

" Richardson's jager, whole plumage, brown ; two middle tail-feathers, abruptly acuminated ; tarsi, black, twenty-two lines long.

"This specimen appears to us to be in full and mature plumage ; we cannot,
therefore, view it as the young, or even as the female, of the *Lestris Buffonii* of
Bojè, which we only know from the characters assigned to it by the Prince of
Musignano. According to this account, the *L. Buffonii* has the bill an inch and
a quarter long from the front ; ours is only an inch : the tarsi are described as
almost smooth, whereas in ours they are particularly rough. The adult, as
figured on plate 762 of the *Pl. Enl.,* has the chin, throat, and sides of the neck
quite white ; but, in our bird, these parts are of the same pure and decided tint
as that of the body, except that the ear-feathers, and a few lower down the neck,

it consists of twelve feathers, and its coverts reach within an inch and a half of its tip ; the wings extend two inches beyond the tail ; a delicate blush is perceivable on the breast and belly ; length of tarsus, two inches.

The head of the female is of a dark dusky slate colour ; in other respects she resembles the male.

In some individuals, the crown is of a dusky grey ; the upper part and sides of the neck of a lead colour ; the bill and legs of a dirty, dark purplish brown. Others have not the white spots above and below the eyes ; these are young birds.

The changes of plumage to which birds of this genus are subject have tended not a little to confound the naturalist ; and a considerable collision of opinion, arising from an imperfect acquaintance with the living subjects, has been the result. To investigate thoroughly their history, it is obviously necessary that the ornithologist should frequently explore their native haunts ; and, to determine the species of periodical or occasional visitors, an accurate comparative examination of many specimens, either alive or recently killed, is indispensable. Less confusion would arise among authors if they would occasionally abandon their accustomed walks—their studies and their museums—and seek correct knowledge in the only place where it is to be obtained—in the grand temple of Nature. As respects, in particular, the tribe under review, the zealous

have a slight tinge of ochre.* The tarsi also, in both the plates cited by the Prince, are coloured yellow. These differences, with the more important one exhibited in the feet, will not permit us to join these birds under one name. Another distinction, which must not be overlooked, is in the colour of the feet. Edwards expressly says of his ‘Arctic Bird’ (pl. 149, which much more resembles ours than that figured on the plate immediately preceding), that ‘the legs and toes are all yellow ;’ whereas, in our bird, these members are of a deep and shining black, while the hinder parts of the tarsi, toes, and connecting membrane, are particularly rough.”—Sw.

This jager breeds in considerable numbers in the barren grounds at a distance from the coast. It feeds on shelly molluscæ, which are plentiful in the small lakes of the Fur Countries, and it harasses the gulls in the same way with others of the genus.—Ed.

---

* The pure colour or uniform tint of the lower parts will not stand as characters ; in our native species they vary constantly.—Ed.

inquirer would find himself amply compensated for all his toil by observing these neat and clean birds coursing along the rivers and coast, enlivening the prospect by their airy movements, now skimming closely over the watery element, watching the motions of the surges, and now rising into the higher regions sporting with the winds, while he inhaled the invigorating breezes of the ocean, and listened to the soothing murmurs of its billows.

The laughing gull, known in America by the name of the black-headed gull, is one of the most beautiful and most sociable of its genus. They make their appearance on the coast of New Jersey in the latter part of April, and do not fail to give notice of their arrival by their familiarity and loquacity. The inhabitants treat them with the same indifference that they manifest towards all those harmless birds which do not minister either to their appetite or their avarice ; and hence the black-heads may be seen in companies around the farmhouse, coursing along the river-shores, gleaning up the refuse of the fisherman, and the animal substances left by the tide, or scattered over the marshes and newly-ploughed fields, regaling on the worms, insects, and their larvæ, which, in the vernal season, the bounty of Nature provides for the sustenance of myriads of the feathered race.

On the Jersey side of the Delaware Bay, in the neighbourhood of Fishing Creek, about the middle of May, the black-headed gulls assemble in great multitudes, to feed upon the remains of the king-crabs which the hogs have left, or upon the spawn which those curious animals deposit in the sand, and which is scattered along the shore by the waves. At such times, if any one approach to disturb them, the gulls will rise up in clouds, every individual squalling so loud, that the roar may be heard at the distance of two or three miles.

It is an interesting spectacle to behold this species when about recommencing their migrations. If the weather be calm, they will rise up in the air spirally, chattering all the while to each other in the most sprightly manner, their notes at such times resembling the singing of a hen, but far louder, changing often into a *haw, ha ha ha haw!* the last syllable lengthened out like the excessive laugh of a negro. When mounting and mingling together, like motes in the sunbeams, their black heads and wing-tips and snow-white plumage give them a very beautiful appearance. After gaining an immense height, they all move off, with one consent, in a direct line towards the point of their destination.

This bird breeds in the marshes. The eggs are three in number, of a dun clay colour, thinly marked with small irregular touches of a pale purple and pale brown ; some are of a deeper dun, with larger marks, and less tapering than others; the egg measures two inches and a quarter by one inch and a half.

The black-heads frequently penetrate into the interior, especially as far as Philadelphia ; but they seem to prefer the neighbourhood of the coast for the purpose of breeding. They retire southward early in autumn.

LITTLE GUILLEMOT. (*Uria alle.*)

PLATE LXXIV.—Fig. 5.

Uria alle, *Temm. Man. d'Orn.* p. 928.—Alca alle, *Linn. Syst.* ed. 12, tom. i. p. 211, 5.—*Gmel. Syst.* i. p. 554, 5.—*Ind. Orn.* p. 795, 10.—Uria minor, *Briss.* vi. p. 73, 2.—Le Petit Guillemot Femelle, *Pl. enl.* 917.—Small Black and White Diver, *Edwards*, pl. 91.—Little Auk, *Lath. Gen. Syn.* iii. p. 327.— *Penn. Arct. Zool.* No. 429.—*Bewick,* ii. p. 158.—*Peale's Museum,* No. 2978.

MERGULUS MELANOLEUCOS.—Ray.*

Mergulus melanoleucos, *Ray, Synop.* p. 125.—*Flem. Brit. Anim.* p. 135.—Uria (subgen. Mergulus) alle, *Bonap. Synop.* p. 425.—Little Auk, *Mont. Orn. Dict. and Supp.*—*Selby, Illust.* pl. lxxxi.—Uria alle, *North. Zool.* ii. p. 479.

Of the history of this little stranger, but few particulars are known. With us it is a very rare bird, and, when seen, it is

---

* I have chosen the name of Ray for this species, as both appropriate, and, as far as my inquiries have led me, entitled to the priority—and the difference in form from the *guillemots* fully entitles it to the rank of a subgenus. It is the only bird allied in any way to the auks, puffins,

generally in the vicinity of the sea. The specimen described was killed at Great Egg Harbour, in the month of December

&c., which has been figured by Wilson, though several forms occur in the northern seas, and have been pointed out by him, which may be now mentioned, but which will be hereafter figured from the remaining volumes of the " Continuation," by the Prince of Musignano, now in the press. I have therefore only added an enumeration from the " Synopsis" of that ornithologist, commencing with the guillemots, for which the genus *Uria* has been adopted ; by some the black guillemot is separated, on account of straightness of the mandibles, whereas in the common they are both bent at the tip. In our present state of knowledge I prefer retaining them together.

Uria, Briss.

1. *U. troile.*—Foolish Guillemot.—Common to both continents, and found during winter on the coasts of the United States.
2. *U. brunichii*, Sab.—Inhabits both continents, and is common in Davis' Straits, Baffin's Bay, &c. It has been said to have occurred once or twice on the British coasts.
3. *U. grylle.*—Black Guillemot.—Common to both continents, and found during winter along the coasts of the United States. A few pairs breed annually on the rocky islands on the Firth of Forth. I have repeatedly found them on the Isle of May.
4. *U. marmorata*, Lath.—Brown, undulated with chestnut ; beneath, dusky, spotted with white ; feet, orange ; bill, black, one inch long. Inhabits the north-western coasts of America, and the opposite shore of Asia.

These are the characters given by Bonaparte to the last bird. Will it not be the immature state of some other species ?

Phaleris, Temm.

1. *P. psittacula*, Temm.—Perrequet Auk.—Inhabits the north-western coasts of America, and the opposite ones of Asia. Common in Kamtschatka.
2. *P. christatella*, Temm.—Crested Auk.—The Prince of Musignano is only of opinion that this may be found on the western shores of America ; it is known in the Japan seas and the north-eastern coast of Africa. He thinks also that the *Alca antiqua* of Latham may prove a third North American species of *Phaleris*.

Another bird (*Phaleris cerorhinca*), entering formerly into this genus, has been separated by the Prince of Musignano, and placed in a subgenus, *Cerorhinca*, to be figured in his fourth volume.

Cerorhinca, Bonap.

1. *C. occidentalis*, Bonap.—Inhabits the western coasts of North America.

Mormon, Illig.

1. *M. cirrhatus*, Temm.—Tufted Auk, Lath.—Inhabits the sea between

1811, and was sent to Wilson as a great curiosity. It measured nine inches in length, and fourteen in extent ; the bill, upper part of the head, back, wings, and tail, were black ; the upper part of the breast and hind head were grey, or white mixed with ash ; the sides of the neck, whole lower parts, and tips of secondaries, were pure white ; feet and legs, black ; shins, pale flesh colour ; above each eye there was a small spot of white ;* the lower scapulars, streaked slightly with the same.

The little guillemot is said to be but a rare visitant of the British Isles. It is met with in various parts of the north, even as far as Spitzbergen ; is common in Greenland, in company with the black-billed auk, and feeds upon the same kind of food. The Greenlanders call it the ice-bird, from the circumstance of its being the harbinger of ice. It lays two bluish white eggs, larger than those of the pigeon. It flies quick, and dives well ; and is always dipping its bill into the water while swimming, or at rest on that element ; walks better on

> North America and Kamtschatka ; often seen on the western coasts of the United States in winter.
>
> 2. *M. glacialis*, Leach.—Puffin of Wilson's List.—Inhabits the arctic parts of both continents ; not uncommon in winter on the coasts of the United States.

This species has of late been looked for on the coasts of Britain, but yet, I believe, without success. The chief and easiest detected difference is in the size and form of the bill Mr Pennant observed a difference in the bills of several species from different parts, and Dr Fleming puts the question—"Have we two species ? " I think it more than probable that this bird has been overlooked from its near alliance, and that, though comparatively rare, it will be yet found to occur on our own coasts.

> 3. *Mormon arcticus.*—Puffin of Wilson's List.—The common puffin of Europe, and migratory to the temperate shores of the United States.

Alca, Linn.

> 1. *A. torda.*—Razorbill of Wilson's List.—Common in winter along the coasts of the United States.
>
> 2. *A. impennis.*—Great Auk.—Inhabits the arctic seas of both continents, where it is almost constantly resident.

---

* In Peale's Museum there is an excellent specimen of this species, which has likewise a smaller spot below each eye.

Head of Turkey Buzzard.

Head of Black Vulture

Drawn from Nature by J.Nelson.

Engraved by W.H.Lizars.

1. Turkey Buzzard. 2. Black Vulture. 3. Raven.

75.

land than others of the genus. It grows fat in the stormy season, from the waves bringing plenty of crabs and small fish within its reach. It is not a very crafty bird, and may be easily taken. It varies to quite white, and sometimes is found with a reddish breast.*

To the anatomist, the internal organisation of this species is deserving attention : it is so constructed as to be capable of contracting or dilating itself at pleasure. We know not what Nature intends by this conformation, unless it be to facilitate diving, for which the compressed form is well adapted ; and likewise the body, when expanded, will be rendered more buoyant, and fit for the purpose of swimming upon the surface of the water.

TURKEY VULTURE, OR TURKEY BUZZARD.
(*Vultur aura.*)

PLATE LXXV.—Fig. 1.

Vultur aura, *Linn. Syst.* ed. 10, tom. i. p. 86, 4.—*Ind. Orn.* p. 4, No. 8.—
    Vieillot, *Ois. de l'Am. Sep.* i. p. 25, pl. 2, vis.—Carrion Crow, *Sloane, Jam.* ii.
    p. 294, fol. 254.—Carrion Vulture, *Lath. Gen. Syn.* i. p. 9.—Le Vautour du
    Brésil, *Briss.* i. p. 468.—Turkey Buzzard, *Catesby, Car.* i. p. 6.—*Bartram's
    Travels*, p. 289.—Cozca quantitti, *Clavigero, Hist. Mex.* i. p. 47, *Eng. trans.*
    —American Vulture, *Shaw, Gen. Zool.* vii. p. 36.—*Peale's Museum*, No. 11,
    m. ; 12, fem.
CATHARTES AURA.—Illiger.†

Cathartes aura, *Illig. Prod.—Bonap. Synop.* p. 33.—*North. Zool.* ii. p. 4.

THIS species is well known throughout the United States, but is most numerous in the southern section of the Union. In

---

* Latham ; Pennant.

† The vultures are comparatively a limited race, and exist in every quarter of the world, New Holland excepted ; * but their range is chiefly in the warm latitudes.

Those of the New World seem to be contained in two genera, *Sarco-*

---

* I have said "New Holland excepted," because we have yet no well authen-ticated instance of anything approaching this form from that very interesting country. The New Holland vulture of Latham rests, to a certain extent, on dubious authority, and cannot now be referred to. I have no doubt that some representing group will be ultimately discovered, which may perhaps elucidate

the northern and middle States it is partially migratory, the greater part retiring to the south on the approach of cold

ramphus of Dumeril, and *Cathartes* of Illiger ; the one containing the condor and Californian vultures ; the other, the turkey buzzards, &c., of Wilson. They are, perhaps, generally, the most unseemly and disgusting of the whole feathered race, of loose and ill-kept plumage, of sluggish habits when not urged on by hunger, feeding on any animal food which they can easily tear to pieces, but often upon the most putrid and loathsome carrion. They have been introduced by the ancients in their beautiful but wild conceptions and imagery, and have been embodied in the tales of fiction and poems of the modern day as all that is lurid, disgusting, and horrible. They are the largest of the feathered race, if we except the *Struthionidæ*, or that group to which the ostrich, cassowary, and bustards belong, and have long been celebrated on account of their great strength. Many fabulous stories are recorded of the formidable condor carrying off men, bullocks, and even elephants.

They have been called the scavengers of nature ; and in warm climates, where all animal matter so soon decays, they are no doubt useful in clearing off what would soon fill the air with noxious miasmata. In many parts of Spain and southern Europe, the *Neophron percnopterus*, or Egyptian vulture of Savigny, and in America, the native species, are allowed to roam unmolested through the towns, and are kept in the market-places, as storks are in Holland, to clear away the refuse and offal ; and a high penalty is attached to the destruction of any of them. In this state they become very familiar and independent. Mr Audubon compares them to a garrisoned half-pay soldier : " To move is for them a hardship ; and nothing but extreme hunger will make them fly down from the roof of the kitchen into the yard. At Natchez, the number of these expecting parasites is so great, that all the refuse within their reach is insufficient to maintain them." They appear also to have been used for a most revolting purpose among barbarous nations, or at least, in conjunction with wild animals, were depended upon to assist in destroying and clearing away the dead, which were purposely exposed to their ravages. Some, however, are elegant and graceful in their form and plumage, and vie with the eagles in strength and activity. Such is the *Vultur barbatus* of Edwards, the lammergeyer of the European Alps.

Independent of the species mentioned by our author, three others

the principal forms wanting to the *Raptores*, and I know that Mr Swainson possesses a New Holland bird, whose station he has been unable to decide, whether it will enter here, or range with the gallinaceous birds. I trust that that gentleman will, ere long, work out its affinities as far as possible, and give it to the public.—Ed.

weather; but numbers remain all the winter in Maryland, Delaware, and New Jersey, particularly in the vicinity of the

have been described as natives of this continent, *Sarcoramphus gryphus* and *Californianus* of Dumeril, and the *Cathartes papa* of Illiger; the former supposed to be the celebrated *Roc* of Sinbad, the no less noted condor of moderns. They are found on the north-west chain of the Andes, frequenting, and not indeed generally met with until near, the limits of eternal snow, where they may be seen perched on the summit of a projecting rock, or sweeping round on the approach of an intruder, in expectation of prey, and looking, when opposed to a clear sky, of double magnitude.

> Moving athwart the evening sky,
> Seem forms of giant height.

The stories of their destructive propensities are, to a certain extent, unfounded. No instance is recorded by any late travellers of children being carried off, and all their inquiries proved the reverse. It is a much followed occupation by the peasantry at the base of the Andes to ascend in search of ice for the luxury of the towns, and their children, at a very tender age, carried with them, are frequently left at considerable distances unprotected; they always remain in security. The *S. Californianus* was first known from a specimen in the British Museum, brought from California. Mr Douglas found it more lately in the woody districts of that country; and I have transcribed his interesting account of its manners, &c. "These gigantic birds, which represent the condor in the northern hemisphere, are common along the coast of California, but are never seen beyond the woody parts of the country. I have met with them as far to the north as 49° north lat. in the summer and autumn months, but nowhere so abundantly as in the Columbian valley, between the Grand Rapids and the sea. They build their nests in the most secret and impenetrable parts of the pine forests, invariably selecting the loftiest trees that overhang precipices, on the deepest and least accessible parts of the mountain valleys. The nest is large, composed of strong thorny twigs and grass, in every way similar to that of the eagle tribe, but more slovenly constructed. The same pair resort for several years to the same nest, bestowing little trouble or attention in repairing it. Eggs, two, nearly spherical, about the size of those of a goose, jet black. Period of incubation twenty-nine or thirty-one days. They hatch generally about the 1st of June. The young are covered with thick whitish down, and are incapable of leaving the nest until the fifth or sixth week. Their food is carrion, dead fish, or other dead animal substance; in no instance will they attack any living animal, unless it be wounded and unable to walk. Their senses of smelling and seeing are remarkably keen. In searching for prey, they soar to a very

large rivers and the ocean, which afford a supply of food at all seasons.

In New Jersey,* the turkey buzzard hatches in May, the deep recesses of the solitary swamps of that State affording situations well suited to the purpose. The female is at no pains to form a nest with materials; but having chosen a suitable place, which is either a truncated hollow tree, an excavated stump, or log, she lays on the rotten wood from two to four eggs, of a dull dirty white or pale cream colour, splashed

great altitude, and when they discover a wounded deer or other animal, they follow its track, and when it sinks, precipitately descend on their object. Although only one is at first seen occupying the carcass, few minutes elapse before the prey is surrounded by great numbers; and it is then devoured to a skeleton within an hour, even though it be one of the larger animals—*Cervus elaphus,* for instance—or a horse. Their voracity is almost insatiable, and they are extremely ungenerous, suffering no other animal to approach them while feeding. After eating, they become so sluggish and indolent, as to remain in the same place until urged by hunger to go in quest of another repast. At such times they perch on decayed trees, with their heads so much retracted, as to be with difficulty observed through the long, loose, lanceolate feathers of the collar. The wings, at the same time, hang down over the feet. This position they invariably preserve in dewy mornings or after rains."

The third species, *C. papa,* not mentioned by Wilson, is introduced in the "Synopsis of Birds of the United States" by the Prince of Musignano, who mentions its occurrence only in the warmer parts of North America; it appears occasionally in Florida during summer. The other two are of much more frequent occurrence, and are of less noble dispositions, more sluggish, very easily intimidated, and dirty in the extreme. Truly clearing away all animal matter, they assemble in vast troops upon the discovery of some dead or nearly dying animal, and exhibit at their feasts scenes of the utmost gluttony and filth. Their power of scenting their quarry from afar has been proved erroneous by the well-managed experiments of Mr Audubon; and, indeed, I never was inclined to think that any birds were endowed with any remarkable development of this particular sense.—Ed.

* Mr Ord mentions New Jersey in particular, as in that State he has visited the breeding-places of the turkey buzzard, and can therefore speak with certainty of the fact. Pennsylvania, it is more than probable, affords situations equally attractive, which are also tenanted by this vulture for hatching and rearing its young.

all over with chocolate, mingled with blackish touches, the blotches largest and thickest towards the great end ; the form something like the egg of a goose, but blunter at the small end ; length, two inches and three-quarters ; breadth, two inches. The male watches often while the female is sitting ; and, if not disturbed, they will occupy the same breeding-place for several years. The young are clothed with a whitish down, similar to that which covers young goslings. If any person approach the nest and attempt to handle them, they will immediately vomit such offensive matter, as to compel the intruder to a precipitate retreat.

The turkey buzzards are gregarious, peaceable, and harmless, never offering any violence to a living animal, or, like the plunderers of the *Falco* tribe, depriving the husbandman of his stock. Hence, though, in consequence of their filthy habits, they are not beloved, yet they are respected for their usefulness ; and in the southern States, where they are most needed, they, as well as the black vultures, are protected by a law which imposes a fine on those who wilfully deprive them of life. They generally roost in flocks, on the limbs of large trees ; and they may be seen on a summer morning, spreading out their wings to the rising sun, and remaining in that posture for a considerable time. Pennant conjectures that this is "to purify their bodies, which are most offensively fetid." But is it reasonable to suppose that *that* effluvia can be offensive to them which arises from food perfectly adapted to their nature, and which is constantly the object of their desires ? Many birds, and particularly those of the granivorous kind, have a similar habit, which doubtless is attended with the same exhilarating effects as an exposure to the pure air of the morning has on the frame of one just risen from repose.

These birds, unless when rising from the earth, seldom flap their wings, but sweep along in ogees, and dipping and rising lines, and move with great rapidity. They are often seen in companies, soaring at an immense height, particularly previous to a thunderstorm. Their wings are not spread horizontally, but form a slight angle with the body upwards, the tips having an upward curve. Their sense of smelling is astonishingly exquisite, and they never fail to discover carrion, even when at the distance of several miles from it. When once they have found a carcass, if not molested, they will not leave the place until the whole is devoured. At such times they eat so immoderately, that frequently they are incapable of rising, and may be caught without much difficulty; but few that are acquainted with them will have the temerity to undertake the task. A man in the State of Delaware, a few years since, observing some turkey buzzards regaling themselves upon the carcass of a horse which was in a highly putrid state, conceived the design of making a captive of one, to take home for the amusement of his children. He cautiously approached, and springing upon the unsuspicious group, grasped a fine plump fellow in his arms, and was bearing off his prize in triumph ; when lo ! the indignant vulture disgorged such a torrent of filth in the face of our hero, that it produced all the effects of the most powerful emetic, and for ever cured him of his inclination for turkey buzzards.

On the continent of America, this species inhabits a vast range of territory, being common,[*] it is said, from Nova Scotia to Tierra del Fuego.[†] How far to the northward of North California[‡] they are found, we are not informed; but it is probable that they extend their migrations to the Columbia,

[*] In the northern States of our Union, the turkey buzzard is only occasionally seen. It is considered a rare bird by the inhabitants.

[†] " Great numbers of a species of vulture, commonly called carrion crow by the sailors (*Vultur aura*), were seen upon this island (New-Year's Island near Cape Horn, lat. 55 S. 67 W.), and probably feed on young seal-cubs, which either die in the birth, or which they take an opportunity to seize upon." Cook calls them turkey buzzards. Forster's Voyage, ii. p. 516, 4to, London, 1777. We strongly suspect that the sailors were correct, and that these were black vultures, or carrion crows.

[‡] Pérouse saw a bird, which he calls the black vulture, probably the *Vultur aura*, at Monterey Bay, North California.—*Voyage*, ii. p. 203.

allured thither by the quantity of dead salmon which, at certain seasons, line the shores of that river.

They are numerous in the West India islands, where they are said to be "far inferior in size to those of North America."* This leads us to the inquiry whether or not the present species has been confounded by all the naturalists of Europe with the black vulture or carrion crow, which is so common in the southern parts of our continent. If not, why has the latter been totally overlooked in the numerous ornithologies and nomenclatures with which the world has been favoured, when it is so conspicuous and remarkable that no stranger visits South Carolina, Georgia, or the Spanish provinces, but is immediately struck with the novelty of its appearance? We can find no cause for the turkey buzzards of the islands† being smaller than ours, and must conclude that the carrion crow, which is of less size, has been mistaken for the former. In the history which follows, we shall endeavour to make it evident that the species described by Ulloa as being so numerous in South America is no other than the black vulture. The ornithologists of Europe, not aware of the existence of a new species, have without investigation contented themselves with the opinion that the bird called by the above-mentioned traveller the *gallinazo* was the *Vultur aura*, the subject of our present history. This is the more inexcusable, as we expect in naturalists a precision of a different character from that which distinguishes vulgar observation. If the Europeans

---

* Pennant ; Arctic Zoology.

† The vulture which Sir Hans Sloane has figured and described, and which he says is common in Jamaica, is undoubtedly the *Vultur aura.* " The head, and an inch in the neck, are bare, and without feathers, of a flesh colour, covered with a thin membrane, like that of turkeys, with which the most part of the bill is covered likewise ; bill, below the membrane, more than an inch long, whitish at the point ; tail, broad, and nine inches long ; legs and feet, three inches long ; it flies exactly like a kite, and preys on nothing living ; but when dead, it devours their carcasses, whence they are not molested."—*Sloane, Natural History of Jamaica*, vol. ii. p. 294, folio.

had not the opportunity of comparing living specimens of the two species, they at least had preserved subjects in their extensive and valuable museums, from which a correct judgment might have been formed. The figure in the *Planches Enluminées*, though wretchedly drawn and coloured, was evidently taken from a stuffed specimen of the black vulture.

Pennant observes, that the turkey vultures " are not found in the northern regions of Europe or Asia, at least in those latitudes which might give them a pretence of appearing there. I cannot find them," he continues, " in our quarter of the globe higher than the Grison Alps,* or Silesia,† or at farthest Kalish, in Great Poland."‡

Kolben, in his account of the Cape of Good Hope, mentions a vulture which he represents as very voracious and noxious. " I have seen," says he, " many carcasses of cows, oxen, and other tame creatures, which the eagles had slain. I say carcasses, but they were rather skeletons, the flesh and entrails being all devoured, and nothing remaining but the skin and bones. But the skin and bones being in their natural places, the flesh being, as it were, scooped out, and the wound by which the eagles enter the body being ever in the belly, you would not, till you had come up to the skeleton, have had the least suspicion that any such matter had happened. The Dutch at the Cape frequently call those eagles, on account of their tearing out the entrails of beasts, *strunt-vogels—i.e.*, dung-birds. It frequently happens that an ox that is freed from the plough, and left to find his way home, lies down to rest himself by the way ; and if he does so, it is a great chance but the eagles fall upon him and devour him. They attack an ox or cow in a body, consisting of an hundred and upwards."§

Buffon conjectures that this murderous vulture is the turkey buzzard, and concludes his history of the latter with the

* Willoughby, Ornithology, p. 67. † Schwenckfeldt, Av. Silesia, 375.
‡ Rzaczynski, Hist. Nat. Poland, 298.
§ Medley's Kolben, vol. ii. p. 135.

following invective against the whole fraternity:—" In every part of the globe they are voracious, slothful, offensive, and hateful, and, like the wolves, are as noxious during their life as useless after their death."

If Kolben's account of the ferocity of his eagle,* or vulture, be just, we do not hesitate to maintain that that vulture is not the turkey buzzard, as, amongst the whole feathered creation, there is none, perhaps, more innoxious than this species ; and that it is beneficial to the inhabitants of our southern continent, even Buffon himself, on the authority of Desmarchais, asserts. But we doubt the truth of Kolben's story ; and, in this place, must express our regret that enlightened naturalists should so readily lend an ear to the romances of travellers, who, to excite astonishment, freely give currency to every ridiculous tale which the designing or the credulous impose upon them. We will add further, that the turkey buzzard seldom begins upon a carcass until invited to the banquet by that odour which in no ordinary degree renders it an object of delight.

The turkey vulture is two feet and a half in length, and six feet two inches in breadth ; the bill from the corner of the mouth is almost two inches and a half long, of a dark horn colour for somewhat more than an inch from the tip, the nostril a remarkably wide slit or opening through it ; the tongue is greatly concave, cartilaginous, and finely serrated on its edges ; ears inclining to oval ; eyes dark, in some specimens reddish hazel ; the head and neck, for about an inch and a half below the ears, are furnished with a reddish wrinkled skin, beset with short black hairs, which also cover the bill as far as the interior angle of the nostril, the neck not so much caruncled as that of the black vulture ; from the hind head to the neck-feathers

---

* These bloodthirsty eagles, we conjecture, are black vultures, they being in the habit of mining into the bellies of dead animals to feast upon the contents. With respect to their attacking those that are living, as the vultures of America are not so heroic, it is a fair inference that the same species elsewhere is possessed of a similar disposition.

the space is covered with down of a sooty black colour; the fore part of the neck is bare as far as the breast-bone, the skin on the lower part, or pouch, very much wrinkled; this naked skin is not discernible without removing the plumage which arches over it; the whole lower parts, lining of the wings, rump, and tail-coverts, are of a sooty brown, the feathers of the belly and vent hairy; the plumage of the neck is large and tumid, and, with that of the back and shoulders, black; the scapulars and secondaries are black on their outer webs, skirted with tawny brown, the latter slightly tipped with white; primaries and their coverts, plain brown, the former pointed, third primary the longest; coverts of the secondaries and lesser coverts, tawny brown centred with black—some of the feathers at their extremities slightly edged with white; the tail is twelve inches long, rounded, of a brownish black, and composed of twelve feathers, which are broad at their extremities; inside of wings and tail, light ash; the wings reach to the end of the tail; the whole body and neck beneath the plumage are thickly clothed with a white down, which feels like cotton; the shafts of the primaries are yellowish white above, and those of the tail brown, both pure white below; the plumage of the neck, back, shoulders, scapulars, and secondaries is glossed with green and bronze, and has purple reflections; the thighs are feathered to the knees; feet considerably webbed; middle toe three inches and a half in length, and about an inch and a half longer than the outer one, which is the next longest; the sole of the foot is hard and rough; claws, dark horn colour; the legs are of a pale flesh colour, and three inches long. The claws are larger, but the feet slenderer than those of the carrion crow. The bill of the male is pure white; in some specimens the upper mandible is tipped with black. There is little or no perceptible difference between the sexes.

The bird from which the foregoing description was taken was shot for this work, at Great Egg Harbour, on the 30th of January. It was a female, in perfect plumage, excessively fat, and weighed five pounds one ounce avoirdupois. On dissection it emitted a slight musky odour.

The vulture is included in the catalogue of those fowls declared unclean and an abomination by the Levitical law, and which the Israelites were interdicted eating.* We presume that this prohibition was religiously observed, so far at least as it related to the vulture, from whose flesh there arises such an unsavoury odour, that we question if all the sweetening processes ever invented could render it palatable to Jew, Pagan, or Christian.

Since the above has been ready for the press, we have seen the History of the expedition under the command of Lewis and Clark, and find our conjecture with respect to the migration of the turkey buzzard verified, several of this species having been observed at Brant Island, near the Falls of the Columbia.†

BLACK VULTURE, OR CARRION CROW.
(*Vultur jota.*)

PLATE LXXV.—FIG. 2.

Vultur jota, *Gmel. Syst.* i. p. 247.—*Molina, Hist. Chili,* i. p. 185, *Am. trans.* —Zalipot, *Clavigero, Hist. Mex.* i. p. 47, *Eng. trans.*—Gallinazo, *Ulloa, Voy.* i. p. 52, *Amsterdam ed.*—Vultur atratus, *Bartram,* p. 289.—Vautour du Brésil, *Pl. enl.* 189.—Vultur aura, *B. Lath. Ind. Orn.* p. 5.—Le Vautour urubu, *Vieil. Ois. de l'Am. Sep.* i. p. 23, pl. 2.—*Peale's Museum,* No. 13.

VULTUR JOTA.—Bonaparte.‡

Vultur jota, *Bonap. Synop.* p. 23.—Cathartes atratus, *North. Zool.* ii. p. 6.

ALTHOUGH an account of this vulture was published more than twenty years ago by Mr William Bartram, wherein it

---

* Leviticus xi. 14 ; Deuteronomy xiv. 13.

† History of the Expedition, vol. ii. p. 233.

‡ Mr Swainson, in a note to the description of this bird in the "Northern Zoology," remarks, as a reason for changing the name given by Bonaparte—" We have not considered it expedient to apply to this bird the scientific name of *Iota*, given by Molina to a black vulture of Chili,

was distinctly specified as a different species from the preceding, yet it excites our surprise that the ornithologists should have persisted in confounding it with the turkey buzzard ; an error which can hardly admit of extenuation when it is considered what a respectable authority they had for a different opinion.

The habits of this species are singular. In the towns and villages of the southern States, particularly Charleston and Georgetown, South Carolina, and in Savannah, Georgia, the carrion crows may be seen either sauntering about the streets, sunning themselves on the roofs of the houses and the fences, or, if the feather be cold, cowering along the tops of the chimneys, to enjoy the benefit of the heat, which to them is a peculiar gratification. They are protected by a law or usage, and may be said to be completely domesticated, being as common as the domestic poultry, and equally familiar. The inhabitants generally are disgusted with their filthy, voracious habits ; but notwithstanding, being viewed as contributive to the removal of the dead animal matter, which, if permitted to putrify during the hot season, would render the atmosphere impure, they have a respect paid them as scavengers, whose labours are subservient to the public good. It sometimes happens that, after having gorged themselves, these birds vomit down the chimneys, which must be intolerably disgusting, and must provoke the ill-will of those whose hospitality is thus requited.

The black vultures are indolent, and may be observed in companies loitering for hours together in one place. They do not associate with the turkey buzzards ; and are much darker in their plumage than the latter. Their mode of flight also varies from that of the turkey buzzard. The black vulture flaps its wings five or six times rapidly, then sails with them extended nearly horizontally ; the turkey buzzard seldom flaps its wings, and when sailing, they form an angle with the body

because there is no evidence to prove that it is the turkey buzzard of North America." Neither is there present proof that it is not, therefore we retain Bonaparte's name.—ED.

upwards. The latter, though found in the vicinity of towns, rarely ventures within them, and then always appearing cautious of the near approach of any one. It is not so impatient of cold as the former, and is likewise less lazy. The black vulture, on the ground, hops along very awkwardly; the turkey buzzard, though seemingly inactive, moves with an even gait. The latter, unless pressed by hunger, will not eat of a carcass until it becomes putrid; the former is not so fastidious, but devours animal food without distinction.

It is said that the black vultures sometimes attack young pigs, and eat off their ears and tails; and we have even heard stories of their assaulting feeble calves, and picking out their eyes. But these instances are rare: if otherwise, they would not receive that countenance or protection which is so universally extended to them in the States of South Carolina and Georgia, where they abound.

" This undescribed species," says Mr Bartram, " is a native of the maritime parts of Georgia and of the Floridas, where they are called carrion crows. They flock together, and feed upon carrion, but do not mix with the turkey buzzard (*Vultur aura*). Their wings are broad, and round at their extremities. Their tail, which they spread like a fan when on the wing, is remarkably short. They have a heavy, laborious flight, flapping their wings, and sailing alternately. The whole plumage is of a sable or mourning colour." *

In one of Mr Wilson's journals, I find an interesting detail of the greedy and disgusting habits of this species; and shall give the passage entire, in the same unadorned manner in which it is written.

" *February* 21, 1809.—Went out to Hampstead † this forenoon. A horse had dropped down in the street in convulsions, and dying, it was dragged out to Hampstead, and skinned. The ground, for a hundred yards around it, was black with carrion crows; many sat on the tops of sheds, fences, and houses

* MS. in the possession of Mr Ord.
† Near Charleston, South Carolina.

within sight; sixty or eighty on the opposite side of a small run. I counted at one time two hundred and thirty-seven, but I believe there were more, besides several in the air over my head and at a distance. I ventured cautiously within thirty yards of the carcass, where three or four dogs and twenty or thirty vultures were busily tearing and devouring. Seeing them take no notice, I ventured nearer, till I was within ten yards, and sat down on the bank. Still they paid little attention to me. The dogs, being sometimes accidentally flapped with the wings of the vultures, would growl and snap at them, which would occasion them to spring up for a moment, but they immediately gathered in again. I remarked the vultures frequently attack each other, fighting with their claws or heels, striking like a cock, with open wings, and fixing their claws in each other's head. The females, and, I believe, the males likewise, made a hissing sound, with open mouth, exactly resembling that produced by thrusting a red-hot poker into water; and frequently a snuffling, like a dog clearing his nostrils, as I suppose they were theirs. On observing that they did not heed me, I stole so close that my feet were within one yard of the horse's legs, and again sat down. They all slid aloof a few feet; but, seeing me quiet, they soon returned as before. As they were often disturbed by the dogs, I ordered the latter home: my voice gave no alarm to the vultures. As soon as the dogs departed, the vultures crowded in such numbers, that I counted at one time thirty-seven on and around the carcass, with several within; so that scarcely an inch of it was visible. Sometimes one would come out with a large piece of the entrails, which in a moment was surrounded by several others, who tore it in fragments, and it soon disappeared. They kept up the hissing occasionally. Some of them, having their whole legs and heads covered with blood, presented a most savage aspect. Still as the dogs advanced, I would order them away, which seemed to gratify the vultures; and one would pursue another to within a foot or two of the spot where I was sitting. Sometimes I observed them stretching their necks along the ground, as if to press the food downwards."

The carrion crow is seldom found on the Atlantic to the northward of Newbern, North Carolina,* but inhabits the whole continent to the southward, as far as Cape Horn. Don Ulloa, in noticing the birds of Carthagena, gives an account of a vulture, which we shall quote, in order to establish the opinion, advanced in the preceding history, that it is the present species. We shall afterwards subjoin other testimony in confirmation of this opinion. With respect to the marvellous tale of their attacking the cattle in the pastures, it is too improbable to merit a serious refutation.

"It would be too great an undertaking to describe all the extraordinary birds that inhabit this country; but I cannot refrain from noticing that to which they give the name of *gallinazo*, from the resemblance it has to the turkey-hen. This bird is of the size of a pea-hen, but its head and neck are something larger. From the crop to the base of the bill it has no feathers: this space is surrounded with a wrinkled, glandulous, and rough skin, which forms numerous warts and other similar inequalities. This skin is black, as is the plumage of the bird, but usually of a brownish black. The bill is well proportioned, strong, and a little hooked. These birds are familiar in Carthagena; the tops of the houses are covered with them; it is they which cleanse the city of all its animal impurities. There are few animals killed whereof they do not obtain the offals; and when this food is wanting, they have recourse to other filth. Their sense of smelling is so acute, that it enables them to trace carrion at the distance of three or four leagues, which they do not abandon until there remains nothing but the skeleton.

"The great number of these birds found in such hot climates is an excellent provision of nature, as otherwise the

---

* Since writing the above, I have been informed by a gentleman who resides at Detroit, on Lake Erie, that the carrion crow is common at that place.

putrefaction caused by the constant and excessive heat would render the air insupportable to human life. When first they take wing, they fly heavily ; but afterwards, they rise so high as to be entirely invisible. On the ground they walk sluggishly. Their legs are well proportioned ; they have three toes forward, turning inwards, and one on the inside, inclining a little backwards, so that, the feet interfering, they cannot walk with any agility, but are obliged to hop : each toe is furnished with a long and stout claw.

" When the gallinazos are deprived of carrion or food in the city, they are driven by hunger among the cattle of the pastures. If they see a beast with a sore on the back, they alight on it, and attack the part affected ; and it avails not that the poor animal *throws itself upon the ground*, and endeavours to intimidate them by its bellowing : *they do not quit their hold !* and by means of their bill they so soon enlarge the wound, that the animal finally becomes their prey." *

The account, from the same author, of the beneficial effects resulting from the fondness of the vultures for the eggs of the alligator merits attention :—

" The gallinazos are the most inveterate enemies of the alligators, or rather, they are extremely fond of their eggs, and employ much stratagem to obtain them. During the summer, these birds make it their business to watch the female alligators ; for it is in that season that they deposit their eggs in the sand of the shores of the rivers, which are not then overflowed. The gallinazo conceals itself among the branches and leaves of a tree, so as to be unperceived by the alligator ; and permits the eggs quietly to be laid, not even interrupting the precautions that she takes to conceal them. But she is no sooner under the water than the gallinazo darts upon the nest ; and, with its bill, claws, and wings, uncovers the eggs, and

* Voyage Historique de l'Amerique Meridionale, par Don George Juan et Don Antoine de Ulloa, liv. i. chap. viii. p. 52, à Amsterdam et à Leipzig, 1752, 4to.

gobbles them down, leaving nothing but the shells. This
banquet would, indeed, richly reward its patience, did not a
multitude of gallinazos join the fortunate discoverer, and share
in the spoil.

"How admirable the wisdom of that Providence which
hath given to the male alligator an inclination to devour its
own offspring, and to the gallinazo a taste for the eggs of the
female! Indeed, neither the rivers nor the neighbouring
fields would otherwise be sufficient to contain the multitudes
that are hatched; for, notwithstanding the ravages of both
these insatiable enemies, one can hardly imagine the numbers
that remain." *

The Abbé Clavigero, in his "History of Mexico," has clearly
indicated the present species as distinguished from the turkey
buzzard :—

"The business of clearing the fields of Mexico is reserved
principally for the *zopilots*, known in South America by the
name of *gallinazzi*, in other places by that of *aure*, and
in some places, though very improperly, by that of *ravens*.
There are two very different species of these birds : the one,
the zopilot, properly so called ; the other, called the cozcaquauhtli: they are both bigger than the raven. These two
species resemble each other in their hooked bill and crooked
claws, and by having upon their head, instead of feathers, a
wrinkled membrane with some curling hairs. They fly so
high, that, although they are pretty large, they are lost to the
sight ; and especially before a hailstorm they will be seen
wheeling in vast numbers under the loftiest clouds, till they
entirely disappear. They feed upon carrion, which they discover, by the acuteness of their sight and smell, from the
greatest height, and descend upon it with a majestic flight,
in a great spiral course. They are both almost mute. The
two species are distinguished, however, by their size, their
colour, their numbers, and some other peculiarities. The
zopilots, properly so called, have black feathers, with a brown

* Liv. iv. chap. ix. p. 172.

head, bill, and feet ; they go often in flocks, and roost together upon trees. This species is very numerous, and is to be found in all the different climates ; while, on the contrary, the coz-caquauhtli is far from numerous, and is peculiar to the warmer climates alone.* The latter bird is larger than the zopilot, has a red head and feet, with a beak of a deep red colour, except towards its extremity, which is white. Its feathers are brown, except upon the neck and parts about the breast, which are of a reddish black. The wings are of an ash colour upon the inside, and upon the outside are varie-gated with black and tawny.

"The cozcaquauhtli is called by the Mexicans *king of the zopilots ;* † and they say that when these two species happen to meet together about the same carrion, the zopilot never begins to eat till the cozcaquauhtli has tasted it. The zopilot is a most useful bird to that country, for it not only clears the fields, but attends the crocodiles, and destroys the eggs which the females of those dreadful amphibious animals leave in the sand to be hatched by the heat of the sun. The de-struction of such a bird ought to be prohibited under severe penalties." ‡

We are almost afraid of trespassing upon the patience of the reader by the length of our quotations ; but as we are very anxious that the subject of this article should enjoy that right to which it is fairly entitled, of being ranked as an in-dependent species, we are tempted to add one testimony more, which we find in the "History of Chili," by the Abbé Molina.

"The *jota* (*Vultur jota*) resembles much the *aura,* a species of vulture of which there is, perhaps, but one variety. It is distinguished, however, by the beak, which is grey, with a black point. Notwithstanding the size of this bird, which is nearly that of the turkey, and its strong and crooked talons,

* This is a mistake.

† This is the *Vultur aura.* The bird which now goes by the name of *king of the zopilots* in New Spain is the *Vultur papa* of Linnæus.

‡ Clavigero's Mexico, translated by Cullen, vol. i. p. 47, London.

it attacks no other, but feeds principally upon carcasses and reptiles. It is extremely indolent, and will frequently remain for a long time almost motionless, with its wings extended, sunning itself upon the rocks or the roofs of the houses. When in pain, which is the only time that it is known to make any noise, it utters a sharp cry like that of a rat; and usually disgorges what it has eaten. The flesh of this bird emits a fetid smell that is highly offensive. The manner in which it builds its nest is perfectly correspondent to its natural indolence: it carelessly places between rocks, or even upon the ground, a few dry leaves or feathers, upon which it lays two eggs of a dirty white." *

The black vulture is twenty-six inches in length, and four feet four inches in extent; the bill is two inches and a half long, of a dark horn colour as far as near an inch; the remainder, the head, and a part of the neck, are covered with a black, wrinkled, caruncled skin, beset with short black hairs, and downy behind; nostril, an oblong slit; irides, reddish hazel; the throat is dashed with yellow ochre; the general colour of the plumage is of a dull black, except the primaries, which are whitish on the inside, and have four of their broadened edges below of a drab or dark cream colour, extending two inches, which is seen only when the wing is unfolded; the shafts of the feathers white on both sides; the rest of the wing-feathers dark on both sides; the wings, when folded, are about the length of the tail, the fifth feather being the longest; the secondaries are two inches shorter than the tail, which is slightly forked; the exterior feathers, three-quarters of an inch longer than the rest; the legs are limy, three inches and a half in length, and, with the feet, are thick and strong; the middle toe is four inches long, side toes, two inches, and considerably webbed, inner toe rather the shortest; claws strong, but not sharp, like those of the *Falco* genus; middle claw, three-quarters of an inch long; the stomach is not lined with

* Hist. Chili, Am. Trans. i. p. 185.

hair, as reported. When opened, this bird smells strongly of musk.

Mr Abbot informs me that the carrion crow builds its nest in the large trees of the low wet swamps, to which places they retire every evening to roost. "They frequent," says he, "that part of the town of Savannah where the hog-butchers reside, and walk about the streets in great numbers, like domestic fowls. It is diverting to see, when the entrails and offals of the hogs are thrown to them, with what greediness they scramble for the food, seizing upon it, and pulling one against another, until the strongest prevails. The turkey buzzard is accused of killing young lambs and pigs by picking out their eyes; but I believe that the carrion crow is not guilty of the like practices. The two species do not associate."

RAVEN. (*Corvus corax.*)

PLATE LXXV.—Fig. 3.

Gmel. Syst. i. p. 364.—*Ind. Orn.* p. 150.—Le Corbeau, *Briss.* ii. p. 8, et var.—*Buff. Ois.* iii. p. 13.—*Pl. enl.* 495.—*Temm. Man. d'Orn.* p. 107.—Raven, *Lath. Gen. Syn.* i. p. 367.—*Id. Supp.* p. 74.—*Penn. Brit. Zool.* No. 74.—*Arct. Zool.* No. 134.—*Shaw, Gen. Zool.* vii. p. 341.—*Bewick,* i. p. 100.—*Low, Fauna Orcadensis,* p. 45.—*Peale's Museum,* No. 175.

CORVUS CORAX.—Linnæus.

Corvus corax, *Bonap. Synop.* p. 56.—*Flem. Br. Anim.* p. 87.—Raven, *Mont. Orn. Dict. and Supp.* p. 67.—*Selby, Illust. Br. Orn.* pl. 27.

A KNOWLEDGE of this celebrated bird has been handed down to us from the earliest ages; and its history is almost coeval with that of man. In the best and most ancient of all books, we learn that, at the end of forty days after the great flood had covered the earth, Noah, wishing to ascertain whether or no the waters had abated, sent forth a raven, which did not return into the ark.* This is the first notice that is taken of this species. Though the raven was declared unclean by the law of Moses, yet we are informed that when the prophet

* Genesis viii. 7.

Elijah provoked the enmity of Ahab by prophesying against him, and hid himself by the brook Cherith, the ravens were appointed by Heaven to bring him his daily food.* The colour of the raven has given rise to a similitude, in one of the most beautiful of eclogues, which has been perpetuated in all subsequent ages, and which is not less pleasing for being trite or proverbial. The favourite of the royal lover of Jerusalem, in the enthusiasm of affection, thus describes the object of her adoration, in reply to the following question :—

> What is thy beloved more than another beloved,
> O thou fairest among women?
> My beloved is white and ruddy, the chiefest among
> Ten thousand. His head is as the most fine gold;
> His locks are bushy, and black as a raven! †

The above-mentioned circumstances taken into consideration, one would suppose that the lot of the subject of this chapter would have been of a different complexion from what history and tradition inform us is the fact. But in every country we are told the raven is considered an ominous bird, whose croakings foretell approaching evil; and many a crooked beldam has given interpretation to these oracles, of a nature to infuse terror into a whole community. Hence this ill-fated bird, from time immemorial, has been the innocent subject of vulgar obloquy and detestation.

Augury, or the art of foretelling future events by the flight, cries, or motions of birds, descended from the Chaldeans to the Greeks, thence to the Etrurians, and from them it was transmitted to the Romans.‡ The crafty legislators of those cele-

* 1 Kings xvii. 5, 6.
† Song of Solomon v. 9, 10, 11.
‡ That the science of augury is very ancient we learn from the Hebrew lawgiver, who prohibits it, as well as every other kind of divination—Deut. chap. xviii. The Romans derived their knowledge of augury chiefly from the Tuscans or Etrurians, who practised it in the earliest times. This art was known in Italy before the time of Romulus, since that prince did not commence the building of Rome till he had taken the auguries. The successors of Romulus, from a conviction of the usefulness of the science, and at the same time not to render it contemptible by becoming too familiar, employed the most skilful augurs

brated nations, from a deep knowledge of human nature, made superstition a principal feature of their religious ceremonies, well knowing that it required a more than ordinary policy to govern a multitude ever liable to the fatal influences of passion, and who, without some timely restraints, would burst forth like a torrent, whose course is marked by wide-spreading desolation.   Hence to the purposes of polity the raven was made subservient; and the Romans having consecrated it to Apollo, as to the god of divination, its flight was observed with the greatest solemnity, and its tones and inflections of voice were noted with a precision which intimated a belief in its infallible prescience.

But the ancients have not been the only people infected with this species of superstition; the moderns, even though favoured with the light of Christianity, have exhibited as much folly, through the impious curiosity of prying into futurity, as the Romans themselves.   It is true that modern nations have not instituted their sacred colleges or sacerdotal orders for the purposes of divination, but in all countries there have been self-constituted augurs, whose interpretations of omens have been received with religious respect by the credulous multitude.   Even at this moment, in some parts of the world, if a raven alight on a village church, the whole fraternity is in an uproar, and Heaven is importuned, in all the ardour of devotion, to avert the impending calamity.

The poets have taken advantage of this weakness of human nature, and, in their hands, the raven is a fit instrument of terror.   Shakespeare puts the following malediction into the mouth of his Caliban :—

> As wicked dew as e'er my mother brushed
> With raven's feather from unwholesome fen,
> Drop on you both ! *

----

from Etruria to introduce the practice of it into their religious ceremonies; and by a decree of the senate, some of the youth of the best families in Rome were annually sent into Tuscany to be instructed in this art.— *Vide Ciceron. de Divin.;* also Calmet and the Abbé Banier.

* Tempest, act i. scene 2.

The ferocious wife of Macbeth, on being advised of the approach of Duncan, whose death she had conspired, thus exclaims—

> ———The raven himself is hoarse
> That croaks the fatal entrance of Duncan
> Under my battlements.*

The Moor of Venice says—

> ———It comes o'er my memory,
> As doth the raven o'er the infected house,
> Boding to all.†

The last quotation alludes to the supposed habit of this bird's flying over those houses which contain the sick, whose dissolution is at hand, and thereby announced. Thus Marlowe in the " Jew of Malta," as cited by Malone—

> ———The sad presaging raven tolls
> The sick man's passport in her hollow beak ;
> And, in the shadow of the silent night,
> Doth shake contagion from her sable wing.

But it is the province of philosophy to dispel these illusions which bewilder the mind, by pointing out the simple truths which Nature has been at no pains to conceal, but which the folly of mankind has shrouded in all the obscurity of mystery.

The raven is a general inhabitant of the United States, but is more common in the interior. On the lakes, and particularly in the neighbourhood of the Falls of the Niagara river, they are numerous ; and it is a remarkable fact, that where they so abound, the common crow (*C. corone*) seldom makes its appearance ; being intimidated, it is conjectured, by the superior size and strength of the former, or by an antipathy which the two species manifest towards each other. This I had an opportunity of observing myself, in a journey during the months of August and September along the Lakes Erie and Ontario. The ravens were seen every day prowling

* Macbeth, act i. scene 5. † Othello, act. iv. scene 1.

about in search of the dead fish which the waves are continually casting ashore, and which afford them an abundance of a
favourite food ; but I did not see or hear a single crow within
several miles of the lakes, and but very few through the whole
of the Genesee country.

The food of this species is dead animal matter of all kinds,
not excepting the most putrid carrion, which it devours in
common with the vultures ; worms, grubs, reptiles, and shellfish, the last of which, in the manner of the crow, it drops
from a considerable height in the air on the rocks, in order
to break the shells ; it is fond of birds' eggs, and is often observed sneaking around the farmhouse in search of the eggs of
the domestic poultry, which it sucks with eagerness ; it is likewise charged with destroying young ducks and chickens, and
lambs which have been yeaned in a sickly state.   The raven,
it is said, follows the hunters of deer for the purpose of falling
heir to the offal ; * and the huntsmen are obliged to cover
their game, when it is left in the woods, with their hunting
frocks, to protect it from this thievish connoisseur, who, if he
have an opportunity, will attack the region of the kidneys,
and mangle the saddle without ceremony.

Buffon says that " the raven *plucks out the eyes of buffaloes,*
and then, *fixing on the back, it tears off the flesh deliberately ;*
and what renders the ferocity more detestable, it is not incited
by the cravings of hunger, but by the appetite for carnage ;
for it can subsist on fruits, seeds of all kinds, and indeed may
be considered as an omnivorous animal."   This is mere fable,
and of a piece with many other absurdities of the same roman-
cing author.

This species is found almost all over the habitable globe.
We trace it in the north from Norway to Greenland, and hear
of it in Kamtschatka.   It is common everywhere in Russia

---

* This is the case in those parts of the United States where the deer
are hunted without dogs ; where these are employed, they are generally
rewarded with the offal.

and Siberia, except within the arctic circle ; * and all through
Europe. Kolben enumerates the raven among the birds of the
Cape of Good Hope ; † De Grandpré represents it as numerous
in Bengal, where they are said to be protected for their usefulness ; ‡ and the unfortunate La Pérouse saw them at Baie
de Castries, on the east coast of Tartary, likewise at Port des
Francois, 58° 37′ north latitude, and 139° 50′ west longitude;
and at Monterey Bay, North California.§ The English cir-
cumnavigators met with them at Nootka Sound, ‖ and at the
Sandwich Islands, two being seen in the village of Kakooa ;
also at Owhyhee, and supposed to be adored there, as they
were called Eatooas. ¶ Our intrepid American travellers,
under the command of Lewis and Clark, shortly after they
embarked on the Columbia river, saw abundance of ravens,
which were attracted thither by the immense quantity of dead
salmon which lined the shores.** They are found at all seasons at Hudson's Bay ; †† are frequent in Mexico ; ‡‡ and it is
more than probable that they inhabit the whole continent of
America.

The raven measures, from the tip of the bill to the end of
the tail, twenty-six inches, and is four feet in extent ; the bill
is large and strong, of a shining black, notched near the tip,
and three inches long ; the setaceous feathers which cover the
nostrils extend half its length ; the eyes are black ; the general colour is a deep glossy black, with steel-blue reflections ;
the lower parts are less glossy ; the tail is rounded, and extends about two inches beyond the wings ; the legs are two
inches and a half in length, and, with the feet, are strong and
black ; the claws are long.

This bird is said to attain to a great age ; and its plumage

---

* Latham.   † Medley's Kolben, ii. p. 136.
‡ Voy. in the Indian Ocean, p. 148.
§ Voy. par J. F. G. de la Pérouse, ii. p. 129, 203, 443.
‖ Cook's Last Voy., ii. p. 236, Am. ed.
¶ Idem, iii. p. 329.   ** Gass's Journal, p. 153.
†† Charlevoix ; Kalm ; Hearne's Journey.   ‡‡ Fernandez.

to be subject to change from the influence of years and of climate. It is found in Iceland and Greenland entirely white.

The raven was the constant attendant of Lewis and Clark's party in their long and toilsome journey. During the winter, at Fort Mandan, they were observed in immense numbers, notwithstanding the cold was so excessive that, on the 17th December 1804, the thermometer stood at 45° below 0.

Like the crow, this species may be easily domesticated, and in that state would afford amusement by its familiarity, frolics, and sagacity. But such noisy and mischievous pets, in common with parrots and monkeys, are not held in high estimation in this quarter of the globe, and are generally overlooked for those universal favourites, which either gratify the eye by the neatness or brilliancy of their plumage, or gladden the ear by the simplicity or variety of their song.

GREAT-FOOTED HAWK.   (*Falco peregrinus.*)*

PLATE LXXVI.—FEMALE.

Falco peregrinus, *Gmel. Syst.* i. p. 272, 88.—*Briss.* i. p. 341, 6, and var. A.—*Ind. Orn.* p. 33, No. 72.—Falco barbarus, *Linn. Syst.* ed. 10, tom. i. p. 88, No. 6.— *Gmel. Syst.* i. p. 272, 8.—*Ind. Orn.* p. 33, No. 71.—Falco hornotinus, *Briss.* i. p. 324, A.—Falco niger, *Id.* p. 327, E.—Falco maculatus, *Id.* p. 329, F.— Peregrine Falcon, *Lath. Syn.* i. p. 73, No. 52.—*Id. Sup.* p. 18.—*Penn. Brit. Zool.* No. 48, pl. 20.—*Arct. Zool.* No. 97.—*Shaw, Gen. Zool.* vol. vii. p. 128.— *Montagu, Orn. Dict. and Supp.*—*Low, Fauna Orcadensis*, p. 150.—Common Falcon, *Lath. Syn.* i. p. 65, No. 49, var. A; p. 67, var. E; p. 68, var. F.— Spotted Hawk or Falcon, *Edwards*, i. p. 3.—Black Hawk or Falcon, *Id.* pl. 4, both from Hudson's Bay.—Le Lannier, *Pl. enl.* 430, old male.—Le Faucon Noir et Passager, *Id.* 469, young female?—Le Faucon Hors, *Id.* p. 470, yearly.— Faucon pèlerin, *Temm. Man. d'Orn.* p. 22.—*Peale's Museum*, No. 386, female.

FALCO PEREGRINUS.—LINNÆUS.†

Falco peregrinus, *Bonap. Synop.* p. 27.—*North. Zool.* ii. p. 23.—Peregrine Falcon, *Selby, Illust. Br. Orn.* pl. 15, p. 37.—*Flem. Br. Anim.* p. 49.—Falco peregrinus? *Anst. Birds in Coll. Linn. Soc. by Vig. and Horsf.* vol. xv. p. 183.

THIS noble bird had excited our curiosity for a long time. Every visit which we made to the coast was rendered doubly

---

* It is also a European species.

† Among the *Falconidæ* this bird will present that form best adapted

Great-Footed Hawk.
76.

interesting by the wonderful stories which we heard of its exploits in fowling, and of its daring enterprise. There was not

for seizing the prey in an open manner by the exercise of their own organs. Noble in bearing and carriage, the falcons are as much distinguished from the vultures by their graceful proportions, " as those of the lion place him in the ranks of creation above the gaunt, ravenous, grisly, yet dastard wolf." Placed by their strong and powerful frames far beyond them in all rapacious powers, they feed nearly exclusively on living prey, despising all upon which they have not themselves acted as executioners, and particularly any carrion which has the least savour of beginning putrescence. For these purposes they are possessed with a compactly-formed body, the neck comparatively short, and supported by muscles of more than ordinary strength ; the feet and thighs remarkably powerful, and the wings of that true hirundine form and texture which points out the greater development of their power. The prey is generally struck while upon the wing with a rapid sweep, and is at once borne off, unless completely above the weight of the assailer, when it is struck to the ground, and despatched more at leisure.

The peregrine falcon has a considerable geographical range, extending over the whole of temperate Europe, North America, and New Holland. The specimens from the latter country, I may remark, are all smaller in size, but hardly any other distinction can be fixed upon. In Britain, it is abundant on all the rocky coasts towards the north, breeding and frequenting the precipitous headlands ; in many districts inland it is also frequent, but the choice of them is more arbitrary and local. The vale of Moffat, in Dumfriesshire, is one of the most favourite stations I am aware of ; many pairs breed there, and on the confines of Selkirkshire, choosing their eyries among the precipitous cliffs and streams of that mountainous district. They return to the same rock year after year, and often fix upon the same nest for their breeding-place. When either of the birds are killed, a mate is speedily found by the survivor, and returns with him to the old abode, and some of the eyries there have been known, and handed down to recollection, as far as the annals of the district extend. The Bass Rock and Isle of May in the Firth of Forth each possess a pair long renowned in deeds of falconry, and the Isle of Man can boast of many a noble bird, whose ancestors have joined in that now nearly forgotten sport. I am aware of no instance in this country where the peregrine builds on trees, as mentioned by Ord in America ; nor does it seem its true habit there. Dr Richardson remarks, that it is a rare bird in the wooded districts of the Fur Countries, and the greater part of the specimens which have reached this country have been procured upon the coast.

To the American falcons may be added the merlin *F. esalon*, which

a gunner along the shore but knew it well, and each could relate something of it which bordered on the marvellous. It was described as darting with the rapidity of an arrow on the ducks when on the wing, and striking them down with the projecting bone of its breast. Even the wild geese were said to be in danger from its attacks, it having been known to sacrifice them to its rapacity.

To behold this hero, the terror of the wildfowl and the wonder of the sportsmen, was the chief object of our wishes. Day after day did we traverse the salt marshes, and explore the ponds and estuaries which the web-footed tribes frequent in immense multitudes, in the hope of obtaining the imperial depredator ; even all the gunners of the district were summoned to our aid, with the assurance of a great reward if they procured him, but without success. At length, in the month of December 1812, to the unspeakable joy of Mr Wilson, he received from Egg Harbour a fine specimen of the far-famed duck-hawk, which was discovered, contrary to his expectations, to be of a species which he had never before beheld.

If we were to repeat all the anecdotes which have been related to us of the achievements of the duck-hawk, they would swell our pages at the expense, probably, of our reputation. Naturalists should be always on their guard when they find themselves compelled to resort to the observations of others, and record nothing as fact which has not been submitted to the temperate deliberations of reason. The reverse of this procedure has been a principal cause why errors and absurdities have so frequently deformed the pages of works of science, which, like a plain mirror, ought to reflect only the genuine images of nature.

From the best sources of information, we learn that this

was met with by Dr Richardson, who thinks it has been there confounded, from its similarity in some states, with the pigeon-hawk. We may also mention a bird described by Mr Audubon as new under the name of *F. temerarius*, but which appears nothing more than the adult plumage of *F. columbarius*.—ED.

species is uncommonly bold and powerful ; that it darts on its prey with astonishing velocity ; and that it strikes with its formidable feet, permitting the duck to fall previously to securing it. The circumstance of the hawk's never carrying the duck off on striking it has given rise to the belief of that service being performed by means of the breast, which vulgar opinion has armed with a projecting bone adapted to the purpose. But this cannot be the fact, as the breast-bone of this bird does not differ from that of others of the same tribe, which would not admit of so violent a concussion.

When the waterfowl perceive the approach of their enemy, a universal alarm pervades their ranks ; even man himself, with his engine of destruction, is not more terrible. But the effect is different. When the latter is beheld, the whole atmosphere is enlivened with the whistling of wings ; when the former is recognised, not a duck is to be seen in the air : they all speed to the water, and there remain until the hawk has passed them, diving the moment he comes near them. It is worthy of remark, that he will seldom, if ever, strike over the water, unless it be frozen, well knowing that it will be difficult to secure his quarry. This is something more than instinct.

When the sportsmen perceive the hawk knock down a duck, they frequently disappoint him of it, by being first to secure it. And as one evil turn, according to the maxim of the multitude, deserves another, our hero takes ample revenge on them, at every opportunity, by robbing them of their game, the hard-earned fruits of their labour.

The duck-hawk, it is said, often follows the steps of the gunner, knowing that the ducks will be aroused on the wing, which will afford it an almost certain chance of success.

We have been informed that those ducks which are struck down have their backs lacerated from the rump to the neck. If this be the fact, it is a proof that the hawk employs only its talons, which are long and stout, in the operation. One

respectable inhabitant of Cape May told us that he has seen
the hawk strike from below.

This species has been long known in Europe, and in the
age of falconry was greatly valued for those qualifications
which rendered it estimable to the lovers and followers of that
princely amusement. But we have strong objections to its
specific appellation. The epithet *peregrine* is certainly not
applicable to our hawk, which is not migratory, as far as our
most diligent inquiries can ascertain; and, as additional evidence of the fact, we ourselves have seen it prowling near the
coast of New Jersey in the month of May, and heard its
screams, which resemble somewhat those of the bald eagle,
in the swamps wherein it is said to breed. We have therefore taken the liberty of changing its English name for one
which will at once express a characteristic designation, or
which will indicate the species without the labour of investigation.*

"This species," says Pennant, "breeds on the rocks of
Llandidno, in Caernarvonshire, Wales.† That promontory has
been long famed for producing a generous kind, as appears
by a letter extant in Gloddaeth Library from the Lord Trea-
surer Burleigh to an ancestor of Sir Roger Mostyn, in which
his lordship thanks him for a present of a fine cast of hawks,
taken on those rocks, which belong to the family. They are
also very common in the north of Scotland, and are sometimes
trained for falconry by some few gentlemen who still take
delight in this amusement in that part of Great Britain.
Their flight is amazingly rapid; one, that was reclaimed by a
gentleman in the shire of Angus, a county on the east side of
Scotland, eloped from its master with two heavy bells attached

* "Specific names, to be perfect, ought to express some peculiarity
common to no other of the genus."—*Am. Orn.*, i. p. 65.

† We suspect that Pennant is mistaken; its name denotes that it is
not indigenous in Great Britain. Bewick says, "The peregrine or
passenger falcon is *rarely* met with in Britain, and consequently is but
little known with us."—*British Birds*, part i.

to each foot on the 24th September 1772, and was killed on the morning of the 26th, near Mostyn, Flintshire."*

The same naturalist in another place observes, that "*the American species is larger than the European.*† They are subject to vary. The black falcon, and the spotted falcon of Edwards, are of this kind; each preserves a specific mark in the black stroke which drops from beneath the eyes down towards the neck.

"Inhabits different parts of North America, from Hudson's Bay as low as Carolina; in Asia, is found on the highest parts of the Uralian and Siberian chain; wanders in summer to the very arctic circle; is common in Kamtschatka."‡

In the breeding season, the duck-hawk retires to the recesses of the gloomy cedar swamps, on the tall trees of which it constructs its nest, and rears its young secure from all molestation. In those wilds, which present obstacles almost insuperable to the foot of man, the screams of this bird, occasionally mingled with the hoarse tones of the heron, and the hooting of the great horned owl, echoing through the dreary solitude, arouse in the imagination all the frightful imagery of desolation. Mr Wilson and the writer of this article explored two of these swamps in the month of May 1813, in pursuit of the great heron and the subject of this chapter; and although they were successful in obtaining the former, yet the latter eluded their research.

The great-footed hawk is twenty inches in length, and three feet eight inches in extent; the bill is inflated, short and strong, of a light blue colour, ending in black; the upper mandible with a tooth-like process, the lower, with a corresponding notch, and truncate; nostrils round, with a central point like the pistil of a flower; the eye is large and dark, surrounded with

---

* British Zoology.

† If we were to adopt the mode of philosophising of the *sapient* Count de Buffon, we should infer that the European species is *a variety of our more generous race, degenerated by the influence of food and climate!*

‡ Arctic Zoology.

a broad bare yellowish skin, the cartilage over it yellow and prominent; frontlet, whitish; the head above, cheeks running off like moustaches, and back, are black; the wings and scapulars are brownish black, each feather edged with paler, the former long and pointed, reaching almost to the end of the tail; the primaries and secondaries are marked transversely on the inner vanes with large oblong spots of ferruginous white; the exterior edge of the tip of the secondaries curiously scalloped, as if a piece had been cut out; the tertials incline to ash colour; the lining of the wings is beautifully barred with black and white, and tinged with ferruginous; on a close examination, the scapulars and tertials are found to be barred with faint ash; all the shafts are black; the rump and tail-coverts are light ash, marked with large dusky bars; the tail is rounding, black, tipped with reddish white, and crossed with eight narrow bars of very faint ash; the chin and breast, encircling the black moustaches, are of a pale buff colour; breast below and lower parts, reddish buff or pale cinnamon, handsomely marked with roundish or heart-shaped spots of black; sides, broadly barred with black; the femorals are elegantly ornamented with herring-bones of black, on a buff ground; the vent is pale buff, marked as the femorals, though with less numerous spots; the feet and legs are of a corn-yellow, the latter short and stout, feathered a little below the knees, the bare part one inch in length; span of the foot, five inches, with a large protuberant sole; the claws are large and black, hind claw the largest. Whether the cere is yellow or flesh-coloured, we were uncertain, as the bird had been some time killed when received; supposed the former.

The most striking characters of this species are the broad patch of black dropping below the eye, and the uncommonly large feet. It is stout, heavy, and firmly put together.

The bird from which the above description was taken was shot in a cedar swamp in Cape May county, New Jersey. It was a female, and contained the remains of small birds, among which were discovered the legs of the sanderling plover.

CONTINUATION

OF

WILSON'S
AMERICAN ORNITHOLOGY.

BY

CHARLES LUCIAN BONAPARTE.

1. Fork tailed Flycatcher. 2. Rocky Mountain Anteater. 3. Female Golden winged Warbler.

Muscicapa Savana. Myiothera Obsoleta. Sylvia Chrysoptera.

CONTINUATION

OF

WILSON'S

AMERICAN ORNITHOLOGY.

FORK-TAILED FLYCATCHER. (*Muscicapa savanna.*)

PLATE I.—Fig. 1.

Muscicapa tyrannus, *Linn.* Syst. i. p. 325, sp. 4.—*Gmel.* Syst. i. p. 931, sp. 4.—
Lath. Ind. p. 484, sp. 69.—Tyrannus savanna, *Vieill. Ois. de l'Am. Sept.* i. p.
72, pl. 43 (a South American specimen).—*Vieill. Nouv. Dict. d'Hist. Nat.*
xxxv. p. 87.—Muscicapa tyrannus cauda bifurca, *Briss. Av.* ii. p. 395, sp. 20,
pl. 39, fig. 3.—Le Moucherolle Savanna, *Buff.* iv. p. 557, pl. 26.—Le Tyran
à queue fourchue de Cayenne, *Buff. Pl. enl.* 571, fig. 2.—Fork-tailed Flycatcher, *Penn. Arct. Zool.* sp. 265.—*Lath. Syn.* ii. part 1, p. 355, sp. 59.—
Phil. Museum, No. 6620.

MILVULUS SAVANNA.—SWAINSON.*

Tyrannus savanna, *Brand's Journ.* No. lx. p. 282.—Muscicapa savanna, *Bonap.*
Synop. p. 67.

THOUGH Brisson, Linné, and Pennant have stated the fork-
tailed flycatcher to inhabit this region as far north as Canada,

---

* The latter part of the description of this bird will show that more
than one species has been confounded with it ; and I am not sure that
those very closely allied are yet properly unravelled. The present
species has more decidedly marked habits, and will most probably be
found entirely restricted to the northern parts of the south continent,
and extending slightly, and in small numbers, into the north.

The peculiar form of the tail, and the extreme development of the
organs of flight, induced Mr Swainson to use these birds as typical of

still the fact seemed more than doubtful, since this bird
escaped the researches of Vieillot, and, what is more extraordinary, those of the indefatigable Wilson. It is, therefore,
a very gratifying circumstance that we are able to introduce
this fine bird with certainty into the ornithology of the
United States, and, by the individual represented in the annexed plate, to remove all doubt on the subject. The specimen from which our drawing was made is a beautiful male
in full plumage; it was shot near Bridgetown, New Jersey,
at the extraordinary season of the first week in December,
and was presented by Mr J. Woodcraft, of that town, to Mr
Titian Peale, who favoured me with the opportunity of examining it.

Brisson published the first account of this bird. That we
have rejected the name given by Linné may appear contrary
to our principles; but, in this instance, we certainly have no
option, inasmuch as the same name has been very properly
retained by Wilson, agreeably to Brisson, for the *Lanius
tyrannus* of Linné. Had Linné himself included them both
in the same genus, he would doubtless have retained that
specific name for the kingbird, which is unquestionably a
Muscicapa, and not a *Lanius*. As the kingbird is a very
abundant species, known to every zoological reader by the
name of *Tyrannus*, it is obvious that less inconvenience will
be produced by changing the name of an almost unknown
species, than would result from altering that of one with
which we are so familiar. We have, therefore, adopted
Vieillot's specific name of *savanna*, taken by that author
from Montbeillard, who in Buffon's work thereby endeavoured
to commemorate this bird's habit of frequenting inundated

a section among the tyrants, to which he has given the characters of
" *Alæ longæ remigum pogoniis internis emarginatis; cauda longissima forficata;* " * and afterwards to give it the rank of a subgenus, under the
title of *Milvulus*, with which will also range the swallow-tailed flycatcher
of next plate, *Milvulus longipennis*, &c.; making this curious group to
contain at least six or seven known species.—ED.

---

* See Monographical Sketch, Brand's Journal, No. LX. p. 282.

savannas. Naturalists who separate *Tyrannus* from *Musci-capa* generically, disagree with respect to the arrangement of this species. For ourselves, we consider the former as a sub-genus of *Muscicapa*, including the larger species, among which our fork-tailed flycatcher must be placed.

This species is fourteen inches long, its tail measuring nearly ten ; the extent, from the tip of one wing to that of the other, is fourteen inches. The bill is somewhat more slender and depressed at base than that of the kingbird, and, as well as the feet, is black. The irides are brown. The upper part of the head, including the cheeks and superior origin of the neck, is velvet black. The feathers of the crown are some-what slender, elevated, and of a yellow orange colour at base, constituting a fine spot, not visible when they are in a state of repose ; the remaining part of the neck above and the back are greyish ash ; the rump is of a much darker greyish ash, and gradually passes into black, which is the colour of the superior tail-coverts ; the inferior surface of the body, from the base of the bill, as well as the under wing and under tail-coverts, is pure white. The wings are dusky, the coverts being somewhat lighter at tip and on the exterior side ; the first primary is edged with whitish on the exterior web, and is equal in length to the fourth ; the second primary is longest ; the three outer ones have a very extraordinary and profound sinus or notch on their inner webs, near the tip, so as to ter-minate in a slender process. The tail is very profoundly forked, the two exterior feathers measuring nearly ten inches in perfect individuals, whilst the two succeeding are but five inches long, and the other feathers become gradually and pro-portionally shorter, until those in the middle are scarcely two inches in length ; the tail is, in fact, so deeply divided, that if the two exterior feathers were removed, it would still exhibit a very forked appearance. All the tail feathers are black, the exterior one on each side being white on the remarkably narrow outer web, and on the shaft beneath, for nearly three-fourths of its length.

L

I cannot agree with those who say that the female is distinguished from the other sex by wanting the orange spot on the head, as I think we may safely conclude, from analogy, that there is hardly any difference between the sexes. The young birds are readily recognised by being destitute of that spot, as well as by having the head cinereous instead of black; the colour of the whole upper part of the body is also darker, the tail considerably shorter, and the exterior feathers not so much elongated as those of the adult. It is proper to remark, that the elongated tail-feathers of the full-grown bird are sometimes very much worn, in consequence of the rapidity with which it passes through the bushes.

Two coloured figures have been given of the fork-tailed flycatcher—the one by Buffon, which is extremely bad, although the rectilinear form of the tail is correctly represented; the other by Vieillot, which has the exterior tail-feathers unnaturally curved, and notwithstanding it is preferable to Buffon's figure, yet it is far from being accurate. This author having been unable to procure a North American specimen, chose nevertheless to introduce the species in his "Natural History of North American Birds," on the authority of former authors, giving a figure from a South American specimen. The error in representing the exterior tail-feathers curved doubtless arose from the manner in which the dried skin was packed for transportation. That our drawing of this graceful bird is far superior to those above mentioned will at once be evident on comparison; this superiority is owing to the circumstance of this drawing, like all the others given in the present work, being made from the recent specimen. Buffon's plain figure is a more faithful representation than that given in his coloured engravings.

From the very great rarity of the fork-tailed flycatcher in this region, and the advanced season in which this individual was killed, it is evident that it must have strayed from its native country under the influence of extraordinary circumstances; and we are unable to believe that its wanderings have ever extended as far as Canada, notwithstanding the statements of authors to the contrary. It may be proper to observe, that the difference indicated by Linné and Latham between the variety which they suppose to inhabit Canada and that of Surinam, appears to have no existence in nature.

Although this bird is so very rare and accidental here, we should be led to suppose it a more regular summer visitant of the southern States, were it not impossible to believe that so showy a bird could have escaped the observation of travellers ; hence we infer that the fork-tailed flycatcher must be included in the catalogue of those species which are mere fortuitous visitors to the United States. As but a single specimen of this bird has been obtained, I cannot give any account of its manners and habits from personal observation.

The native country of the fork-tailed flycatcher is Guiana, where it is rather common, and is improperly called *veuve* (widow), from the great length of its tail, in which character only it resembles the African birds of that name.

The habits of the fork-tailed flycatcher resemble those of other species of the same genus. It is a solitary bird, remaining for a long time perched on the limb of a tree, whence it occasionally darts after passing insects ; or, flying downwards, it alights on the tufts of herbage which appear above the water, affording it a resting-place in the midst of those partially inundated lands called savannas, beyond the limits of which it is not frequently seen. While on the tuft, this bird moves its tail in a manner similar to that of the wagtails. Besides insects, the fork-tailed flycatcher feeds occasionally on vegetable substances, as, on dissection, the stomach of our specimen was found to be filled with pokeberries (*Phytolacca decandra*, L.)

Beyond these particulars, we have no positive knowledge of the manners of our flycatcher, though Vieillot has recorded a history of some length, taken from D'Azara ; but the bird observed by the latter author in Paraguay and Buenos Ayres, though closely allied, appears to be specifically distinct from the one we are describing. Vieillot has since been convinced of this difference, and in the (French) "New Dictionary of Natural History," he has separated the more southern species, under the name of *Tyrannus violentus.* In colour, that bird strongly resembles our *Muscicapa savanna,* but it is considerably smaller, and has different habits, being gregarious; whilst the *savanna,* as we have already stated, is a solitary bird.

Another species for which ours may be readily mistaken is the *Tyrannus bellulus,* Vieill., which, however, is much larger, with a still longer tail, differing also by having a large black collar extending to each corner of the eye, margining the white throat ; and the head of the same bluish grey colour with the other superior parts of the body ; the remaining under parts being of the same colour, with a narrow brown line in the middle of each feather ; and by having a whitish line on each side of the head behind the eye, extending to the occiput. The *Tyrannus bellulus* is a native of Brazil.

ROCKY MOUNTAIN ANTCATCHER.
(*Myiothera obsoleta.*)

PLATE I.—Fig. 2.

Troglodytes obsoleta, *Say, in Long's Expedition to the Rocky Mountains,* vol. ii. p. 4.—*Philadelphia Museum,* No. 2420.

TROGLODYTES OBSOLETA.—Say.*

Myiothera obsoleta, *Bonap. Synop.* p. 73.

This bird is one of those beings which seem created to puzzle the naturalist, and convince him that Nature will never conform to his systems, however perfect his ingenuity may be capable of devising them. This will become sufficiently apparent when we consider in what manner different authors would have arranged it.

We cannot positively decide whether Vieillot and his fol-

* We prefer retaining this bird for the present in *Troglodytes.* The habits, colour, and marking, nest and call, of this bird, bring it nearer to the wrens. There is no question, however, of its being an aberrant form, wherever it may rank.—Ed.

lowers would have referred this species to *Myrmothera*, a name they have substituted for *Myiothera* ; to their genus *Thryothorus*, which we unite to *Troglodytes* ; or to their slender-billed section of *Thamnophilus*, rejected by us from that genus, and of which some recent authors have made a genus called *Formicivora* : yet we have very little hesitation in stating our belief that they would have assigned its place among the species of the latter. According to our classification, it is certainly not a *Thamnophilus*, as we adopt the genus, agreeably to the characters given by Temminck, who, not admitting the genus *Troglodytes*, would undoubtedly have arranged this bird with *Myiothera*, as Illiger would also have done.

The only point, therefore, to be established by us is, whether this bird is a *Myiothera* or a *Troglodytes*. It is, in fact, a link intermediate to both. After a careful examination of its form, especially the unequal length of the mandibles, the notch of the superior mandible, and the length of the tarsus, and after a due consideration of the little that is known relative to its habits, we unhesitatingly place it with *Myiothera*, though, in consequence of its having the bill more slender, long, and arcuated than that of any other species I have seen, it must occupy the last station in the genus, being still more closely allied to *Troglodytes* than those species whose great affinity to that genus has been pointed out by Cuvier. This may be easily ascertained by comparing the annexed representation with the figures given by Buffon and Temminck. The figure which our Rocky Mountain antcatcher resembles most is Buffon's *Pl. Enl.* 823, fig. 1 (*Myiothera lineata*). The colours of our bird are also similar to those of a wren ; but this similitude is likewise observed in other *Myiotherœ*.

The bird now before us was brought from the Arkansaw river, in the neighbourhood of the Rocky Mountains, by Major Long's exploring party, and was described by Say under the name of *Troglodytes obsoleta*, from its close resemblance to the Carolina wren (*Troglodytes Ludovicianus*), which Wilson considered a *Certhia*, and Vieillot a *Thryothorus*.

As the Rocky Mountain antcatcher is the first and only species hitherto discovered in North America, we shall make
some general observations on the peculiarities of a genus thus
introduced into the fauna of the United States.

Buffon first formed a distinct group of the antcatchers,
under the name of *Fourmiliers*, and considered them as allied
to his *Brèves*, now forming the genus *Pitta* of Vieillot, they
having been previously placed in that of *Turdus*. Lacépède
adopted that group as a genus, and applied to it the name
of *Myrmecophaga*. Illiger added such species of the genus
Lanius of Linné and Latham as are destitute of prominent
teeth to the bill, and gave to the genus, thus constituted,
the name of *Myiothera*, rejecting Lacépède's designation, as
already appropriated to a genus of mammalia.

Cuvier perceived that some of the *Fourmiliers* of Buffon
were true thrushes ; but he retained the remainder as *Myio-
theræ*, among which he also included the *Pittæ*. Vieillot,
besides the *Pittæ*, removed some other species, in order to
place them in his new genera *Conopophaga* and *Thamnophilus*,
giving the name of *Myrmothera* to the remaining species, with
the exception of the *Myiothera rex*, for which he formed a
distinct genus, with the name of *Grallaria*. We agree with
Vieillot in respect to the latter bird ; but as regards the
other species, we prefer the arrangement of Temminck, who
has adopted the genus *Myiothera* nearly as constituted by
Illiger, including some of the slender-billed *Thamnophili* of
Vieillot, of which our *Myiothera obsoleta* would probably be
one, as above stated.

The genus thus constituted contains numerous species, which
inhabit the hottest parts of the globe, a greater number of
them existing in South America than elsewhere. For the
sake of convenience, several sections may be formed in this
genus, founded on the characters of the bill, tail, and tarsus ;
but as we have only one species, it does not rest with us to
make divisions ; and we shall merely remark, that our *obsoleta*
is referable to the last section, consisting of those whose bills

are the most slender, elongated, and arcuated, in company with the *Turdus lineatus* of Gmelin.

The antcatchers may justly be enumerated amongst the benefactors of mankind, as they dwell in regions where the ants are so numerous, large, and voracious, that, without their agency, co-operating with that of the *Myrmecophaga jubata*, and a few other ant-eating quadrupeds, the produce of the soil would inevitably be destroyed in those fertile parts of the globe. The anthills of South America are often more than twenty feet in diameter, and many feet in height. These wonderful edifices are thronged with two-hundred-fold more inhabitants, and are proportionally far more numerous, than the small ones with which we are familiar. Breeding in vast numbers, and multiplying with great celerity and profusion, the increase of these insects would soon enable them to swarm over the greatest extent of country, were not their propagation and diffusion limited by the active exertions of that part of the animal creation which continually subsist by their destruction.

The antcatchers run rapidly on the ground, alighting but seldom on trees, and then on the lowest branches; they generally associate in small flocks, feed exclusively on insects, and most commonly frequent the large anthills before mentioned. Several different species of these birds are often observed to live in perfect harmony on the same mound, which, as it supplies an abundance of food for all, removes one of the causes of discord which is most universally operative throughout animated nature. On the same principle, we might explain the comparative mildness of herbivorous animals, as well as the ferocity and solitary habits of carnivorous, and particularly of rapacious animals, which repulse all others from their society, and forbid even their own kind to approach the limits of their sanguinary domain.

The antcatchers never soar high in the air, nor do they extend their flight to any great distance without alighting to rest, in consequence of the shortness of their wings and tail, which, in fact, seem to be seldom employed for any other purpose than to assist them in running along the ground, or in leaping from branch to branch of bushes and low trees,—an exercise in which they display remarkable activity. Some species, like the woodpeckers, climb on the trunks of trees in pursuit of insects; and it would appear, from their restless habits and almost constant motion, that their limited excursions are entirely attributable to the want of more ample provision for flight. The antcatchers are never found in settled districts, where their favourite insects are generally less abundant; but they live in the dense and remote parts of forests, far from the abodes of man and civilisation. They also dislike open and wet countries.

The note of the antcatchers is as various as the species are different; but it is always very remarkable and peculiar. Their flesh is oily, and disagreeable to the taste; and, when the bird is opened, a very offensive odour is diffused from the remains of half-digested ants and other insects contained in the stomach.

The plumage of the antcatchers very probably undergoes considerable changes in colour. The size of the sexes is different, the female being much larger than the male. Such variations may have induced naturalists to consider many as species that really do not exist as such in nature.

The nest of these birds is hemispherical, varying in magnitude according to the size of the species, composed of dried grass rudely interwoven; it is fixed to small trees, or attached by each side to a branch, at the distance of two or three feet from the ground. The eggs are nearly round, and three or four in number.

The discovery of any species of this genus in the Old World is quite recent, and it had previously been believed that the genus was peculiar to South America; and though the existence of ant-destroying birds was suspected in other tropical regions, they were supposed to be generically distinct from those of the corresponding parts of America, as was known to

be the fact in the case of the ant-eating quadrupeds. This
opinion was founded on the admitted axiom that Nature
always varies her groups in remote tropical regions having no
communication with each other. The reverse, however, is the
fact in the case of the ant-catching birds, as we find perfect
analogies between the species residing in those distant parts
of the globe, even throughout the different sections into which
the genus may be divided.

The Rocky Mountain antcatcher is six inches long. The
bill, measured from the corner of the mouth, is more than one
inch in length, being slightly curved almost from the base ;
it is very slender, being nearly two-eighths of an inch in diameter at the base, and only the sixteenth of an inch in the
middle, whence it continues to diminish to the tip ; and is of
a dark horn colour, paler beneath. The feet are dusky, and
the length of the tarsus is seven-eighths of an inch. The irides
are dark brown ; the whole plumage above is of a dusky
brownish, slightly undulated with pale, tinted with dull ferruginous on the top of the head and superior portions of the
back. The sides of the head are dull whitish, with a broad
brown line passing through the eye to the commencement of
the neck. The chin, throat, and breast are whitish, each
feather being marked by a longitudinal line of light brown.
The belly is white, and the flanks are slightly tinged with
ferruginous. The primaries are entirely destitute of undulations or spots ; the tail-coverts are pale, each with four or five
fuscous bands ; the inferior tail-coverts are white, each being
bifasciate with blackish brown. The tail is nearly two inches
long, rounded, broadly tipped with ferruginous yellow, and
having a narrow black band before the tip : the remaining
part of the tail is of the same colour with the wings, and is
obsoletely banded, these bands being more distinct on the two
middle feathers, which are destitute of the black and yellowish
termination ; the exterior feather is dusky at tip, marked by
four yellowish white spots on the exterior, and by two larger
ones on the inner web.

The specimen of the Rocky Mountain antcatcher we are describing is a male, shot in the month of July, and possibly not adult. As it is the only one brought by Major Long's party, we cannot determine the extent or nature of the variations the species may undergo from age, sex, or season.

The note of this bird is peculiar, resembling the harsh voice of the terns. It inhabits the sterile country bordering on the river Arkansaw, in the neighbourhood of the Rocky Mountains, where it is frequently observed hopping on the ground, or flitting among the branches and weatherbeaten, half-reclining trunks of a species of juniper: when it flies among the crooked limbs of this tree, it spreads its tail considerably, but was never seen to climb. They were generally observed in small associations of five or six individuals, perhaps composing single families.

FEMALE GOLDEN-WINGED WARBLER.
(*Sylvia chrysoptera.*)

PLATE I.—FIG. 3.

See *Wilson's American Ornithology,* ii. p. 113, pl. 15, fig. 5, for the male.—Motacilla chrysoptera, *Linn. Syst.* i. p. 333, sp. 20.—*Gmel. Syst.* i. p. 971, sp. 20 (male).—Sylvia chrysoptera, *Lath. Ind.* p. 541, sp. 123.—*Vieill. Ois. de l'Am. Sept.* ii. p. 37, pl. 97 (male).—Motacilla flavifrons, *Gmel. Syst.* i. p. 976, sp. 126 (male).—Sylvia flavifrons, *Lath. Ind.* p. 527, sp. 69 (male).— Ficedula Pennsylvanica Cinerea Gutture Nigro, *Brisson, Av. Supp.* p. 109, sp. 80 (male).—Figuier aux aîles dorées, *Buff. Ois.* v. p. 311 (male).—Goldenwinged Flycatcher, *Edwards, Glean.* ii. p. 189, pl. 299 (male).—Gold-winged Warbler, *Penn. Arct. Zool.* sp. 295.—*Lath. Syn.* ii. part 2, p. 492, sp. 118 (male).—Yellow-fronted Warbler, *Penn. Arct. Zool.* sp. 296.—*Lath. Syn.* ii. part 2, p. 461, sp. 67 (male).—Parus alis aureis, the Golden-winged Flycatcher, *Bartram, Trav.* p. 292 (male).—*Philadelphia Museum,* No. 7010, male ; No. 7011, female.

VERMIVORA CHRYSOPTERA.—Swainson.
Male, vol. i. p. 259.

THE female of this pretty little warbler, hitherto unknown to any naturalist, is now figured and described for the first time. For the opportunity of presenting it to the reader, we are indebted to Mr Titian Peale, who shot it on the 24th of May, near Camden, New Jersey, and, with his usual kindness and zeal for natural history, communicated it to us for this work.

This little warbler differs so materially from its mate as to require a distinct figure and description in order to be recognised ; yet we cannot fail to perceive a kind of family resemblance between the sexes ; and by comparing the two descriptions and accompanying figures, our readers will agree with us that they are but one and the same species, in a different garniture of plumage. The distribution of markings is really similar in both sexes, but in the female the colours are paler, and green prevails on those parts which in the male are of a dark slate colour.

The female of the golden-winged warbler is four and a half inches long. The bill is blackish, straight, entire, rounded, and gradually tapering to a sharp point. The feet are brownish ash ; the irides, dark brown. The front is golden yellow ; the top of the head, bright olive yellow ; the back of the head, and superior parts of the neck and body, are of a pale plumbeous hue, the feathers being tipped with yellow olive, more particularly on the rump ; the superior tail-coverts are pure pale plumbeous. A wide slate-coloured stripe passes through the eye from the bill, and dilates on the cheeks ; this is margined by a white line above the eye, and by a wider one on each side of the throat. The throat is of a pale slate colour, becoming still paler on the breast. The remaining under parts are whitish, occasionally tinged with yellow, and with slate colour on the flanks. The wings are of the same colour as the back, but somewhat darker, and are crossed by two wide bands of bright yellow, formed by the tips of the first and second rows of wing-coverts. The primaries are dusky, margined on the exterior web with pale, and on the inner broadly with white. The secondaries are broadly margined with yellow olive on the outer web, and with white on the inner web. The tail is nearly even at tip, of a dusky plumbeous colour ; the three lateral feathers have a large pure white spot on the inner web.

This last essential character also exists in the male, though Wilson has not mentioned it. As to the manners and habits of the species, he has given us no information, except that it is rare, and remains only a few days in Pennsylvania. He says nothing of the female, and Vieillot never saw it.

We regret that we are unacquainted with the form of its nest, and the peculiarity of its song. We can only state, that during its short stay in Pennsylvania, it is solitary and silent, gleaning amongst the branches of trees, and creeping much after the manner of the titmouse, with its head frequently downwards, in pursuit of larvæ and insects, which constitute exclusively the food of this species.

Wilson was impressed with the opinion that the shape of the bill would justify the formation of a distinct subgenus, which would include this bird, the *Sylvia vermivora*, and some other species. In this opinion Cuvier has coincided, by forming his subgenus *Dacnis*, which he places under his extensive genus *Cassicus*, remarking that they form the passage to *Motacilla*. This subgenus we shall adopt, but we differ from Cuvier by arranging it under *Sylvia ;* it will then form the transition to the more slender-billed *Icteri*. Temminck and Vieillot have arranged them also under *Sylvia ;* the latter author, in the (French) "New Dictionary of Natural History," gives them the name of *pitpits ;* and it is most probably from want of examination that he has not considered the present bird as belonging to that section.*

* The opinion of Wilson, now mentioned by his continuator, shows the accurate perception he had of the generic forms and modifications of birds : the subdivision he mentions' has actually been made by various ornithologists. Holding different views, we certainly also prefer placing it among the *Sylviadæ*, but it may lead off in other directions according to the ideas of the systematist and the mode of analysis he pursues. *Vermivora* is now retained, on account of, as far as we are aware, the more restricted form.—ED.

1. Swallow tailed Flycatcher. 2. Arkansas Flycatcher. 3. Says Flycatcher. 4. Female Golden crested Wren.

Muscicapa Forficata.   Muscicapa Verticalis.   Muscicapa Saya.   Regulus Cristatus.

SWALLOW-TAILED FLYCATCHER. (*Muscicapa forficata.*)

PLATE II.—FIG. 1.

Muscicapa forficata, *Gmel. Syst.* i. p. 931, sp. 22.—*Lath. Ind.* p. 485, sp. 70.—*Vieill. Ois. de l'Am. Sep.* i. p. 71.—*Stephens, Cont. of Shaw's Zool.* xx. p. 413, pl. 3.—Tyrannus forficatus, *Say, in Long's Expedition to the Rocky Mountains,* ii. p. 224.—Moucherolle à queue fourchue du Mexique, *Buff. Ois.* iv. p. 564.—Gobe-mouche à queue fourchue du Mexique, *Buff. Pl. enl.* 677.—Swallow-tailed Flycatcher, *Lath. Syn.* ii. part i. p. 356, sp. 50.—*Philadelphia Museum,* No. 6623.

MILVULUS FORFICATUS.—SWAINSON.

Muscicapa forficata, *Bonap. Synop.* p. 67.

THIS rare and beautiful bird is, I believe, now figured from nature for the second time; and as the plate given by Buffon conveys but an imperfect idea of its characters, the representation in the accompanying engraving will certainly prove the more acceptable to naturalists. That author had the merit of publishing the first account of this species; and the individual he described was received from that part of Louisiana which borders on Mexico. Neither Latham, Gmelin, nor Vieillot seem to have had an opportunity of examining this bird, as they have evidently drawn on Buffon for what they have said relative to it. Hence it appears that the swallow-tailed flycatcher has never been obtained from the time of Buffon to the period of Major Long's expedition to the unexplored region it inhabits. The specimen before us, which is a fine adult male, was shot by Mr Titian Peale, on the 24th of August, on the Canadian fork of the Arkansaw river.

Although this bird is very different from the fork-tailed flycatcher, yet, on account of the form of the tail, and the similarity of the common name, they are apt to be mistaken for each other, and when both are immature, some caution is required to avoid referring them to the same species. Notwithstanding this similarity, some authors have placed the fork-tailed flycatcher in their genus *Tyrannus,* and the present bird in *Muscicapa*, whereas, from an inspection of the bills, it will at once be seen that the latter would be still more properly placed in their genus *Tyrannus*, as the form of its bill is exactly the same with that of the kingbird, the type of the subgenus.

The swallow-tailed flycatcher, when in full plumage, is eleven inches long. The bill and feet are blackish; the irides are brown (red, according to authors). The upper part of the head and neck is of a light grey; the back and scapulars are dark cinereous, tinged with reddish brown; the rump is of the same colour, but strongly tinged with black, and the superior tail-coverts are deep black; the under part of the body is milk-white, the flanks being tinged with red; the inferior tail-coverts are pale rosaceous; the wings are brownish black, the upper coverts and secondaries being margined externally, and at tip with dull whitish; the under wing-coverts are whitish rosaceous; the axillary feathers, above and beneath, are of a vivid scarlet colour. The tail is greatly elongated and excessively forked; it is of a deep velvet black colour, each feather having the terminal margin of a dull whitish tint, and the shafts white at their bases. The three exterior feathers on each side are of a delicate pale rosaceous colour on a considerable part of their length from the base. The external one is five inches and a half long; the second and third gradually decrease in length, but the fourth is disproportionately shorter, and from this feather there is again a gradual decrease to the sixth, which is little more than two inches long.

The female of the swallow-tailed flycatcher is probably very similar to the male, but the colours of the young bird are much less vivid, and the exterior tail-feathers are much shorter than those of the adult.

The swallow-tailed flycatcher is as audacious as the kingbird, attacking with unhesitating intrepidity, and turning the flight of the most powerful of the feathered tribe. Its note consists of a chirping sound, like *tsch, tsch,* much resembling that of the prairie dog (*Arctomys Ludoviciana*, Ord), by which it deceived the members of Long's party into a belief that they were approaching one of the villages of this animal.

"A note like that of the prairie dog," writes Say, "for a moment induced the belief that a village of the marmot was near, but we were soon undeceived by the appearance of the beautiful *Tyrannus forficatus* in full pursuit of a crow. Not at first view recognising the bird, the fine elongated tail-plumes, occasionally diverging in a furcate manner, and again closing together to give direction to the aerial evolutions of the bird, seemed like extraneous processes of dried grass or twigs of a tree adventitiously attached to the tail, and influenced by currents of wind. The feathered warrior flew forward to a tree, whence, at our too near approach, he descended to the earth at a little distance, continuing at intervals his chirping note. This bird seems to be rather rare in this region; and as the very powder within the barrels of our guns was wet, we were obliged to content ourselves with only a distant view of it."

The range of the swallow-tailed flycatcher appears to be limited to the trans-Mississippian territories, lying on the south-western frontier of the United States, more especially frequenting the scanty forests, which, with many partial and often total interruptions, extend along the Arkansaw, Canadian, and Platte rivers, where in some districts they do not seem to be very uncommon.

ARKANSAW FLYCATCHER. (*Muscicapa verticalis.*)

PLATE II.—Fig. 2.

Tyrannus verticalis, *Say, in Long's Expedition to the Rocky Mountains*, ii. p. 60.—*Philadelphia Museum*, No. 6624.

TYRANNUS VERTICALIS.—Say.

Muscicapa verticalis, *Bonap. Synop.* p. 67.

This bird, brought from the Rocky Mountains by Major Long's exploring party, is so closely allied to many imperfectly described species of the extensive genus to which it belongs, that ornithologists, at first sight, may very reasonably doubt its pretensions to rank as a new species. But notwithstanding any doubt that may be produced by its similarity to others, it is certainly an addition to the already numerous catalogue of flycatchers.

The total length of the Arkansaw flycatcher is eight inches. The bill is similar to that of the crested flycatcher, but is more rounded above, and more abruptly inflected at tip, being of a blackish colour, as well as the feet. The head above, and nucha, are pure pale plumbeous ; the crown has a restricted bright orange spot in the middle, invisible when the feathers are at rest; there is a dusky spot between the bill and eyes. The cervix and back are pale plumbeous, tinged with olivaceous, and deepening on the rump almost to blackish, which is the colour of the superior tail-coverts. The chin is whitish ; the throat and upper part of the breast are of the same colour as the head, but paler ; the remaining under surface, including the inferior wing and tail-coverts, is yellow. The wings are brown, the secondaries being margined exteriorly with whitish ; the inner webs of the primaries are whitish towards the base, and near the tips they are narrowed ; the first is remarkably so, being almost falciform. The tail is of a deep brown black colour, and very slightly emarginated ; the exterior feather is white on the outer web, the shaft being white on the exterior half, and brown on the interior.

Say first described and named this bird in the second volume of the work above quoted ; and he remarks that it is allied to the *Tyrannus griseus* and *Tyrannus sulphuratus* of Vieillot. There are many species for which the Arkansaw flycatcher might more readily be mistaken : of these, we may mention the crested flycatcher (*Muscicapa crinita*), so well described and figured by Wilson in his second volume ; and particularly the *Muscicapa ferox* * of Gmelin, a South American bird, the

* This bird has been incorrectly considered by Vieillot, in his "Natural History of North American Birds," as identical with the *Muscicapa crinita*, but afterwards perceiving it to be a distinct species, he named it *Tyrannus ferox*. A specimen is in the Philadelphia Museum, designated

description of which agrees so well with the species we are now considering, that it might be equally applied to either. Our bird differs from the two latter by that striking character, the white exterior web of the outer tail-feather. From the *crinita* it may more especially be known by the spot on the crown, which does not exist in that species; by not having the tail and wing-feathers rufous in any part; and by having the primaries narrowed at tip, while the *crinita* has them quite large, entire, and rounded. On a particular comparison with the *ferox*, we shall perceive that the bill of that bird is flattened, broad, and carinate, whilst in the *verticalis* it is almost rounded above. The general colour of the latter is, besides, much paler, and the tail is less deeply emarginated.

The Arkansaw flycatcher appears to inhabit all the region extending west of the Missouri river. The specimen we have been describing is a male, killed in the beginning of July, on the river Platte, a few days' march from the mountains.

SAY'S FLYCATCHER. (*Muscicapa Saya.*)

PLATE II.—FIG. 3.

Philadelphia Museum, No. 6831.

TYRANNULA SAYA.—SWAINSON.*

Muscicapa Saya, *Bonap. Synop.* p. 67.—Tyrannula pallida, *Sw. Synop. Mex. Phil. Mag.* No. 15.—Tyrannula Saya, *North. Zool.* ii. p. 142.

WE now introduce into the fauna of the United States a species which is either a nondescript, or one that has been improperly named; and I dedicate it to my friend Thomas Say, a naturalist of whom America may justly be proud, and whose talents and knowledge are only equalled by his modesty. The

by the fanciful name of ruby-crowned flycatcher (with this Say compared his *Tyrannus verticalis* before he stated it to be new), and in the New York Museum three specimens are exhibited with the erroneous title of whiskered flycatcher (*Muscicapa barbata*).

* One or two specimens of this apparently uncommon bird were procured by the Overland Arctic Expedition, agreeing generally with the birds now described. Mr Swainson has received it from Mexico.—ED.

specimen now before us is a male, shot by Mr T. Peale, on the 17th of July, near the Arkansaw river, about twenty miles from the Rocky Mountains.

We cannot be perfectly sure that this flycatcher has not heretofore been noticed, since we find in the books two short unessential descriptions which might be supposed to indicate it. One of these is the *Muscicapa obscura* of Latham (dusky flycatcher of his "Synopsis"), from the Sandwich Islands; but, besides the difference of the tail-feathers, described as acute in that bird, the locality decides its identity with ours. The other description is that of a bird from Cayenne, the *Muscicapa obscura* of Vieillot,* given by that author as very distinct from Latham's, although he has applied the same name to it, no doubt inadvertently. This may possibly be our bird; but, even in this case, the name we have chosen will necessarily be retained, as that of *obscura* attaches to Latham's species by the right of priority.

This flycatcher strongly resembles the common pewee (*Muscicapa fusca*), but differs from that familiar bird by the very remarkable form of the bill; by the colour of the plumage, which verges above on cinnamon brown, instead of greenish, and beneath is cinereous and rufous, instead of yellowish ochreous; and by the proportional length of the primary feathers, the first being longer than the sixth in our bird, whereas it is shorter in the pewee.

The total length of Say's flycatcher is seven inches; the bill is long, straight, and remarkably flattened; the upper mandible is blackish, and but very slightly emarginated; the lower mandible is much dilated, and pale horn-colour on the disk. The feet are blackish; the irides are brown. The general colour of the whole upper parts is dull cinnamon brown, darker on the head; the plumage at base is of a lead colour. The throat and breast are of the same dull cinnamon tint, gradually passing into pale rufous towards the belly, which is entirely of the latter colour; the under wing-coverts

* Nouv. Dict. d'Hist. Nat., xxi. p. 451.

are white, slightly tinged with rufous. The primaries are dusky, tinged with cinnamon, and having brown shafts; they are considerably paler beneath. The first primary is a quarter of an inch shorter than the second, which is nearly as long as the third; the third is longest; the fourth and fifth gradually decrease, and the sixth is decidedly shorter than the first. The tail is hardly emarginated, and of a blackish brown colour.

We know nothing of the habits of this flycatcher, except what has been communicated by Mr T. Peale, from his manuscript notes. The bird had a nest in July, the time when it was obtained; its voice is somewhat different from that of the pewee, and first called attention to its nest, which was built on a tree, and consisted chiefly of moss and clay, with a few blades of dried grass occasionally interwoven. The young birds were, at that season, just ready to fly.

FEMALE GOLDEN-CROWNED GOLD-CREST.
(*Regulus cristatus.*)

PLATE II.—FIG. 4.

Wilson's American Ornithology, i. p. 126, pl. 8, fig. 2, for the male.—Motacilla regulus, *Linn. Syst.* i. p. 338, sp. 48.—*Gmel. Syst.* i. p. 995, sp. 48.—Sylvia regulus, *Lath. Ind.* p. 548, sp. 152.—*Temm. Man. d'Orn.* p. 229.—*Ranzani, Elem. di Zool.* iii. part 5, p. 105, pl. 16, fig. 3.—Regulus cristatus, *Ray, Syn.* p. 79, sp. 9.—*Aldr. Orn.* ii. p. 649.—*Will. Orn.* p. 163, pl. 42.—*Vieill. Nouv. Dict. d'Hist. Nat.* xxix. p. 420.—Regulus vulgaris, *Stephens, Cont. of Shaw's Zool.* xx. p. 758, pl. 59.—Parus calendula, Regulus cristatus vulgo dicta, *Briss. Av.* iii. p. 579, sp. 17.—Le Roitelet, *Gerardin, Tabl. Elem. d'Orn.* i. p. 318, sp. 26, pl. 15 (not of *Buff. Ois.* v. pl. 363, pl. 16, fig. 2, nor *Pl. enl.* 651, fig. 3, which represent Sylvia ignicapilla of *Brehm*).—Regolo, Storia degli uccelli, iv. pl. 390.—Gold-crested Wren, *Lath. Syn.* ii. part 2, p. 508, sp. 145.—*Penn. Brit. Zool.* sp. 153.—*Penn. Arct. Zool.* sp. 321.—Golden-crowned Wren, *Edw. Glean.* v. p. 95, pl. 254, lower fig. male.—*Philadelphia Museum*, No. 7246, male; No. 7247, female.

REGULUS REGULOIDES.—JARDINE.[*]
For male and note, see vol. i. p. 127.

Two distinct species of gold-crest have been, until lately, considered by naturalists as but one. Are they both inhabitants

[*] For the distinctions of the American bird and the true *R. cristatus*, see as noted above. The Prince of Musignano is now aware of the distinctions

of this continent? and if not, which is the American species? These questions cannot be readily answered, since we have nothing better than negative evidence to offer relative to the first. The present female, however, is decisive as to which of them inhabits this country, and we have therefore concluded that the faithful representation in the accompanying plate will be acceptable to ornithologists. A slight inspection of this specimen leaves no doubt as to its being the female of the *Regulus cristatus ;* and should the *Regulus ignicapillus,* contrary to our expectations, also prove to be an inhabitant of this country, it will appear, along with its mate, in another volume of this work. All the ornithologists state that the latter is a native of this continent, whilst they take no notice whatever of the *Regulus cristatus,* which, if not the only indigenous, is certainly the more common species. This error seems to have originated with Vieillot, who, considering the two species as but one, probably was not careful in selecting the individual from which his drawing was made ; he may, therefore, have chosen a European bird, and, unluckily, of the other species, as both are found in Europe.

However this may be, his figure is certainly that of the *igni-capillus ;* and it is equally obvious that his short description of the female can only apply to the female of the *cristatus,* which corroborates my opinion. In the (French) " New Dictionary of Natural History," Vieillot distinguishes two varieties of *Regulus cristatus,* and again describes the *ignicapillus* as the one he saw in America. If this observation could be relied upon, we should admit that both species are inhabitants of this country, although the present, which must be by far the most numerous, is certainly not the *ignicapillus.*

I agree with Ray, Vieillot, and other authors, and dissent

between the American and British species, which he will himself detail hereafter. The *R. ignicapillus* has not yet been discovered in America, unless the bird which Mr Audubon has figured as *R. Cuverii* may prove to be it ; but which it is impossible to determine from the plate alone. —ED.

from Linné, Latham, Wilson, and Temminck, respecting the propriety of placing these birds in a separate genus from *Sylvia*, and I have therefore changed the generic name adopted by Wilson. This genus forms a link intermediate to the genera *Sylvia* and *Parus*. It is small, both in the number and size of its species, consisting of the two smallest of the European birds, one of which is the subject of this article; an American species, the ruby-crowned gold-crest (*Regulus calendulus*), so well figured and described by Wilson; and a fourth from Asia.

The most obvious characters that distinguish the genus *Regulus* from *Sylvia* are, the bill remarkably slender throughout, and two small decomposed feathers, directed forwards so as to cover the nostrils.

The habits of the gold-crests resemble, in many respects, those of the titmouse. They delight in cold weather, and then often perch on evergreen trees. They display great activity and agility in search of their food, being almost constantly in motion, hopping from branch to branch, or climbing on trees, frequently with the head downwards, searching the chinks of the bark for their prey. These little birds commonly feed on the smallest insects, which they catch adroitly while on the wing. In the winter they seek them in their retreats, where they lie torpid or dead. They are also very expert at finding larvæ and all sorts of small worms, of which they are so fond as to gorge themselves exceedingly. During summer, they occasionally eat little berries and small grains. In autumn, they are fat, and fit for the table, notwithstanding their very diminutive size. The species we are describing is found in great quantities in the neighbourhood of Nuremberg, in Germany, and sold in the markets of that city, where they command a high price.

Wilson, in his account of the present species, observes that " the very accurate description given by the Count de Buffon agrees, in every respect, with ours." Notwithstanding this observation, Buffon's plate and description designate the *igni-*

capillus beyond the possibility of doubt, whilst those of Wilson are intended for the *cristatus*.

This statement of Wilson, joined to the testimony of Vieillot, would have led us to believe the *ignicapillus* to be an American bird, if Wilson's plate, and more especially his description, as well as the inspection of the very individual he delineated, and a hundred others, had not confirmed our own belief. It may, however, be considered extraordinary that so diminutive a being should extend its range so widely as to participate equally in the bounties of two continents, and that another, so closely allied to it as to be generally mistaken for a mere variety, should be limited in its wanderings by the boundaries of but one.

That the reader may be assured of the specific difference between these two birds, I add a short comparative description. The *Regulus cristatus* has the bill very feeble, and quite subulate, whilst that of the *ignicapillus* is also subulate, but is wider at base. The cheeks of the former are pure cinereous, without any white lines, having only a single blackish one through the eye; those of the latter, in addition to the black line through the eye, have a pure white one above and another below, whence Temminck calls it *Roitelet triple bandeau*. The English name also may be derived from this character, or the bird may rather be called fire-crowned gold-crest, from its Latin name. The crest of the male golden-crowned gold-crest is yellowish orange, that of the fire-crowned is of the most vivid orange; but the most obvious difference is between the females, that of the golden-crowned having a lemon yellow crest, which in the female of its congener is orange, like that of the male, only much less vivid. The cheek-bands of the female fire-crowned are by no means so obvious as in its mate; thus the female of this species resembles the male golden-crowned, than which the colours of its crest are not less brilliant. If to these traits we add that the latter is a little larger, we shall complete the enumeration of their differences.

The two species are also somewhat distinguished by their

manner of living. The golden-crowned gold-crest associates in small bands, consisting of a whole family, whilst the fire-crowned is only observed in pairs. The latter is more shy, and frequents the tops of the highest trees; whereas the former is more generally observed amongst low branches and bushes; the voice of the fire-crowned gold-crest is also stronger. Their nests, however, are both of the same admirable construction, having the entrance on the upper part; but the eggs are different in colour, and those of the fire-crowned are fewer in number.

The female golden-crowned gold-crest is three inches and three-quarters long, and six in extent. The bill is black; the feet dusky; the toes and nails wax colour; the irides are dark brown. The frontlet is dull whitish grey, extending in a line over and beyond the eye; above this is a wide black line, confluent on the front, enclosing on the crown a wide longitudinal space of lemon yellow, erectile, slender feathers, with disunited webs; a dusky line passes through the eye, beneath which is a cinereous line, margined below by a narrow dusky one. The cervix and upper part of the body are dull olive green, tinged with yellowish on the rump. The whole inferior surface is whitish; the feathers, like those of the superior surface, being blackish plumbeous at base. The lesser and middling wing-coverts are dusky, margined with olive green, and tipped with whitish; the greater coverts are dusky, the outer ones immaculate, the inner ones have white tips, which form a band on the wings. The inferior wing-coverts, and all the under surface of the wings, are more or less whitish grey; the primaries are dusky, with a narrow greenish yellow outer margin, wider at base, and attenuated to the tip, where it is obsolete. The secondaries are dusky; on the outer web, they are whitish near the base, then black, then with a greenish yellow margin, extending nearly to the tip; the margin of the inner web is white; the secondaries nearest to the body are, moreover, whitish on the terminal margin. The tail is emarginated; the feathers are dusky olive green on the margin of the outer web ; the inner margins, with the exception of the two middle ones, are whitish.

Until their first moult, the young of both sexes are much like the adult female, except in being destitute of the yellow spot on the crest, which is greenish olive. In this state, however, they are not seen here, as they breed farther to the north, and moult before their arrival in the autumn.

YELLOW-HEADED TROOPIAL.   (*Icterus icterocephalus.*)

PLATE III.—Fig. 1, Male ; Fig. 2, Female.

Oriolus icterocephalus, *Linn. Syst.* i. p. 163, sp. 16.—*Gmel. Syst.* i. p. 392, sp. 16.—*Lath. Ind.* p. 183, sp. 32, male.—Icterus icterocephalus, *Daudin, Orn.* ii. p. 337, sp. 9, male.—Pendulinus icterocephalus, *Vieill. Nouv. Dict. d'Hist. Nat.* v. p. 317, male.—Icterus xanthornus icterocephalus Cayanensis, *Briss. Av.* ii. p. 124, sp. 27, pl. 12, fig. 4, male.—Cornix atra ; capite, collo, pectoreque flavis, *Kœlreuter, Nov. Comm. Ac. Sc. Petrop.* xi. p. 435, pl. 15, fig. 7, male.—Les Coiffes Jaunes, *Buff. Ois.* iii. p. 250, male.—Carouge de Cayenne, *Buff. Pl. enl.* 343, male.—Yellow-headed Starling, *Edw. Glean.* iii. p. 241, pl. 323, male.—Yellow-headed Oriole, *Lath. Syn.* i. part 2, p. 441, sp. 30, male.— *Philadelphia Museum*, No. 1528, male ; No. 1529, female.

AGLAIUS ICTEROCEPHALUS.—Jardine.[*]

Icterus (subgen. Xanthornus) xanthocephalus, *Bonap. Synop.* p. 52.—Aglaius xanthocephalus, *North. Zool.* ii. p. 281.

ALTHOUGH this species has long been known to naturalists as an inhabitant of South America, and its name introduced into all their works, yet they have given us no other information concerning it than that it is black with a yellow head and neck. It was added to the fauna of the United States by the expedition of Major Long to the Rocky Mountains.

---

[*] I have retained what appears to be the old specific name for this bird, and which also seems to be the view of our author. Another has been selected in the "Northern Zoology," where this bird is described from species obtained during the last expedition. It is mentioned as reaching the Saskatchewan about the 20th of May, and being even more numerous than the redwings ; commits great havoc among the cornfields.—Ed.

Drawn from Nature by Titian R. Peale.   Engraved by W.H.Lizars.

1.Yellow-headed Blackbird. 2.Female. 3.Female Cape May Warbler.

Icterus Icterocephalus.   *Sylvia Maritima.*

3.

The female has been hitherto entirely unknown, and all the figures yet given of the male being extremely imperfect, from the circumstance of their having been drawn from wretchedly stuffed specimens, we may safely state that this sex also is for the first time represented with a due degree of accuracy in our plate. The figures published by Edwards and Buffon approach the nearest to the real magnitude; but they are mere masses of black, surmounted by a yellow cap: those of Brisson and others are considerably smaller.

As that striking character, the white spot on the wing, is neither indicated in the figure nor description of any author, we might have been induced to believe that our species is different from the South American, if a close comparison of the two had not proved their identity. Another circumstance might have been equally deceptive: Brisson, who gave the first account of this bird, from a Cayenne specimen sent to Reaumur's Museum, and who seems to have been copied by all subsequent authors, states its length to be less than seven inches, a size considerably inferior to that of the living bird. Had this admeasurement been taken from a recent specimen, we could hardly hesitate to believe our bird distinct; but as he had only a dried skin, and as Buffon's figure represents a nearer approach to the size of nature, we conclude that Brisson's estimate is not to be implicitly relied upon. Vieillot, who never saw the bird, states the length to be six inches and a half, and refers it to his genus *Pendulinus*, but it certainly belongs to his genus *Agelaius*.

The male yellow-headed troopial is ten inches and a half long. The bill is dark horn-colour, and formed exactly like that of the red-winged troopial. The feet are black; the irides, dark brown. The whole head, neck, and breast are brilliant orange yellow, more vivid and sericeous on the head, and terminating in a point on the belly; the feathers around the base of the bill, the chin, and a wide stripe passing from the bill through the eye, are black. The remaining parts, excepting some feathers of the belly, and some of the under tail-coverts, which are yellow at base, are glossy black, very slightly tinged with brownish. Some of the exterior wing-coverts are pure white, with black tips, constituting two very remarkable white spots on the wing, the larger of which is formed by the greater coverts of the primaries, and the smaller one by the middling coverts. The first, second, and third primaries are longest and equal. The tail is four inches long, slightly rounded, the two middle feathers being somewhat shorter than those immediately adjoining. This character Wilson remarked in the red-winged troopial; and as other notable traits are common to both species, we must regard them, not only as congeneric, but as very closely allied species of the same subgenus. They differ, however, in colour; and the yellow-headed troopial is larger, having the bill, feet, and claws consequently stronger, and the first primary longer than the second and third, or at least as long; whereas, in the red-winged, the third is the longest.

The female of our troopial is eight inches and a quarter long, a size remarkably inferior to that of the male, and exactly corresponding with the difference existing between the sexes of the red-winged troopial. The bill and feet are proportionally smaller than those of the male, the feet being blackish; the irides are dark brown. The general colour is uniform dark brown, a shade lighter on the margin of each feather. The frontlet is greyish ferruginous, as well as a line over the eye, confluent on the auricles, with a broad line of the same colour passing beneath the eye, including a blackish space varied with greyish. An abbreviated blackish line proceeds from each side of the lower mandible; the chin and throat are whitish; on the breast is a large rounded patch, of a pretty vivid yellow, occupying nearly all its surface, and extending a little on the neck. On the lower part of the breast and beginning of the belly the feathers are skirted with white. The form of the wings and tail is the same as in the male; the wings are immaculate.

The young of this species are very similar to the female ; the young male gradually changing to the rich adult covering.

The yellow-headed troopials assemble in dense flocks, which, in all their varied movements and evolutions, present appearances similar to those of the red-winged, which have been so well described by Wilson. They are much on the ground, like the cow-troopial (cow-bunting of Wilson) ; on dissection, their stomachs have been found filled with fragments of small insects, which seem to constitute their chief food, though doubtless they also feed on vegetable substances. Their notes resemble those of the red-winged troopial, but are more musical. The range of the yellow-headed troopial is very extensive, as it is found from Cayenne to the river Missouri ; although it passes far north in the western region, yet it does not visit the settled parts of the United States.

The fine specimens represented in our plate were killed near the Pawnee villages, on the river Platte, where they were seen in great numbers about the middle of May. The males and females were sometimes observed in separate flocks.

We adopt the genus *Icterus*, nearly as it was established by Brisson, and accepted by Daudin and Temminck. Authors have variously estimated this genus, both in regard to its denomination and limits. One of Wilson's most important nomenclatural errors consisted in placing one of the species under the genus *Sturnus*, with which it has but little similarity, if we except some of its habits, and particularly its gregarious disposition. Linné considered these birds as *Orioli*, in which he was followed by Gmelin and Latham, notwithstanding the remarkable difference existing between them and the *Oriolus galbula* of Europe, the type of that genus. Illiger, and some other naturalists, considering that bird a *Coracias*, appropriated the name of *Oriolus* to our *Icterus*, and separated from it the largest species, which he called *Cassici*. Linné had declared all generic names previously given to arts, diseases, &c., to be inadmissible in natural history ; Illiger, on that principle, altogether rejected the name *Icterus,* as being pre-occupied by a disease.   This may account for the introduction of new names for genera, one of which, at least, ought to have retained its first appellation.   Vieillot, however, would have caused less confusion if he had adopted the name of *Icterus* (which, with *Saxicola,* and all other names of that class, we do not think objectionable) instead of *Agelaius, Pendulinus,* or *Yphantes,* three of his four genera corresponding to our *Icterus.*   But if the latter name was considered as utterly inadmissible, we see no reason why he did not accept that of *Xanthornus,* applied to this genus by Pallas.

All the species of troopial are peculiar to America.   We divide them into four subgenera, the present bird belonging to the second, to which we apply the name of *Xanthornus.* The species of this subgenus are peculiarly social in their dispositions, and their associations are not liable to interruption from the influence of love itself.   Not only do many individuals of the same family combine and labour in concert, but they also unite with very different species.   Their aspect is animated, and their movements are quick, bold, and vigorous; they fly rapidly, at a good height, and are much attached to the places of their birth.   Their song is a kind of whistling ; they walk with the body nearly erect, with a slightly hurried step, and are seen sitting on the ground, or perched on the branches of trees.   They seek no concealment, and never enter the woods, though they are very careful to construct their nests in a safe situation.   The troopials eat no fruits, but derive their subsistence from insects, worms, grains, and small seeds.   They leave the temperate climates at the approach of winter, and are amongst the first birds of passage that return with the spring.

FEMALE CAPE MAY WARBLER. (*Sylvia maritima.*)

PLATE III.--FIG. 3.

My Collection.

SYLVICOLA MARITIMA.—SWAINSON.

Sylvia maritima, *Bonap. Synop.* p. 79.

I WAS so fortunate as to obtain this undescribed little warbler in a small wood near Bordentown, New Jersey, on the 14th of May, at which season ornithologists would do well to be on the alert to detect the passenger warblers, whose stay in this vicinity is frequently limited to a very few days.

Judging by the analogical rules of our science, this bird is no other than the female of Wilson's Cape May warbler. Its appearance is so different from the male he described, that the specific identity is not recognised at first sight; but, by carefully comparing the two specimens, a correspondence in the least variable characters may readily be perceived, especially in the remarkable slenderness of the bill, which distinguishes the Cape May from all other resembling species of North American warblers.

Wilson has given no information relative to the history and habits of this species, having never procured more than a male specimen; and we have equally to regret that, having obtained but a single female, we are unable to supply the deficiency, even in regard to its song.

The female Cape May warbler is four inches and three-quarters long, and more than eight in extent. The bill is slender, delicate, and slightly curved, being black, as well as the feet. The irides are dark brown; the upper part of the head, olive cinereous, each feather having a small blackish spot on the middle. A yellow line extends from the bill over the eye, and is prolonged in an obsolete trace around the auditory region, thence returning to the corner of the mouth. A blackish line passes through the eye, which is circumscribed by a whitish circle; the cheeks are dull cinereous, with very small pale spots; the upper parts of the neck and of the body are olive cinereous, tinged with more cinereous on the neck, and with yellow olive on the rump. The chin is whitish; the throat, breast, and flanks are whitish, slightly tinged with yellowish, each feather having a blackish spot on the middle; the belly is immaculate; the vent and inferior tail-coverts are shaded in the middle of each feather with dusky. The smaller wing-coverts are dull olive green, blackish in the centre; the middling wing-coverts are black, margined exteriorly, and tipped with pure white; the greater wing-coverts are blackish, margined with olive white; the primaries are dusky, finely edged with bright olive green on the exterior web, obsolete on that of the first primary, which is of the same length as the fourth; the second and third are longest, and but little longer than the fourth. The tail is slightly emarginated, the feathers being dusky, edged with bright olive green on the exterior side, and with white on the interior; the two or three exterior feathers on each side have a pure white spot on their inner webs near the tip.

The female Cape May warbler may be very easily mistaken for an imperfect *Sylvia coronata*, of which four or five nominal species have already been made. The striking resemblance it bears to the young, and to the autumnal condition of the plumage in that species, requires a few comparative observations to prevent their being confounded together. The present bird is smaller than the *coronata*, with a more slender, and rather more elongated bill; it is altogether destitute of the yellow spot on the head, as well as of the yellow on the rump, which is a striking character of the *coronata* in all its states, and gives rise to the English name adopted by Wilson.

The colour of the outer edging of the wing and tail-feathers is a very good distinctive mark; in the *maritima* it is olive green, whilst in the *coronata* it is white. The white spot on the inner webs of the exterior tail-feathers is also four times larger in the *coronata* than in the *maritima*.

Drawn from Nature by John J. Audubon & J. Bird.

Engraved by W.H. Lizars.

1. Great Crow Blackbird. 2. Female.

Quiscalus Major.

GREAT CROW BLACKBIRD. (*Quiscalus major.*)

PLATE IV.—FIG. 1, MALE ; FIG. 2, FEMALE.

Quiscalus major, *Vieill. Nouv. Dict. d'Hist. Nat.* xxviii. p. 487.—Gracula quiscala, *Ord, Journ. Acad. Nat. Sc. Philad.* i. p. 253.—Gracula barita, *Wilson,
Am. Orn.* vi. *Index,* p. 8.—Gracula quiscala, the Purple Jackdaw of the sea-
coast, *Bartram, Travels,* p. 290.—Corvus Mexicanus? *Gmel. Syst.* i. p. 375,
sp. 42.—*Lath. Ind.* p. 164, sp. 36, male.— Corvus zanoe? *Gmel. Syst.* i. p.
375, sp. 44.—*Lath. Ind.* p. 164, sp. 37, female.—Pica Mexicana major? *Briss.
Av.* ii. p. 43, sp. 4, male.—Pica Mexicana minor? *Briss. Av.* ii. p. 44, sp. 5,
female.—L'Hocizana? *Buff. Ois.* iii. p. 103, male.—Le Zanoé? *Buff. Ois.* iii.
p. 106, female.—Mexican Crow? *Lath. Syn.* i. p. 396, sp. 34, male.—Lesser
Mexican Crow? *Lath. Syn.* i. p. 397, sp. 36, female.—Hocitzanatl, seu Magnus
Sturnus? *Hernand. Hist. An. Nov. Hisp.* p. 21, male.—Tzanahoei? *Hernand.
Hist. An. Nov. Hist.* p. 22, female.—Hoitzanatl? *Ray, Syn. Av.* p. 162,
male. —Tzanahoei, seu Pica Mexicana Hernandezii ? *Ray, Syn. Av.* p. 162,
female.—*Philadelphia Museum,* No. 1582, male ; No. 1583, female.

QUISCALUS MAJOR.—VIEILLOT.

Quiscalus major, *Bonap. Synop.* p. 54.

No part of natural history has been more confused than that
relating to North and South American birds of black plumage;
which is by no means surprising, when we recollect that they
are chiefly destitute of coloured markings, and that the
greater number of admitted species are founded on the short
and inexact descriptions of travellers who have neglected to
observe their forms, habits, and characters.   But little aid has
been derived from the wretched plates hitherto given, for they
seem better suited to increase the confusion than to exemplify
the descriptions to which they are annexed, and every succeeding compiler has aggravated, rather than diminished, this
complication of error.   It is, therefore, solely by a studious
attention to nature, that we can extricate these species from
the uncertainty involving them, and place them in a distinct
and cognisable situation.   With these views we now give a
faithful representation of both sexes of the great crow blackbird, drawn by that zealous observer of nature, and skilful
artist, Mr John J. Audubon, and hope thereby to remove all
doubt relative to this interesting species.

For the same purpose we give in the following plate a figure of the female common crow blackbird, which differs so little from its mate (admirably represented in the third volume of Wilson's "Ornithology"), that it would be otherwise unnecessary. This measure, we believe, will be acceptable to ornithologists, as it furnishes them with means of comparing the females of both the species in question, whence the most striking distinctive characters are obtained; that of one species differing considerably in size and colour from the male, while the sexes of the other are very similar in appearance.

Wilson having mentioned this species in his catalogue of land birds, evidently intended to describe and figure it; but this he deferred, probably, in expectation of obtaining better opportunities of examination, which are not so readily presented, as the bird does not inhabit this section of the United States.

It would be difficult to ascertain whether or not Linné and Latham have mentioned this bird in any part of their works, but the reader may perceive our opinion on this point by referring to our synonyms, which, however, are given with much doubt, since we do not hesitate to say that those authors have not published any satisfactory description of this species.

We shall not endeavour to settle the question relative to the species inhabiting South America, or even Mexico and the West Indies; but we may assert that this is the only blackbird found in the United States, besides those of Wilson, which, as is the case with all that his pencil or pen has touched, are established incontestably. He may occasionally have been mistaken as to his genera, or incorrect in a specific name, but by the plate, description, and history he has always determined his bird so obviously as to defy criticism and prevent future mistake.

Mr Ord has published an excellent paper in the *Journal of the Academy of Natural Sciences*, proving the existence, in the United States, of two allied species of crow blackbird, in which he gives new descriptions, indicates stable characters, and adds an account of their respective habits; but in attempting to correct Wilson, he has unfortunately misapplied the names. In this instance he should not have charged Wilson with error, who is certainly correct in regard to the species he published; and even had this been doubtful, he who so well described and figured the common crow blackbird ought to have been followed by ornithologists. Therefore, notwithstanding, Mr Ord's decision, we consider the *quiscala* of Wilson unquestionably the true *quiscala* of authors. This is so obvious, that it is unnecessary to adduce any evidence in support of our opinion, which, indeed, is sufficiently afforded by Mr Ord's paper itself.

It is impossible to decide with certainty what bird authors intended to designate by their *Gracula barita*; but after a careful review of the short and unessential indications, repective synonyms, and habitat given by different writers, we feel assured that they have not referred to one and the same species. Thus the *barita* of Linné is a species not found in the United States, but common in the West Indies, called *Icterus niger* by Brisson, and afterwards *Oriolus niger* by Gmelin and Latham; the *barita* of Latham, his boat-tailed grakle, is evidently the same with the *quiscala*;* Gmelin's *barita* is taken partly from that of Linné, and partly from the boat-tailed grakle of Latham, being compounded from both species : we shall not be at the trouble of deciphering the errors of subsequent compilers.

Ornithologists are all at variance as to the classification of these species. Linné and Latham improperly referred them to *Gracula*; Daudin, with no better reason, placed them under *Sturnus*; Temminck considers them as *Icteri*, Cuvier as *Cassici*, and Vieillot has formed a new genus for their reception. I have no hesitation in agreeing with the latter author, and

---

* It was probably by Latham that Mr Ord was led to misapply the names of the two species ; for, perceiving that the *barita* of that author was the *quiscala*, he inferred that the *quiscala* was the *barita*.

adopt his name of *Quiscalus* ; but I add to the genus, as constituted by him, the *Gracula ferruginea*, which he regarded
as a *Pendulinus*, and which other authors have arranged in
several different genera, making of it a profusion of nominal
species. Wilson judiciously included that species in the same
genus with those above mentioned, although other authors
had placed it in *Turdus, Oriolus*, &c.

The genus *Quiscalus* is peculiar to America, and is composed of four well-ascertained species, three of which are found
in the United States ; these are *Quiscalus major,** versicolor*,
and *ferrugineus ;* the fourth, *Quiscalus baritus*, inhabits the
West Indies, and probably South America.

The species of this genus are gregarious and omnivorous,
their food being composed of insects, corn, and small grains,
thus assisting and plundering the agriculturist at the same
time. When the first European settlements were formed in
North America, the havoc made by these birds and the
troopials in the grain-fields was so great that a premium was
given for their heads. Their destruction was easily effected,
as they are not shy, and are more easily approached as their
numbers decrease ; but the evil which resulted from exter-
minating so many of these birds was as unexpected as irre-
mediable. The corn and pastures were so devoured by worms
and insects, that the inhabitants were obliged to spare the
birds in order to avert a scourge which had been previously
unknown. As population increases, and a greater quantity
of grain is cultivated, the ravages of these birds become less
perceptible, and the injury they cause comparatively trifling.

The great crow blackbird is more than sixteen inches long,
and twenty-two in extent. The bill, from the angle of the
mouth, is one inch and three-quarters, and its colour, like that
of the feet, is black ; the roof of the mouth is furnished with a
slight osseous carina ; the irides are pale yellow. The general

---

* We call the present species *Quiscalus major*, agreeably to Vieillot,
who certainly intended this bird, although his description is a mere indication.

appearance of the bird is black ; the whole head and neck having bluish purple reflections ; the interscapular region, breast, belly, sides, and smaller wing-coverts are glossy steel-blue ; the back, rump, and middling wing-coverts are glossed with copper-green ; the vent, inferior tail-coverts, and thighs are plain black. The undescribed parts of the wings are deep black, slightly glossed with green, as well as the tail, which is cuneiform, capable of assuming a boat-shaped appearance, and measures nearly eight inches in length from its insertion, surpassing the tip of the wings by five inches.

The female is considerably shorter, measuring only twelve and a half inches in length, and seventeen inches and a half in extent. The bill, from the angle of the mouth, is one inch and a half long, and, with the feet, is black ; the irides are of a still paler yellow than those of the male. The head and neck above are light brown, gradually passing into dusky towards the back, which, with the scapulars and lesser wing-coverts, has slight greenish reflections ; a whitish line passes from the nostrils over the eye, to the origin of the neck. The chin, throat, and breast, are dull whitish ; the anterior part of the breast is slightly tinged with brownish ; the flanks are brownish ; the belly brownish white ; and the vent and inferior tail-coverts are blackish brown, each feather being margined with pale. The remaining parts are of a dull brownish black, slightly glossed with greenish ; the secondaries, tail-coverts, and tail-feathers, having a slight banded appearance, which is equally observable in the male.

The young at first resemble the female, but have the irides brown, and the males gradually acquire the brilliant plumage of the adult.

The great and common crow blackbirds are both alike distinguished by the very remarkable boat-like form of the tail, but the great difference of size, appearance of the females, length of the tail, prominence of the osseous carina, and brilliancy of colouring, most obviously prove them to be altogether specifically distinct.

The great crow blackbird inhabits the southern part of the Union, where it is called jackdaw ; Georgia and Florida appear to be its favourite residence. The disposition of this species is extremely social, and they frequently mingle with the common crow blackbird ; vast flocks are seen among the sea islands and neighbouring marshes on the mainland, where they feed at low water on the oyster beds and sand flats.

The chuck of our species is shriller than that of the common crow blackbird, and it has other notes, which resemble the noise made by a watchman's rattle ; their song is only heard in the spring, and though the concert they make is somewhat melancholy, it is not altogether disagreeable. Their nests are built in company, on reeds and bushes, in the neighbourhood of marshes and ponds ; they lay about five eggs, which are whitish, spotted with dark brown, as represented in the plate.

Mr Ord mentions in his paper that the first specimens he saw of this bird were obtained on the 22d of January at Ossabaw Island, when but a few males were seen scattered over the cotton plantations. Advancing towards the south, they became more numerous ; and in the early part of February, the males, unaccompanied by females, were common near the mouth of the river St Juan in Florida. A few days after, the females appeared, and associated by themselves on the borders of fresh-water ponds ; they were very gentle, and allowed themselves to be approached within a few feet, without becoming alarmed. Flocks composed of both sexes were seen about the middle of March.

About the latter end of November, they leave even the warm region of Florida to seek winter quarters farther south, probably in the West Indies. Previous to their departure, they assemble in very large flocks, and detachments are seen every morning moving southward, flying at a great height. The males appear to migrate later than the females, as not more than one female (easily distinguishable, even in the higher regions of the air, by its much smaller size) is observed for a hundred males, in the last flocks.

Drawn from Nature by Titian R. Peale.   Engraved by W.H.Lizars.

1.*Female Crow Blackbird.* 2.*Orange-crowned Warbler.* 3.*Lark Finch.*

Quiscalus Versicolor   *Sylvia Celata.*   *Fringilla Grammaca.*

5.

The great crow blackbird is also very numerous in the West Indies, Mexico, and Louisiana ; but it does not frequent the northern, or even the middle States, like the common crow blackbird. Our opinion, that the *Corvus Mexicanus* of authors is the male of this species, and their *Corvus zanoe* the female, is corroborated by the male and female great crow blackbird being seen in separate flocks.

FEMALE COMMON CROW BLACKBIRD. (*Quiscalus versicolor.*)

PLATE V.—FIG. 1.

Wilson's American Ornithology, iii. p. 44, pl. 21, fig. 4, for the male, and history. —Quiscalus versicolor, *Vieill. Nouv. Dict. d'Hist. Nat.* xxviii. p. 488.—*Nobis, Obs. Nom. Wils. Orn. Journ. Acad. Nat. Sc. Philad.* iii. p. 365.—Gracula quiscala, *Linn. Syst.* i. p. 165, sp. 7.—*Gmel. Syst.* i. p. 397, sp. 7.—*Lath. Ind.* p. 191, sp. 7.—Gracula barita, *Gmel. Syst.* i. p. 396, sp. 4.—*Lath. Ind.* p. 191, sp. 6.—*Ord, Journ. Acad. Nat. Sc. Philad.* i. p. 254 (not of Linn.) —Oriolus Ludovicianus, *Gmel. Syst.* i. p. 387, sp. 31 (pied variety).—Oriolus leucocephalus, *Lath. Ind.* p. 175, sp. 4 (pied variety).—Pica Jamaicensis, *Briss. Av.* ii. p. 41, sp. 3.—Monedula purpurea, the Purple Jackdaw, *Catesby, Carolina,* i. p. 12, pl. 12.—Gracula purpurea, the Lesser Purple Jackdaw, or Crow Blackbird, *Bartram, Trav.* p. 291.—Pie de la Jamaique, *Buffon, Ois.* iii. p. 97.—Cassique de la Louisiane, *Buff. Ois.* iii. p. 242.—*Pl. enl.* 646 (pied variety).—Purple Grakle, *Penn. Arct. Zool.* sp. 153.—*Lath. Syn.* i. part 2, p. 462, sp. 6.—Boat-tailed Grakle, *Penn. Arct. Zool.* sp. 154.—*Lath. Syn.* i. part 2, p. 460, sp. 5.—White-headed Oriole, *Penn. Arct. Zool.* sp. 147.—*Lath. Syn.* i. part 2, p. 422, sp. 4 (pied variety).—*Philadelphia Museum,* No. 1578, male; No. 1579, female; No. 1602, whitish variety.

QUISCALUS VERSICOLOR.—Vieillot.

The female common crow blackbird is figured in the annexed plate, that naturalists may have an opportunity of comparing it with the corresponding sex of the great crow blackbird, and thus receive a distinct idea of the difference between the two species, so well manifested in their females.

The specific name of this bird (*quiscala*) has been changed, in consequence of its having been applied to the genus : we have substituted the name given by Vieillot, which is admirably appropriate. The English name employed by Wilson being now rendered inadmissible by the generic change, we have thought proper to adopt a local appellation.

The female common crow blackbird is eleven inches in length, and sixteen and a half in extent. The bill is nearly an inch and a half long, and, as well as the feet, black; the irides are yellowish white; the whole head, neck, and upper part of the breast, are blackish, with steel-blue, green, and violet reflections, which are not so vivid as in the male. The general colour of the body, wings, and tail is deep sooty brown; the feathers of the back are margined with coppery and purplish; the rump, tail-coverts, and wing-coverts are glossed with purplish; the lower part of the breast and flanks have a coppery reflection; the inferior tail-coverts are obscurely glossed with violet. The tail is cuneiform, but slightly concave in flight, and is five inches long, extending two and a half inches beyond the tip of the wings; the feathers are glossed with very obscure greenish. In the male, the tail is also cuneiform, and greatly concave, exhibiting a singular boat-shaped appearance, as in the preceding species, and even more remarkably so, according to Mr Ord, which induced him to change the name.

We shall not attempt to make any additions to the almost complete and very excellent history of this species given by Wilson; but as the four species of *Quiscalus* are liable to be confounded, we shall proceed to give a few comparative observations, that the student may be enabled to distinguish them from each other.

Amongst other remarkable traits, the *Quiscalus ferrugineus* is at once known in all its various states by its even tail and comparatively smaller bill, which somewhat resembles that of a thrush. In addition to the characters drawn from its dimensions, the *Quiscalus versicolor* can always be distinguished from its congeners by the slight difference in size and colour between the sexes; while, in the other species, the males and females are remarkably dissimilar: the mouth of this species is, moreover, armed with a prominent osseous carina, a quarter of an inch long, which, in the others, is much smaller. That the *Quiscalus major* and *Quiscalus baritus* should have been confounded together, is not a little surprising, as the former is sixteen inches long, the tail being eight inches, and extending five inches beyond the tip of the wings ; whilst the latter is only ten inches, the tail much less cuneiform, four inches and a half long, and extending but two inches beyond the tip of the wings ; the osseous carina is similar in these two species, and the markings of the females are much alike. From this statement it is apparent that the females of the largest and smallest crow blackbirds correspond in the disposition of their colours—a parity that does not exist in the intermediate species. In comparative size, however, they differ considerably : the female of the *baritus*, though smaller, as we have already stated, is, in proportion to its mate, considerably larger than that of the other, being only half an inch, whilst the female of the *major* is nearly four inches, smaller.

The individual represented in the annexed plate is a remarkably fine one, in the most perfect state of plumage. It therefore more strongly resembles the male than is usual with its sex, which are generally much less brilliant in colouring, and more sooty brown. This bird was obtained at Great Egg Harbour on the 21st of May, and was selected as the best female of several pairs, assembled to breed at one of the identical fish-hawk nests, in the interstices of which Wilson mentions having seen them building. One of their nests contained three eggs, and the species had not ceased to lay.

These birds, as we have had occasion personally to observe, like most of the feathered tribes, are subject to become either wholly or partially albinos. From this circumstance, numerous errors have been introduced in the pages of ornithological works.

ORANGE-CROWNED WARBLER. (*Sylvia celata.*)

PLATE V.—FIG. 2.

Sylvia celata, *Say, in Long's Expedition to the Rocky Mountains,* i. p. 169.—
Philadelphia Museum, No. 7013.

VERMIVORA ? CELATA.—JARDINE.

Sylvia celata, *Bonap. Synop.* p. 88.

THIS little bird, discovered, early in May, at Engineer Cantonment on the Missouri river, was first described and named by Say; the species was not uncommon at that season, and appeared to be on its passage farther north. It is more particularly interesting, inasmuch as it enriches the fauna of the United States with another species of the small subgenus *Dacnis,* which may be ascertained by inspecting the bill, represented in the annexed plate.

The orange-crowned warbler is full five inches long, and seven in extent. The bill is dark horn-colour, slender, straight entire, and tapering to an acute point; the base of the inferior mandible is whitish beneath; the legs are dusky; the irides dark brown. The general plumage above is dull greenish olive, the rump and tail-coverts being bright yellowish olive. The head is very slightly and inconspicuously crested; the feathers of the crest are orange at base, constituting a spot on the crown, visible only when they are elevated, being tipped with the common colour. The whole bird beneath is dull olive yellow; the inferior tail-coverts are pure yellow. The wings are destitute of spots or bands; the primaries are dark brown, olive green on the exterior margin, which is much paler on the outer ones; the interior margin is whitish; the four outer primaries are subequal; the fifth is but very little shorter. The tail is even, the feathers being dark brown, edged with olive green on the outer, and with white on the inner web.

The orange-crowned warbler resembles several species of indigenous and foreign warblers; and the females of others, such as that of the *Sylvia trichas*, may also be mistaken for it ; but it may be distinguished from each of them respectively by particular characters, which it is not necessary to detail, as the concealed orange spot of the crown is a peculiarity not possessed by either of the allied species. The Nashville warbler (*Sylvia rubricapilla*) of Wilson seems to be more closely related to the orange-crowned warbler than any other. That bird, also, is evidently a *Dacnis*, and scarcely differs from our species, except in the white belly, the light ash colour of the head and neck, and the deep chestnut colour disposed in small touches on the crown, instead of an uniform orange colour.

The figure given in our plate is that of a male ; and the only difference observable between the sexes is, that the rump of the male is of a brighter colour, approaching in old birds to a pure yellow.

During winter, the orange-crowned warbler is one of the most common birds in the neighbourhood of St Augustin, Florida, almost exclusively frequenting the orange-trees. Their manners resemble those of the kindred species, though they have a remarkable habit of constantly inflecting the tail, like the pewee. The note consists of a chuck and a faint squeak, but little louder than that of a mouse.

LARK FINCH. (*Fringilla grammaca.*)

PLATE V.—Fig. 3.

Fringilla grammaca, *Say, in Long's Expedition*, i. p. 139.—*Phil. Museum*, No. 6288.

PLECTROPHANES? GRAMMACA.—Jardine.

Fringilla grammaca, *Bonap, Synop.* p. 108.

For this very interesting new species, ornithology is again indebted to Long's Expedition, and particularly to Say, who gave it the name we have adopted, and informs us, in his notes, that many of these birds were shot in the month of

June, at Bellefontaine, on the Missouri; and others were observed, the following spring, at Engineer Cantonment, near Council Bluffs.

It seems probable that the range of this bird is limited, in a great measure, by the Mississippi on the east. Like the larks, they frequent the prairies, and very seldom, if ever, alight on trees. They sing sweetly, and often continue their notes while on the wing.

The lark finch is six inches and a half long; its bill, a little notched at tip, is of a pale horn-colour, with a slight elevation on the roof of the upper mandible. The feet are pale flax-colour, tinged with orange; the irides are dark brown. On the top of the head are two dilated lines, blackish on the front, and passing into ferruginous on the crown and hind head, separated from each other by a whitish cinereous line; from the eye to the superior mandible is a black line, which, as well as the eye, is enclosed by a dilated white line, contracted behind the eye; from the angle of the mouth proceeds a black line, which is much dilated into a ferruginous spot on the auricles; below this is a broad white line, margined beneath by a narrow black one, originating at the inferior base of the lower mandible; the chin and throat are pure white. The neck above, the back, and rump, are dull cinereous brown, each feather of the interscapular region having a blackish brown disk; the neck beneath and breast are dull whitish cinereous; a small blackish brown spot is on the middle of the breast; the belly and vent are white. The wings are dusky brown; the lesser wing-coverts are margined with dull cinereous; the exterior primary is equal to the third; both are very little shorter than the second, which is longest; the outer webs of the second, third, and fourth primaries, being whitish near their bases, form a distinct spot on the wing. The tail is rounded, the feathers being blackish brown; the two intermediate ones are immaculate, somewhat paler than the others. The adjoining ones have a small white spot at tip, which, on the lateral feathers, increases in

Drawn from Nature by Titian R. Peale

Engraved by W.H.Lizars

1. Crimson-necked Bullfinch. 2. Female. 3. Arkansaw Siskin. 4. Female American Goldfinch. 5. Lazuli Finch.

Pyrrhula Frontalis. Fringilla Psaltria. Fringilla Tristis. Fringilla Amæna.

6.

size, until, on the exterior one, it occupies half the total length
of the feather, whilst its exterior web is white to the base.

The female is very similar to the male, but the colours are
duller, and the stripes on the head are not so decided ; the
auriculars, moreover, are yellowish brown.

This species has the bill and feet precisely similar to those
of Wilson's black-throated bunting, and those other *Fringillæ*
and supposed *Emberizæ*, of which I have constituted the subgenus *Spiza* in my " Observations on Wilson's Ornithology."
It cannot be mistaken for any other species, being very peculiar
in its markings and manners.

CRIMSON-NECKED BULLFINCH. *(Pyrrhula frontalis.)*

PLATE VI.—Fig. 1, Male ; Fig. 2, Female.

Fringilla frontalis, *Say, in Long's Expedition to the Rocky Mountains,* ii. p. 40.—
Philadelphia Museum, No. 6279, male ; No. 6277, female.

ERYTHROSPIZA FRONTALIS.—Bonaparte.

Erythrospiza, *Bonap. Oss. Sulla,* 2d ed. *Del Regn. Anim. Cuv.* p. 80.—See note,
vol. i. p. 121.

Much confusion exists in the works of naturalists respecting
those finches and bullfinches that are tinged with red ; and,
in fact, their great resemblance to each other, and their intricate synonymy, render them very difficult to elucidate.   The
only species in Wilson's work with which the present may be
confounded is the *Fringilla purpurea,* a bird closely related
to ours, and for the first time well figured and permanently
established by that author.*   But several other allied species

---

* He was rather precipitate in asserting the *Fringilla rosea* and *Loxia
erythrina* to be identical with his bird, as they are actually two very
distinct species, belonging to the genus *Pyrrhula,* and proper to the
old continent, whilst the *purpurea* is a true *Fringilla,* and peculiar to
America.   To those who have not critically investigated the subject, it
may appear somewhat inconsistent to state that the *erythrina* is not an
inhabitant of this continent, when it is a well-known fact that many
authors speak of it as an American bird.   This apparent contradiction

may be mistaken for the crimson-necked bullfinch; two of these, belonging to the genus *Pyrrhula*, present so much analogy with the present species, judging from their descriptions, that we doubted the correctness of giving the latter a separate place, considering it identical with *Pyrrhula erythrina* of Temminck, whose description agrees better with it than that of any other. Yet, in addition to some differences discoverable by comparing the crimson-necked bullfinch with his description, we cannot admit that an arctic bird of the old continent, known to visit even the more northern portion of the temperate climates only during very cold winters, and then not very regularly, should be found in the month of July on the sultry plains of the Arkansaw, and of course breeding there. We therefore conclude that our bird is not the *erythrina*, although we regret our inability to give differential characters, having never seen that species, as our endeavours to obtain a specimen have not been attended with success. The southern residence of our bird might lead us to suppose it the *Loxia* (*Pyrrhula*) *violacea*, which we have not seen, neither do we think the species well established. But if we are to rely on the short description given of it, and on Catesby's figure, we cannot perceive much resemblance between them; their identity, however, would not much surprise us, when we consider that Catesby's figure of the *Pyrrhula violacea* is as much like our bird as his figure of the purple finch is like what it is intended to represent. Having the

may be readily removed by considering what bird those authors alluded to when they stated the *erythrina* to be a native of North America. When Latham expressed a doubt in his "Synopsis" whether the birds in the neighbourhood of New York so much resembling the *erythrina* were not specifically the same, he alluded to the *Fringilla purpurea*: Gmelin, as usual, in his miserable compilation, inserted this doubt of Latham as a certainty. As to the crimson-headed finch of Pennant, it is evidently the *purpurea*, thus excusing, in part, the strange assertion of Wilson. Latham also committed an error in his index, by placing the *Loxia erythrina* of Pallas and Gmelin, his own crimson-head finch, as a variety of *Fringilla rosea*.

authority of Say, we consider it as new, notwithstanding these doubts.

The crimson-necked bullfinch was procured by Long's party near the Rocky Mountains, and Say described it in the journal of that expedition under the name of *Fringilla frontalis*, adopting that genus in the comprehensive limits assigned by Illiger and Cuvier. The specific name given by Say is preoccupied in that genus by an African species, but as we consider our bird a *Pyrrhula*, we think proper to retain his name.

The crimson-necked bullfinch is five inches and a half long. The bill and feet are horn colour; the lower mandible is paler; the irides are dark brown; the head, neck beneath, and superior portion of the breast, are brilliant crimson, most intense near the bill and over the eye; the space between the bill and the eye is cinereous grey, as well as the cheeks, and the small feathers immediately around the bill; the crimson feathers are brown at base, being red only at tip; the occiput, and the neck above and on each side, are brown, with a reddish cast, the feathers being margined with pale; the back is dusky brownish; the rump and superior tail-coverts are crimson, but less vivid than that of the head; the inferior portion of the breast, the belly, and vent, are whitish, each feather having a broad fuscous line; the general plumage is lead colour at base. The wings are blackish brown, the primaries being broadly margined within, towards the base, with whitish, and exteriorly edged with greyish; the coverts and secondaries are edged with dull greyish; the tail is blackish brown, hardly emarginated; the lateral feathers are edged on the inner side with whitish.

Such is the description of our male specimen; but as it was procured when summer was far advanced, a season in which the plumage begins to fade, it is proper to observe that the colouring of this bird is probably much more brilliant in its full spring dress, the crimson extending much farther down on the back, &c. As the season advances, the tips of the feathers, which are the only parts of a crimson colour, being gradually worn off, the bird as gradually loses its brilliancy, and in the autumnal and winter plumage exhibits the humble appearance of the female.

The female is altogether destitute of the brilliant colour, being dusky brown above, the feathers margined on each side with dull whitish; the whole inferior surface is whitish, each feather having a brown longitudinal line in the middle, obsolete on the vent, which is almost pure white.

A change similar to that above mentioned takes place in the purple finch, whose habits also much resemble those of the crimson-necked bullfinch; but the form of its bill is certainly that of a finch, and will always distinguish it from the species we are describing, the bill of which is unequivocally of the bullfinch form. The different tints of red adorning these birds will also at once strike the eye of the least expert in discriminating species; in the present bird the tint is vivid crimson, whilst in the purple finch it is rosaceous. In addition to these characters, the latter is a somewhat larger bird, with a pure white belly, inferior tail-coverts, and a deeply emarginated tail; whilst the former has a nearly even tail, and its belly and inferior tail-coverts are striped with dusky.

Some persons, without doubt, may think it highly improper to separate generically two birds so closely allied as the present species and the purple finch, which may be mistaken for the same species; but we may remark, that they stand at the extreme limit of their respective genera, and form the links of union between *Pyrrhula* and *Fringilla*. It is true that the intimate alliance of these two groups would seem to justify Illiger, Meyer, and others in uniting them under the same genus; but as *Fringilla* is so vast in the number of its species, and *Pyrrhula* has a few distinctive characters, we choose to follow Temminck, Vieillot, and other naturalists, by arranging them generically separate. The closeness of affinity between these two birds, when thus properly disposed, affords no good reason for the unity of their genera; for, if we proceed to the abolition of all artificial distinctions between genera united by almost imperceptible gradations, *Sylvia* would be joined to *Turdus*, *Myiothera* to *Troglodytes*, *Lanius* to *Muscicapa*, the whole of these would be confused together, and, in fact, orders and classes would be considered as genera ; and even the vast groups thus formed would be still observed to unite inseparably at their extremes, and we should finally be compelled to consider all living bodies, both animal and vegetable, as belonging to one genus. This argument, however, may not convince every naturalist of the propriety of our arrangement, and they must, therefore, place the two species strictly according to nature in one genus, and consider the present as a *Fringilla ;* but how unnatural will then be the situation of *Pyrrhula vulgaris*, and *Pyrrhula enucleator !*

The inflated form of the bill, the curvature of both mandibles, very apparent in the superior one, as well as the compression of both at tip, are obvious characters which distinguish the species of *Pyrrhula* from the *Fringillæ*, in which both mandibles are nearly straight, and present a conic form on every side.

Berries and seeds which they extract from the pericarp, buds and young shoots of different plants, constitute the food of the bullfinches. They generally frequent forests and bushy places, building their nests on small trees or low branches of large ones. The females lay four or five eggs. The greater number of the species moult twice a year ; the sexes differ considerably in appearance. They reside in cold and temperate climates, with the exception of a few species that inhabit Africa and South America.

The crimson-necked bullfinch is found in the district of country extending along the base of the Rocky Mountains, near the Arkansaw river, and has not been observed elsewhere. In the month of July, when our specimens were obtained, these birds occur in small scattered flocks, keeping mostly on the tops of the cotton-wood trees, on whose buds they partially feed. Their voice considerably resembles that of their relative the *Fringilla purpurea.*

ARKANSAW SISKIN. (*Fringilla psaltaria.*)

PLATE VI.—FIG. 3.

Fringilla psaltaria, *Say, in Long's Exped.* ii. p. 40.—*Phil. Museum,* No. 6278.
See note, vol. i. pp. 12 and 15.

CARDUELIS PSALTARIA.—BONAPARTE.

Fringilla (subgen. Carduelis) psaltaria, *Bonap. Synop.* p. 111.

"A VERY pretty little bird," writes Say, in his precious zoological notes to the Journal of Long's Expedition, "was frequently seen hopping about in the low trees or bushes, singing sweetly, somewhat in the manner of the American goldfinch or hempbird, *Fringilla tristis.* The tints, and the distribution of the colours of its plumage, resemble in a considerable degree those of the autumnal and less brilliant vesture of that well-known species. It may, however, be distinguished, in addition to other differences, by the black tip of its tail-feathers and the white wing spot."

The Arkansaw siskin inhabits the country near the base of the Rocky Mountains, south of the river Platte, and probably is also to be found in Mexico. The only specimen brought by the party was shot on the 16th of July, near Boiling Spring Creek : on the annexed plate, it is figured in company with the American goldfinch in autumnal plumage, for the sake of comparison.

The Arkansaw siskin is four inches and a quarter long; the bill is yellowish, tipped with blackish ; the feet are flesh colour ; the irides, burnt umber. The top of the head is blue black ; the cheeks are dusky olivaceous ; the neck above, and half its side, the back and rump, are olivaceous, more or less intermixed with dusky and yellowish, particularly on the rump ; the superior tail-coverts are black, varied with olivaceous ; all the under parts, from the very base of the bill to the under tail-coverts inclusively, are of a pure bright yellow. The wings are brownish black, the smaller wing-coverts being very slightly tinged with blue, and edged with olivaceous ; the greater wing-coverts are tipt with white, which forms a narrow band across the wing; the primaries, excepting the exterior one, are slightly edged with white; the third, fourth, fifth, sixth, and seventh are white towards the base, so as to exhibit a white spot beyond the wing-coverts; the first four primaries are nearly equal in length, the fifth is a quarter of an inch shorter; the secondaries are broadly margined with white exteriorly towards their tips. The tail is slightly emarginated, the feathers being blackish, slightly edged with dull whitish; the three exterior ones are widely pure white on the middle of their inner webs.

The specimen we have just described is a male, evidently in perfect plumage; the female, and state of imperfect plumage, are unknown; but, without risking any great deviation from the truth, we may state, from analogy, that the young resemble the female, which must be destitute of the black cap, and have the colours less vivid and less pure.

The Arkansaw siskin certainly resembles the American goldfinch in its winter dress; but a still more striking similarity exists in some other birds, such as the European siskin (*Fringilla spinus*), and the Olivarez (*Fringilla Magellanica*, Vieill.) of South America; and it is so similar to the European, that it might, with a much greater degree of propriety, be considered as a variety, than those regarded as such by authors. They can, however, be easily distinguished by the following comparative characters :—All the under parts of the Arkansaw siskin are bright yellow, whilst the corresponding parts of the European siskin are tinged with greenish, the throat being black, and the belly, vent, and flanks, whitish, spotted longitudinally with black; the margins and spots of the wing and tail-feathers are white in our bird, and yellow in the European siskin; the white spots on the tail of the Arkansaw siskin are confined to the three outer feathers, whilst, in the foreign bird, all the feathers, excepting the two middle ones, are marked with yellow; the bill of our species is also a little shorter, less compressed, and less acuminated; finally, we may notice another trifling difference, which consists in the proportional length of the primaries, the four first being nearly equal in the American bird, and the three first only in the European, the fourth being almost a quarter of an inch shorter. The other approximate species, *Fringilla Magellanica,* Vieill., considered by Gmelin and Latham as a variety of the European siskin, is readily distinguishable by having the head entirely black.

Though the Mexican siskin (*Fringilla Mexicana,* Gmel.) may prove to be the female of our bird, or the male in an imperfect state of plumage (and, from the locality, we should possibly have referred it to that name, had the classification of it fallen to our lot), yet, as nothing positive can be drawn from so unessential an indication as that of the Mexican siskin, we have no hesitation in following the same course with Say, who considers it as entirely new, and have retained his elegant name of *Fringilla psaltaria.* It is very possible that not only the *Fringilla Mexicana,* but also the black Mexican siskin (*Fringilla catotol,* Gmel.), may be the same bird as our *Fringilla psaltaria ;* but how can we determine from the vague descriptions that have been given of those species? They are equally applicable to the American goldfinch in its dull state of plumage ; and Wilson expresses a doubt whether or not the black Mexican siskin is the same as his new species, *Fringilla pinus.*

All these pretty little birds belong to the subgenus *Carduelis,* having a more slender, acute, and elongated bill than other *Fringillæ.*

FEMALE AMERICAN GOLDFINCH. (*Fringilla tristis.*)

PLATE VI.—FIG. 4.

Wilson's American Ornithology, i. p. 20, pl. 1, fig. 2, for the male and history.—
Fringilla tristis, *Linn. Syst.* i. p. 320, sp. 12, male.—*Gmel. Syst.* i. p. 907, sp.
12.—*Lath. Ind.* p. 452, sp. 64.—*Vieill. Nouv. Dict. d'Hist. Nat.* xii. p. 167.
—Fringilla spinus, var. γ, *Gmel. Syst.* i. p. 914, sp. 25, male in winter
plumage.—Carduelis Americana, *Briss. Av.* iii. p. 64, sp. 3.—Carduelis
Americanus, the American Goldfinch, *Catesby, Carolina,* i. p. 43, pl. 43, male
in spring dress.—*Bart. Trav.* p. 291.—Chardonneret jaune, *Buff. Ois.* iv. p.
212.—Chardonneret du Canada, *Buff. Pl. enl.* 202, fig. 2, male in spring
dress.—Tarin de la Nouvelle York, *Buff. Ois.* iv. p. 231.—*Pl. enl.* 292,
fig. 1, male changing; fig. 2, male in winter dress.—Golden Finch, *Penn.
Arct. Zool.* sp. 242.—American Goldfinch, *Edwards, Glean.* ii. p. 133, pl.
274, male and female.—*Lath. Syn.* ii. part 1, p. 288, sp. 57; *Id.* 1st Suppl.
p. 196.—Siskin, var. B, *Lath. Syn.* ii. part 1, p. 291, sp. 58, male changing.—
Philadelphia Museum, No. 6344, male; No. 6345, female; No. 6346, albino.

CARDUELIS AMERICANA.—EDWARDS.

Male and note, see vol. i. pp. 11, 15.

WE have been induced, by the analogy existing between the
preceding new species and this common bird, to figure them
as companions on the same plate, that they may be immediately and readily compared. To give the present figure more
interest, we have chosen the female, though we might, with
equal propriety, have selected the male in winter plumage, as
the latter differs but slightly from its mate during that season.
The very great dissimilarity between the sexes in their spring
dress will justify the reappearance of a bird already given by
Wilson, more especially as it has, in this state, been mistaken
for a distinct species, and most unaccountably arranged in the
systems as a variety of the European siskin.

The history of this bird, which so completely resembles the
goldfinch of Europe in song and habits, being nearly completed by the golden pen of Wilson, we shall not attempt to
add any observations of our own, but shall refer the reader to
his volume, quoted above, for its biography. As we cannot
but observe that his description is short and somewhat imper-

fect, probably owing to the opinion he at first entertained, but afterwards judiciously relinquished, that a minute description of common birds is superfluous, we shall proceed to describe the species in all its different states.

The male American goldfinch in summer dress, represented by Wilson in his first plate, is four and a half inches long, and eight in extent. The bill resembles that of the European goldfinch, and, as well as the feet, is of a reddish cinnamon colour ; the irides are dark brown ; the front and vertex are glossy black ; the remaining part of the head, and all the body, rich lemon yellow; the superior and inferior tail-coverts are white, as well as the thighs ; the wings and tail are black, the small coverts of the wings being yellow externally, and white on the inner side and at tip ; the greater coverts are tipt with white, an arrangement which exhibits two white bands across the wings ; the first and third primaries are equal, hardly shorter than the second, which is the longest, the fourth being nearly as long as the third ; the secondaries are margined with white. The tail is emarginated, the feathers being black, slightly edged with white, and having a large pure white spot on the inner web at tip.

The female, as is usual in this family of birds, is rather smaller than the male, and is widely different from that sex in the colours of its plumage. The bill and feet are brownish ; the lower mandible is whitish at base ; the head has no appearance of black, and, with the neck, the back, and rump, is brownish olive, the latter part being of a lighter shade than the preceding portions ; the upper tail-coverts are greenish white ; the frontlet, cheeks, sides of the neck, throat, and upper part of the breast, are pale greenish yellow ; the lower portion of the breast, belly, vent, flanks, under wing and under tail-coverts, are whitish. The wings and tail, which always afford the most constant specific characters, are like those of the male, except that the black colour is less intense, and the white is less pure, being slightly tinged with rufous.

In this state of plumage, the bird closely resembles the

Fringilla citrinella of the south of Europe, which, however, can always be distinguished from it by several characters, but more particularly by its greenish yellow rump, and by being destitute of the whitish spot at the tip of the inner web of the tail-feathers. The young are so like the females as to be distinguished with difficulty ; their colours, however, are still less lively ; they assume the adult livery in the spring, but do not exhibit all the brilliancy of the perfect bird until the third moult.

The American goldfinch moults twice a year, in the seasons of spring and autumn. At the spring moult, the males obtain their vivid colouring, which is lost at the autumnal change, and replaced by a more humble dress, similar to that of the female, from which sex they cannot then be readily distinguished. The black of the wings is, however, somewhat more intense ; the white of the wings and of the tail is dull and dirty, and a yellowish tint prevails around the eyes, as well as on the neck. From this statement it follows, that Wilson's figure represents the adult male in that brilliant dress in which it appears for the space of four or five months only; whilst the figure in the annexed plate exhibits the invariable colours of the female and young, as well as the appearance of the male for the remaining seven months in the year.

As the season advances, the plumage of the adult male gradually changes, but not simultaneously in the different individuals,'so that in the spring and autumn we rarely find two that are alike; some being more or less yellow, having a rudiment of black on the head, &c., according as the moulting process is more or less advanced.

A remarkable variety is exhibited in a changing male, which I shot near Philadelphia in the month of April, and which is therefore considerably advanced towards perfect plumage. All the primaries are pure white on the outer web towards the base, thus constituting, in the most obvious manner, that white spot beyond the wing-coverts assigned by Say as a good discriminating mark between this species and the preceding. The fact we have related diminishes the value of this character, which is nevertheless a very good one ; but as many other distinctions are observable, we need not rely exclusively upon it. The deviation we have here mentioned is the more remarkable, as the greater number of species allied to this bird have that spot, either white or yellow.

Since writing the above, I obtained, from one of the large flocks in which these birds congregate in the autumn, several specimens of both sexes, more or less distinguished by the marking above stated as peculiar to the variety.

LAZULI FINCH. (*Fringilla amœna.*)

PLATE VI.—Fɪɢ. 5.

Emberiza amœna, *Say, in Long's Expedition to the Rocky Mountains,* ii. p. 47.— *Philadelphia Museum,* No. 5919.

SPIZA AMŒNA.—Bᴏɴᴀᴘᴀʀᴛᴇ.*

Fringilla (subgen. Spiza), *Bonap. Synop.* p. 106.

Tʜᴇ genus *Emberiza*, though very natural, and distinguished by well-marked characters, has, notwithstanding these advantages, been often misunderstood ; and authors, without consulting the boundaries assigned to it by themselves, have recorded a copious list of species, whilst in nature its limits are much restricted. We are not therefore surprised that so acute a zoologist as Say should have arranged his bird in that genus, particularly as it is more closely allied to *Emberiza* than many of those, not only of Wilson, but even of Linné and Latham.

---

* In the Prince of Musignano's " Synopsis of the Birds of the United States," in the "Annals of the Lyceum of Natural History, New York," he has instituted a subgenus under the name of *Spiza*, to contain a portion of the *Fringillidæ*, dividing it into two sections. The first contains our present bird, with the *F. cyanea* and *ciris* of Wilson, and to which we think the subgenus should be restricted. Those which form the second section run much more into the *Emberizæ*, and although it may " form the passage to the buntings," it is of sufficient importance to constitute a small *sub-group.*—Eᴅ.

This bird, which we have no hesitation in pronouncing one of the most beautiful of its tribe, would be placed by Vieillot in his genus *Passerina ;* but, according to my classification, it belongs to the genus *Fringilla,* and to that American subgenus lately established in my "Observations on the Nomenclature of Wilson's Ornithology," under the name of *Spiza.* As a species, it is more intimately allied to *Fringilla ciris* and *Fringilla cyanea,** which I stated in that paper to differ so much from their congeners, particularly in the greater curvature of the upper mandible, as to deserve, perhaps, a separation into a small subgenus by themselves. This would unite *Fringilla* to *Tanagra,* as *Spiza,* on the other hand, shows its transition to *Emberiza.*

The lazuli finch is five inches and three-quarters long ; the bill is formed like that of the indigo-bird (*Fringilla cyanea* Wilson), but is emarginated near the tip, being horn colour, as well as the feet ; the irides are dark brown ; the whole head and neck are brilliant verdigris-blue ; the back is brownish black, intermixed with blue, and a little ferruginous brown ; the rump is pure verdigris-blue ; the superior portion of the breast is pale ferruginous ; the lower part of the breast, the belly, and inferior tail-coverts are white ; the smaller wing-coverts are blue ; the middling coverts are blackish at base, and broadly tipped with white, forming a wide band across the wing ; the greater wing-coverts are blackish, obscurely margined with blue, and slightly tipt with white on the exterior web, constituting a second band across the wings parallel to the first, but much narrower ; the primaries and secondaries are blackish, obscurely margined with blue on the outer web ; the under wing-coverts are whitish, a little intermixed with blue ; the tail is slightly emarginated, the feathers being blackish, edged with blue on the outer web, and with white on the inner web at tip.

The above description of this handsome bird is taken from

---

* Its relation to *Fringilla cyanea,* considered as *Emberiza,* probably induced Say to place it under that genus.

a male in summer plumage, the only specimen brought by
Long's exploring party ; hence we are unable to give any positive information relative to the female and young, though,
from analogy, we must believe them in great part destitute of
the blue colour, and otherwise less brilliantly adorned.

This species appears to be rather rare ; it is found along
the Arkansaw river, near the base of the Rocky Mountains,
during the summer months ; they frequent the bushy valleys,
keeping much in the grass, and seldom alight on shrubs or
trees.   In this respect also they resemble the indigo-bird, and
probably their habits are the same, although the note is entirely
dissimilar.

FULVOUS OR CLIFF SWALLOW.   (*Hirundo fulva.*)

PLATE VII.—Fig. 1.

Hirundo fulva, *Vieill. Ois. de l'Am. Sept.* i. p. 62, pl. 32.—*Stephens, Cont. of
    Shaw's Zool.* x. part i. p. 126.—*De Witt Clinton, Ann. Lyceum Nat. Hist. N.
    Y.* i. p. 156.—Hirundo lunifrons, *Say, in Long's Exp. to the Rocky Mountains,*
    ii. p. 47.—*Philadelphia Museum,* No. 7624.

HIRUNDO FULVA.—Vieillot?

Hirundo fulva, *Bonap. Synop.* p. 64.—The Republican, or Cliff Swallow, *Aud.* pl.
    68, m. and y.—*Orn. Biog.* i. p. 353.—Hirundo lunifrons, *North. Zool.* ii. p.
    331.

With the exception of a very imperfect description, little
was known relative to this interesting bird anterior to Long's
expedition to the Rocky Mountains.   One of the notes annexed
to the account of that journey contains an excellent description of this swallow, with a notice of its habits, and remarkable
manner of building.   Mr De Witt Clinton has recently published a paper on the same subject, accompanied by some
observations from Mr Audubon.   Combining what these gentlemen have made known with the information previously
given by Vieillot and Say, we can present a tolerably complete history of the cliff swallow.

Some doubts having been entertained whether the *Hirundo
lunifrons* of the Rocky Mountains be the same species as the

1. *Fulvous or Cliff Swallow.* 2.*Burrowing Owl.*

Hirundo Fulva.   *Strix Cunicularia.*

Hirundo fulva of the western part of New York, I was desirous of deciding the question by comparing the specimens ; this I accomplished through the politeness of Dr Dekay of New York, who, with the kindness and liberality distinctive of those who cultivate science for its own sake, sent me the specimen and nest deposited by Mr Clinton in the Cabinet of the Lyceum. Thus being possessed of the individuals in question, we are enabled to place their specific identity beyond the reach of future uncertainty.*

* There can be nothing more annoying than being in a manner obliged to give an opinion regarding a disputed point from descriptions and plates, without the actual comparison of the birds themselves. The authors of the " Northern Zoology " consider the *H. lunifrons* of Say different from the *H. fulva* of Vieillot, on account of the pure white front and slightly forked tail of the former ; but the Prince of Musignano makes them identical, from actual comparison with authentic specimens of *H. fulva.* The alternative, therefore, must be, that the specimens brought by the Northern Expedition are distinct from either, and yet unnamed. Audubon's figure, however, is very nearly pure white in the frontlet, and he insists upon that colour even in the young ;—the tail is square, a decided mark of our author. There are either two species confused in these, or inattention has been paid to the drawing and colouring of those parts where distinction chiefly is insisted on.

On the precipitous coast of the Firth of Forth, near Tantallon Castle, in Haddingtonshire, there was, in 1826, and for several years previous, a colony of fifty or sixty pairs of *H. urbica,* building their nests in the usual form, but in the same manner, under a huge projecting cliff, as represented of the *H. fulva.* They struck me at the time of first seeing them as a remarkable situation for the species, and the plate here immediately recalled them to my memory. Mr Audubon's description of their manner of building may add to that of our author:—

" About daybreak they flew down to the shore of the river, one hundred yards distant, for the muddy sand of which the nests were constructed, and worked with great assiduity until near the middle of the day, as if it were that the heat of the sun was necessary to dry and harden their moist tenements. They then ceased from labour for a few hours amongst themselves, by performing aerial evolutions, courted and caressed their mates with much affection, and snapped at flies and other insects on the wing. They often examined their nests to see if they were sufficiently dry ; and as soon as these appeared to have acquired the requisite firmness, they renewed their labours. Until the females began to sit, they

That Say considered his *Hirundo lunifrons* as a new bird, is entirely attributable to the incorrectness of Vieillot's figure, which is one of those better suited to mislead than to assist the naturalist in his researches. The most striking characteristic of the *Hirundo fulva* is its even tail; yet Vieillot has represented this part as forked. We are therefore not surprised that our learned zoologist, who had no opportunity of consulting the coloured plate, should not have even thought of comparing his bird with that of Vieillot, who probably figured it with a forked tail merely because it was a swallow. The characters of the cliff swallow are so remarkable, and its manner of building is so peculiar, that, when these are accurately delineated, it cannot be mistaken for any other species.

The cliff swallow is five and a half inches long. The bill is black, and the feet dusky; the irides are dark brown. A narrow black line extends over the bill to each eye; the front is pale rufous, and the remaining part of the crown, black violaceous; the chin, throat, and cheeks are dark ferruginous, extending in a narrow band on the hind head; the upper part of the body is black, glossed with violaceous; the inferior part of the rump, and some of the tail-coverts, are pale ferruginous;

all roosted in the hollow limbs of the sycamores (*Platanus occidentalis*) growing on the banks of the Licking river, but when incubation commenced, the males alone resorted to the trees. A second party arrived, and were so hard pressed for time, that they betook themselves to the holes in the wall where bricks had been left out for the scaffolding. These they fitted with projecting necks, similar to those of the complete nests of the others. Their eggs were deposited on a few bits of straw, and great caution was necessary in attempting to procure them, as the slightest touch crumbled their frail tenement into dust. By means of a table-spoon, I was enabled to procure many of them. Each nest contained four eggs, which were white, with dusky spots. Only one brood is raised in a season. The energy with which they defended their nests was truly astonishing. Although I had taken the precaution to visit them at sunset, when I supposed they would all have been on the sycamores, yet a single female happened to be sitting, and gave the alarm, which immediately called out the whole tribe. They snapped at my hat, body, and legs, passed between me and the nests within an inch of my face, twittering their rage and sorrow."—Ed.

the breast is of a pale rufous ash colour, and the remaining under parts are whitish, tinged with brownish ferruginous; the wings and tail are blackish, the small wing-coverts being glossed with violaceous; the inferior wing-coverts are ashy brown; the tail is nearly entire, somewhat shorter than the tips of the wings; the exterior tail-feather is slightly edged with whitish on the inner vane; the wing and tail-feathers have their shafts black above and white beneath.

This description is taken from our finest male, which is also represented in the plate; no difference exists between the sexes, and the young, even during early age, can scarcely be distinguished from the parents, except by having the front white, instead of rufous. We are informed by Vieillot that some individuals have all the inferior surface of the body tinged with the same colour as that of the throat: these are probably very old males.

A very singular trait distinguishes the migrations of this bird. While the European, or white variety of the human race, is rapidly spreading over this continent, from its eastern borders to the remotest plains beyond the Mississippi, the cliff swallow advances from the extreme western regions, annually invading a new territory farther to the eastward, and induces us to conclude, that a few more summers will find it sporting in this immediate vicinity, and familiarly established along the Atlantic shores.

Like all other North American swallows, this species passes the winter in tropical America, whence in the spring it migrates northward for the purpose of breeding. It appears to be merely a spring passenger in the West Indies, remaining there but a few days according to Vieillot, who, not seeing any in the United States, and observing some while at sea, in August, in the latitude of Nova Scotia, supposed that they propagated in a still more northern region. As we have not received any account of their inhabiting the well-explored countries around Hudson's Bay, we are led to the conclusion that the western wilds of the United States have hitherto been their summer resort, and that not until recently have they ventured within the domains of civilised man. Be this as it may, they were observed in great numbers by Major Long's party near the Rocky Mountains in the month of July, and a few were also seen on the banks of the Missouri river. Within ten or twelve years they have become familiar in different localities of Ohio, Kentucky, &c., whence they are extending very rapidly, and have recently appeared in the western part of New York. In order to show the rapid progress of this little stranger, we quote the following passage from Mr Clinton's interesting paper :—

The fulvous swallow "first made its appearance at Winchell's Tavern, on the highroad, about five miles south of Whitehall, near Lake Champlain, and erected its nest under the eaves of an outhouse, where it was covered by the projection of a roof. This was in 1817, and in this year there was but one nest ; the second year, seven ; the third, twenty-eight ; the fourth, forty ; and in 1822 there were seventy, and the number has since continued to increase.

"It appeared in 1822 at Whitehall on the 5th of June, and departed on the 25th of July ; and these are the usual times of its arrival and disappearance."

This active little bird is, like its congeners, almost continually on the wing, and feeds on flies and other insects, while performing its aerial evolutions. Its note is different from that of other swallows, and may be well imitated by rubbing a moistened cork around in the neck of a bottle. The species arrive in the west from the south early in April, and immediately begin to construct their symmetrical nests, which are perfected by their united and industrious efforts. At the dawn of day they commence their labours by collecting the necessary mud from the borders of the river or ponds adjacent, and they persevere in their work until near midday, when they relinquish it for some hours, and amuse themselves by sporting in the air, pursuing insects, &c. As soon as the nest acquires the requisite firmness, it is completed, and the female begins to deposit her eggs, which are four in number, white, spotted with dusky brown. The nests are extremely friable, and will readily crumble to pieces : they are assembled in communities, as represented in the background of our plate. In unsettled countries, these birds select a sheltered situation under a projecting ledge of rock ; and in civilised districts, they have already evinced a predilection for the abodes of man, by building against the walls of houses, immediately under the eaves of the roof, though they have not in the least changed their style of architecture. A nest from the latter situation is now before me ; it is hemispherical, five inches wide at its truncated place of attachment to the wall, from which it projects six inches, and consists exclusively of a mixture of sand and clay, lined on the inside with straw and dried grass, negligently disposed for the reception of the eggs. The whole external surface is roughened by the projection of the various little pellets of earth which compose its substance. The entrance is near the top, rounded, projecting, and turning downward, so that the nest may be compared to a chemist's retort, flattened on the side applied to the wall, and with the principal part of the neck broken off.

So great is the industry of these interesting little architects, that this massive and commodious structure is sometimes completed in the course of three days. About the middle of July, some nests found near the Rocky Mountains contained young ones, while in others the process of incubation had not terminated. It is probable that the cliff swallows rear two broods in that region, though in Kentucky and Ohio, agreeably to Mr Audubon, they have but one in the year. During the first few days of August they assemble in flocks, and, after several attempts to commence their migration, they finally succeed in obtaining a unanimity of purpose, and they disappear as suddenly as they came.

BURROWING OWL.   (*Strix cunicularia.*)

PLATE VII.—Fɪɢ. 2.

Strix cunicularia, *Molina, Hist. Chili* (Am. ed.) i. p. 184.—*Gmel. Syst.* i. p. 292,
    sp. 28.—*Lath. Ind.* p. 63, sp. 38.—*Vieill. Ois. de. l'Am. Sept.* i. p. 48.—*Say,
    in Long's Expedition to the Rocky Mountains,* ii. p. 36 and 200.—Ulula
    cunicularia, *Feuillée, Journ. Obs. Phys.* p. 562.—Noctua Coquimbana, *Briss.
    Av.* i. p. 525, sp. 11.—Coquimbo Owl, *Lath. Syn.* i. p. 145, sp. 33.—*Phila-
    delphia Museum,* No. 472.

ULULA CUNICULARIA.—Feuillée.*
Strix (subgen. Surnia) cunicularia, *Bonap. Synop.* p. 36.

Venerable ruins, crumbling under the influence of time and
vicissitudes of season, are habitually associated with our recollections of the owl ; or he is considered as the tenant of
sombre forests, whose nocturnal gloom is rendered deeper and
more awful by the harsh dissonance of his voice.   In poetry
he has long been regarded as the appropriate concomitant of
darkness and horror ; and when heard screaming from the
topmost fragments of some mouldering wall, whose rugged-
ness is but slightly softened by the mellowing moonlight,
imagination loves to view him as a malignant spirit, hooting
triumphantly over the surrounding desolation !   But we are
now to make the reader acquainted with an owl to which none
of these associations can belong—a bird that, so far from seeking refuge in the ruined habitations of man, fixes its residence
within the earth, and, instead of concealing itself in solitary
recesses of the forest, delights to dwell on open plains, in
company with animals remarkable for their social disposition,
neatness, and order.   Instead of sailing heavily forth in the

---

* I am far from being satisfied with the generic appellation I have
now provisionally bestowed on this bird.   It is completely a *day owl*
in its habits, but in many parts of its structure resembles the nocturnal
species.   Wherever it may be hereafter placed by a strict analysis, it
will prove a most interesting form, and perhaps show some connections
that we do not at present anticipate.   The Prince of Musignano has
properly incorporated Mr Say's remarks with his description.—Eᴅ.

obscurity of the evening or morning twilight, and then retreating to mope away the intervening hours, our owl enjoys the broadest glare of the noonday sun, and flying rapidly along, searches for food or pleasure during the cheerful light of day.

The votaries of natural science must always feel indebted to the learned and indefatigable Say for the rich collection of facts he has made whenever opportunities have been presented, but more especially in the instance of this very singular bird, whose places of resort in this country are too far distant to allow many the pleasure of examining for themselves. We feel doubly disposed to rejoice that the materials for the history of our bird are drawn from his ample store, both on account of their intrinsic excellence, and because it affords us an opportunity of evincing our admiration of the zeal, talents, and integrity which have raised this man to the most honourable and enviable eminence as a naturalist.

In the trans-Mississippian territories of the United States, the burrowing owl resides exclusively in the villages of the marmot or prairie dog, whose excavations are so commodious, as to render it unnecessary that our bird should dig for himself, as he is said to do in other parts of the world, where no burrowing animals exist. These villages are very numerous, and variable in their extent, sometimes covering only a few acres, and at others spreading over the surface of the country for miles together. They are composed of slightly elevated mounds, having the form of a truncated cone, about two feet in width at base, and seldom rising as high as eighteen inches above the surface of the soil. The entrance is placed either at the top or on the side, and the whole mound is beaten down externally, especially at the summit, resembling a much-used footpath.

From the entrance, the passage into the mound descends vertically for one or two feet, and is thence continued obliquely downwards, until it terminates in an apartment, within which the industrious marmot constructs, on the approach of the cold season, a comfortable cell for his winter's sleep. This cell, which is composed of fine dry grass, is globular in form, with an opening at top capable of admitting the finger ; and the whole is so firmly compacted, that it might, without injury, be rolled over the floor.

It is delightful, during fine weather, to see these lively little creatures sporting about the entrance of their burrows, which are always kept in the neatest repair, and are often inhabited by several individuals. When alarmed, they immediately take refuge in their subterranean chambers, or, if the dreaded danger be not immediately impending, they stand near the brink of the entrance, bravely barking and flourishing their tails, or else sit erect to reconnoitre the movements of the enemy.

The mounds thrown up by the marmot in the neighbourhood of the Rocky Mountains have an appearance of greater antiquity than those observed on the far distant plains. They sometimes extend to several yards in diameter, although their elevation is trifling, and, except immediately surrounding the entrance, are clothed with a scanty herbage which always distinguishes the area of these villages. Sometimes several villages have been observed almost entirely destitute of vegetation, and recollecting that the marmot feeds exclusively on grasses and herbaceous plants, it seems singular that this animal should always choose the most barren spot for the place of his abode. However this may be accounted for, it at least affords an opportunity of beholding the approach of his enemies, and allows him to seek, within the bosom of the earth, that security which he has neither strength nor arms to command.

In all these prairie-dog villages, the burrowing owl is seen moving briskly about, or else in small flocks scattered among the mounds, and, at a distance, it may be mistaken for the marmot itself when sitting erect. They manifest but little timidity, and allow themselves to be approached sufficiently close for shooting ; but if alarmed, some or all of them soar away, and settle down again at a short distance ; if further disturbed, their flight is continued until they are no longer in view, or they descend into their dwellings, whence they are difficult to dislodge.

The burrows into which these owls have been seen to descend, on the plains of the river Platte, where they are most numerous, were evidently excavated by the marmot ; whence it has been inferred by Say, that they were either common though unfriendly residents of the same habitation, or that our owl was the sole occupant of a burrow acquired by the right of conquest. The evidence of this was clearly presented by the ruinous condition of the burrows tenanted by the owl, which were frequently caved in, and their sides channelled by the rains, while the neat and well-preserved mansion of the marmot showed the active care of a skilful and industrious owner. We have no evidence that the owl and marmot habitually resort to one burrow, yet we are well assured by Pike and others that a common danger often drives them into the same excavation, where lizards and rattlesnakes also enter for concealment and safety.

The owl observed by Vieillot in St Domingo digs itself a burrow two feet in depth, at the bottom of which its eggs are deposited on a bed of moss, herb stalks, and dried roots. These eggs are two in number, of a very pure white, nearly spheroidal, and about as large as those of the dove. When the young are only covered with down, they frequently ascend to the entrance to enjoy the warmth of the sun, but as soon as they are approached, they quickly retire into the burrow.

The note of our bird is strikingly similar to the cry of the marmot, which sounds like *cheh, cheh,* pronounced several times in rapid succession ; and were it not that the burrowing owls of the West Indies, where no marmots exist, utter the same sound, it might be inferred that the marmot was the unintentional tutor to the young owl : this cry is only uttered as the bird begins its flight. Vieillot states that the burrowing owl inhabiting St Domingo sometimes alights on farmhouses at night, and produces a note which resembles that of the syllables *hoo, hoo, oo, oo ;* but has he not mistaken a nocturnal species for it in this case ?

The food of the bird we are describing appears to consist entirely of insects, as, on examination of its stomach, nothing but parts of their hard wing-cases were found. The authors we have quoted inform us that in Chili and St Domingo the burrowing owls also feed on rats, mice, and reptiles, which we cannot suppose to be the case with the bird found in the United States, as our explorers never could discover the slightest reason for believing that they preyed on the marmots whose dwellings they invade.

Throughout the region traversed by the American expedition, the marmot was unquestionably the artificer of the burrow inhabited by the owl, while the testimony of Vieillot is equally conclusive that the owl digs for himself when he finds no burrow to suit his purpose ; but, preferring one already made, his fondness for the prairie-dog villages is readily explained.

Whether only a single species of burrowing owl inhabits the vast continent of North and South America, or whether that of Chili mentioned by Molina, that of St Domingo described by Vieillot, and the owl of the Western American territory, be distinct though closely allied species, can only be determined by accurate comparisons.* When we consider the extraordinary habits attributed to all those, as well as their correspondence in form and colours noted in the several descriptions, we are strongly inclined to believe that they are all of the same species ; nevertheless, Vieillot states his bird to be somewhat different from that of Molina, and the eggs of the burrowing owl of the latter are spotted with yellow, whilst those of the former are immaculate. We have to regret that no figure has hitherto been published, and we cannot well understand why Vieillot did not thus exemplify so interesting a bird. Our

---

* Should they prove to be different species, new appellations must be given ; and as that of *Strix cunicularia* will, by right of priority, be exclusively retained for the Coquimbo owl, we would propose for the present bird the name of *Strix hypugæa.*

figure will be the more acceptable to ornithologists, as it is the first which has been given of the burrowing owl: in the distance we have introduced a view of the prairie-dog village.

The peculiar subgenus of this bird has not hitherto been determined, owing to the neglect with which naturalists have treated the arrangement of extra-European owls. Like all diurnal owls, our bird belongs to the subgenus *Noctua* of Savigny, having small oval openings to the ears, which are destitute of operculum, the facial disk of slender feathers, small and incomplete, and the outer edges of the primaries not recurved; but it differs from them in not having the tarsus and toes covered by long thick feathers.

The burrowing owl is nine inches and a half long, and two feet in extent. The bill is horn colour, paler on the margin, and yellow on the ridges of both mandibles; the inferior mandible is strongly notched on each side : the capistrum before the eyes terminates in black rigid bristles, as long as the bill : the irides are bright yellow. The general colour of the plumage is a light burnt-umber, spotted with whitish, paler on the head and upper part of the neck; the lower part of the breast and belly are whitish, the feathers of the former being banded with brown : the inferior tail-coverts are white immaculate. The wings are darker than the body, the feathers being much spotted and banded with whitish ; the primaries are five or six banded, each band being more or less widely interrupted near the shaft, and margined with blackish, which colour predominates towards the tip ; the extreme tip is dull whitish ; the shafts are brown above and white beneath : the exterior primary is finely serrated, and equal in length to the fifth, the second and fourth being hardly shorter than the third, which is the longest. The tail is very short, slightly rounded, having its feathers of the same colour as the primaries, and, like them, five or six banded, but more purely white at tip. The feet are dusky, and remarkably granulated, extending, when stretched backwards, an inch and a half beyond the tail; the tarsi are slender, much elongated, covered before and on each side with loose webbed feathers, which are more thickly set near the base, and become less crowded towards the toes, where they assume the form of short bristles ; those on the toes being altogether setaceous, and rather scattered. The lobes beneath the toes are large, and much granulated ; the nails are black, and rather small, the posterior one having no groove beneath.

The individual we have described is a male, and no difference is observable in several other specimens : the female differs in nothing except that her eyes are of a pale yellow colour.

YOUNG YELLOW-BELLIED WOODPECKER.
(*Picus varius.*)

PLATE VIII.—Fig. 1, 2.

See *Wilson's American Ornithology*, i. p. 147, pl. 9, fig. 2, for the adult and history.—Picus varius, *Linn. Syst.* i. 176, sp. 20.—*Gmel. Syst.* i. p. 438, sp. 20.—*Lath. Ind.* p. 232, sp. 21.—*Vieill. Ois. de. l'Am. Sept.* ii. p. 63, pl. 118, adult male ; pl. 119, very young.—Picus varius Carolinensis, *Briss. Av.* iv. p. 62, sp. 24.—Picus varius minor, Ventre luteo, the Yellow-bellied Woodpecker, *Catesby, Carolina*, i. p. 21, pl. 21, left figure, adult male.—*Bartram, Trav.* p. 291.—Epeiche ou Pic varié de la Caroline, *Buff. Ois.* vii. p. 77, *Pl. enl.* 785, adult male.—Yellow-bellied Woodpecker, *Penn. Arct. Zool.* sp. 166.—*Latham, Syn.* i. p. 574, sp. 20.—*Philadelphia Museum*, 2004, adult male ; No. 2005, adult female.—*My Collection*, young and variety.

PICUS VARIUS.—Bonaparte.

Picus varius, *Bonap. Synop.* p. 45.

As Wilson's history of this well-known woodpecker is complete, and his description obviously discriminates the sexes and young, we shall refer the reader entirely to him for information on those points. The present bird is introduced on account of its anomalous plumage ; for although the colour of the head is but slightly advanced towards its red tint, having only two or three reddish points visible on the forehead, yet the patch on the breast is quite as obvious as it is found in the adult state. In young birds of the first and second years,

1.and 2.Young Yellow-bellied Woodpeckers. 3.Band-tailed Pigeon.

Picus Varius.　　　　　Columba Fasciata.

8

this patch is usually obsolete, the breast being chiefly dusky grey, although the crown is entirely red.

The specimen before us, possibly exhibiting one of the periodical states of plumage of this changeable bird, is the only one we have been able to procure, amongst a great number of the young of both sexes in the ordinary dress.   The well-marked patch on the breast might induce the belief that this individual is an adult female, and that this sex, as several writers have erroneously remarked, is destitute of the red crown ; but, in addition to the fact that our specimen proved, on dissection, to be a male, we obtained, almost every day during the month of November, young birds of both sexes, with the crown entirely red, or more or less sprinkled with that colour, the intermixture arising altogether from age or advanced plumage, and not from sex.   We are unable to state, with any degree of certainty, at what period the bird assumes the plumage now represented ; and we rather incline to the opinion that it is an accidental variety.

For the purpose of comparison, we have added, on the same plate, the most interesting portion of a young bird, as it usually appears in November of the first year ; and though the sexes are then alike in plumage, we had the figure taken from a young male, in order to complete the iconography of that sex.

Vieillot's figure represents the young before the first moult, when, like our anomalous specimen, they have no red on the crown ; differing, however, in not having the head of a glossy black, but of a dull yellowish grey, and the patch on the breast also of a dull grey tint.

BAND-TAILED PIGEON. (*Columba fasciata.*)

PLATE VIII.—Fig. 3.

Columba fasciata, *Say, in Long's Expedition to the Rocky Mountains*, ii. p. 10.—
Philadelphia Museum, No. 4938.

COLUMBA FASCIATA.—Say.*

Columba fasciata, *Bonap. Synop.* p. 119.—*Wagl. Syst. Av. Columba*, No. 47 ?

THIS bird, which is a male, was shot in July, by Mr Titian
Peale, at a saline spring on a small tributary of the river Platte,
within the first range of the Rocky Mountains ; it was accompanied by another individual, probably its mate, which escaped.
As no other specimens have been discovered, the reader will
not be surprised that our specific description is unaccompanied
by a general history of their manners.

The band-tailed pigeon is thirteen inches long ; the bill is
yellow, black at tip, and somewhat gibbous behind the nostrils ; the feet are yellow, and the nails black ; the irides are
blackish ; the head is of a purple cinereous colour ; the neck,
at its junction with the head, has a white semiband, beneath

---

* We have already passed two distinct forms among the *Columbidæ*
in the passenger and Carolina pigeons, of long and slender form and
wedge-shaped tails ; and the diminutive ground doves, whose size and
strength sometimes hardly exceed that of a sparrow. In the bird now
described with the *leucocephala*, figured in the present volume, we see a
third form, and perhaps that to which the title *Columba* should be
restricted, including, as more familiar examples, the common tame pigeon
and the cushat of Europe. Some of the other forms in this beautiful
group seem more restricted in their distribution. Thus the ground doves
and passenger pigeons will nearly claim America; *Vinago* will claim India
and different parts of the Asiatic continent ; and that lovely group, with
feathered tarsi, known under *Ptilonopus*, Swain., takes India, New Holland, and the range of the South Pacific, while those of the present division
will extend over the world. Their form is strongly made, with highly
developed means of a powerful flight ; plumage remarkably dense and
strong. They are gregarious, except during the breeding season, easily
domesticated, and their flesh generally good ; breed more than once
during the season, and feed on grain or on the leaves and soft parts of
vegetables, according to circumstances. In disposition they are timid
and watchful, but rather pugnacious among themselves.—ED.

which its back and sides are brilliant golden green, the feathers being brownish purple at base; the under part of the neck is pale vinaceous purplish, this colour becoming paler as it approaches the vent, which, together with the inferior tail-coverts, is white. The anterior portion of the back, the wing-coverts, and scapulars, are brownish ash; the primaries are dark brown, edged with whitish on the exterior webs; the lower part of the back, the rump, tail-coverts, inferior wing-coverts, and sides, are bluish ash, brighter beneath the wings. The shafts of the body feathers and tail-coverts are remarkably robust, tapering rather suddenly near the tip. The tail, which consists of twelve feathers, is slightly rounded at tip, with a definite blackish band at two-thirds the length from the base, visible on both sides; before this band the colour is bluish ash, and behind dirty greyish; the tail is much lighter on the inferior surface.

This species is closely allied to *Columba caribœa* of Gmelin, with which Say stated its analogy, and also to *Columba leucocephala* of Linné. In fact, it possesses some characters in common with each of these species, such as the band on the tail of the former, and an indication of white on the head of the latter. This character may induce some naturalists to suppose it the young of the *leucocephala*; but by a careful comparison all doubt will be removed, and it will be admitted to the rank of a distinct species.

The *caribœa* may readily be distinguished from the present species by its superior size, and by being destitute of the white band on the neck; by having a reddish bill, tipt with yellow, and dark red feet. The *leucocephala*, in the adult state, has the whole head white above; but as it is destitute of this distinction when young, acquiring it gradually as it advances in age, other discriminating characters must be employed; the tail is without a band, the bill is red, with a white tip, and the feet are red.

WILD TURKEY.   (*Meleagris gallopavo.*)

PLATE IX.—MALE AND FEMALE.

Meleagris gallopavo, *Linn. Syst.* i. p. 268, sp. 1.—*Gmel. Syst.* i. p. 732. sp. 1.—
    Lath. Ind. p. 618, sp. 1.—*Temm. Hist. Nat. des Pig. et Gall.* Index, iii. p.
    676. —*Wilson, Am. Orn.* vi. Index, p. xvii.—*Stephens, Cont. of Shaw's Zool.*
    xi. part i. p. 156, pl. 8.—*Ranzani, Elem. di Zool.* iii. part i. p. 154.—Melea-
    gris sylvestris, *Vieill. Nouv. Dict. d'Hist. Nat.* ix. p. 447.—Gallopavo,
    Aldrovandi, Orn. ii. p. 35, fig. on p. 39, domestic variety, male; fig. on p.
    40, *id.* female.— *Willoughby, Orn.* p. 113, pl. 27, fig. 4, dom. var. male.—
    Johnston, Theatrum Universale de Avibus, p. 55, pl. 24, fig. 1, dom. var. male;
    fig. 2, *id.* female.—*Briss. Av.* i. p. 158, pl. 16, dom. var. male.—Gallus
    Indicus, Welscher Han, *Johnston, Th. Av.* p. 83, pl. 29, fig. 1, dom. var.
    male.—Gallopavo sylvestris Novæ-Angliæ, a New England Wild Turkey, *Ray,*
    Syn. p. 51, sp. 3.—Gallopavo sylvestris, *Catesby, Carolina,* i. App. p. xliv.—
    Meleagris Americanus, the Wild Turkey, *Bartr. Trav.* p. 290.—Gallo Pavo,
    Gallo Pavone volgarmente Pollo d'India, *Storia degli Uccelli,* ii. pl. 222,
    dom. crested var. male; pl. 223, dom. white, black spotted var. young; pl.
    224, dom. white, yellowish spotted var. young; pl. 225. dom. black, var.
    young; pl. 226, dom. black, white spotted var. young.—Coc d'Inde, *Belon,*
    Histoire de la Nature des Oiseaux, p. 248, with fig. dom. var. male.—Dindon,
    Buff. Ois. ii. p. 132, pl. 3.—*Pl. enl.* 97, dom. whitish, var. male.—*Temm.*
    Hist. Nat. des Pig. et Gall. ii. p. 374.—*Gerardin, Tabl. Elem. d'Orn.* ii. p.
    103, pl. 21, fig. 2. dom. var. male.—Turkie, *Josselyn, Voyages to New*
    England, p. 99.—*New England's Rarities,* p. 8.—Wild Turkey, *Clayton,*
    Virginia Phil. Trans. xvii. p. 992.—*Id. Abridg.* iii. p. 590.—*Lawson,*
    Carolina, p. 149.—*Penn. Phil. Trans.* lxxi. p. 67.—*Arct. Zool.* sp. 178.—
    American Turkey, *Lath. Syn.* ii. part ii. p. 676, sp. 1.—Domestic Turkey,
    Penn. Brit. Zool. i. sp. 97.

MELEAGRIS GALLOPAVO.—LINNÆUS.[*]

Meleagris gallopavo, *Bonap. Synop.* p. 123.—The Wild Turkey, *Aud.* m. pl. 1,
    fem. and young, p. vi.—*Orn. Biog.* i. pp. 1 and 33.—*Gard. and Menag. of*
    Zool. Soc. Birds, p. 209.

THE native country of the wild turkey extends from the
north-western territory of the United States to the Isthmus

---

[*] The natural history of the turkey is so well and fully detailed by
our author, that almost nothing can be added, even from the later
observations of Audubon.   From the evidence that has been collected, there seems to be little doubt that Great Britain is indebted, in
a secondary way, for the introduction of these valuable domestic birds;
and I have added the observations of Mr Bennet on the subject of its

Drawn from Nature by Titian R. Peale.   Engraved by W.H. Lizars.

Wild Turkey. Male & Female.
Meleagris Gallopavo.

9.

of Panama, south of which it is not to be found, notwithstanding the statements of authors, who have mistaken the curassow

original introduction from the Gardens and Menagerie of the Zoological Society.

" The turkey domesticated by the Spaniards seems to have found its way to England almost immediately. This fact may be easily accounted for by the extensive intercourse subsisting between the two great maritime nations at that early period ; but it is somewhat singular that no traces of its transmission from Spain should remain, either in the name of the bird, or in popular tradition. On the other hand, it is barely possible that it may have been brought directly from America to England by Chabot, who made such extensive discoveries on the coast of the newly found continent. According to a popular rhyme, quoted by Baker in his Chronicle—

> 'Turkeys, carps, hoppes, pinaret, and beer,
> Came into England all in one year ; '

which remarkable year is said to have been about the 15th of King Henry the Eighth, or 1524. Barnaby Googe, an old writer on husbandry, who published in 1614, speaking of ' those outlandish birds called ginny-cocks and turkey-cocks,' says that ' before the yeare of our Lord 1530, they were not seene with us ; ' but in this he merely translates from Heresbach, a German author, whose treatise forms the basis of his work. A more positive authority is Hakluyt, who, in certain instructions given by him to a friend at Constantinople, bearing date in 1582, mentions, among other valuable things introduced into England from foreign parts, ' turkey cocks and hennes,' as having been brought in ' about fifty years past.' We may therefore fairly conclude that they became known in this country about the year 1530. Why they were denominated turkeys, an appellation which bears no resemblance to their name in any other language, we have no probable grounds even for conjecture. Willoughby supposes the name to be derived from a notion that they were brought from Turkey. Such an erroneous opinion may possibly have arisen from that confusion which appears to have at first existed between them and the guinea-fowls, the latter being probably commonly obtained from the Levant ; and being also, in the sixteenth century, exceedingly rare in England.

" The turkey, on the country, speedily became a common inhabitant of our poultry yards, and a standing dish at all festivals. So early as the year 1541, we find it mentioned in a constitution of Archbishop Cranmer, published in Leland's *Collectanea*, by which it was ordered, that of such large fowls as cranes, swans, and turkey-cocks, ' there should be but one in a dish.' The serjeants-at-law created in 1555 provided,

for it. In Canada, and the now densely peopled parts of the United States, wild turkeys were formerly very abundant; but, like the Indian and buffalo, they have been compelled to yield to the destructive ingenuity of the white settlers, often wantonly exercised, and seek refuge in the remotest parts of the interior. Although they relinquish their native soil with slow and reluctant steps, yet such is the rapidity with which settlements are extended and condensed over the surface of this country, that we may anticipate a day, at no dis-

according to Dugdale, in his 'Origines Juridicales,' for their inauguration dinner, among other delicacies, two turkeys and four turkey chicks, which, as they were rated at only four shillings each, while swans and cranes were charged ten shillings, and capons half-a-crown, could not have been esteemed very great rarities. Indeed, they had become so plentiful in 1573, that honest Tusser, in his 'Five Hundred Points of Good Husbandry,' enumerates them among the usual Christmas fare at a farmer's table, and speaks of them as 'ill neighbors' both to 'peason' and to hops.

"A Frenchman, named Pierre Gilles, has the credit of having first described the turkey in this quarter of the globe, in his editions to a Latin translation of Ælian, published by him in 1535. His description is so true to nature, as to have been almost wholly relied on by every subsequent writer down to Willoughby. He speaks of it as a bird that he had seen; and he had not then been farther from his native country than Venice; and states it to have been brought from the New World. That turkeys were known in France at this period is further proved by a passage in Champier's treatise 'De re Cibaria,' published in 1560, and said to have been written thirty years before. This author also speaks of them as having been brought but a few years back from the newly discovered Indian islands. From this time forward, their origin seems to have been entirely forgotten; and for the next two centuries we meet with little else in the writings of ornithologists concerning them than an accumulation of citations from the ancients, which bear no manner of relation to them. In the year 1566, a present of twelve turkeys was thought not unworthy of being offered by the municipality of Amiens to their King, at whose marriage, in 1570, Anderson states in his 'History of Commerce,' but we know not on what authority, they were first eaten in France. Heresbach, as we have before seen, asserts that they were introduced into Germany about 1530; and a sumptuary law made at Venice, 1557, quoted by Zanoni, particularises the tables at which they were permitted to be served."—ED.

tant period, when the hunter will seek the wild turkey in vain.

We have neglected no means of obtaining information from various parts of the Union relative to this interesting bird; and having been assisted by the zeal and politeness of several individuals, who, in different degrees, have contributed to our stock of knowledge on this subject, we return them our best thanks. We have particular satisfaction in acknowledging the kindness of Mr John J. Audubon, from whom we have received a copious narrative, containing a considerable portion of the valuable notes collected by him on this bird during twenty years that he has been engaged in studying ornithology, in the only book free from error and contradiction, the great book of Nature. His observations, principally made in Kentucky and Louisiana, proved the more interesting, as we had received no information from those States; we have, in consequence, been enabled to enrich the present article with several new details of the manners and habits of the wild turkey.

The wooded parts of Arkansaw, Louisiana, Tennessee, and Alabama; the unsettled portions of the States of Ohio, Kentucky, Indiana, and Illinois; the vast expanse of territory north-west of these States on the Mississippi and Missouri, as far as the forests extend, are more abundantly supplied than any other parts of the Union with this valuable game, which forms an important part of the subsistence of the hunter and traveller in the wilderness. It is not probable that the range of this bird extends to, or beyond, the Rocky Mountains; the Mandan Indians, who, a few years ago, visited the city of Washington, considered the turkey one of the greatest curiosities they had seen, and prepared a skin of one, to carry home for exhibition.

The wild turkey is not very plenty in Florida, Georgia, and the Carolinas; is still less frequently found in the western parts of Virginia and Pennsylvania; and is extremely rare, if indeed it exists at all, in the remaining northern and eastern

parts of the United States; in New England, it even appears
to have been already destroyed one hundred and fifty years
back.   I am, however, credibly informed that wild turkeys
are yet to be found in the mountainous districts of Sussex
county, New Jersey.   The most eastern part of Pennsylvania
now inhabited by them appears to be Lancaster county; and
they are often observed in the oak woods near Philipsburg,
Clearfield county.   Those occasionally brought to the Philadelphia and New York markets are chiefly obtained in
Pennsylvania and New Jersey.

The wild turkeys do not confine themselves to any particular food; they eat maize; all sorts of berries, fruits, grasses,
beetles, and even tadpoles, young frogs, and lizards, are
occasionally found in their crops; but where the becan-nut is
plenty, they prefer that fruit to any other nourishment; their
more general predilection is, however, for the acorn, on which
they rapidly fatten.   When an unusually profuse crop of
acorns is produced in a particular section of country, great
numbers of turkeys are enticed from their ordinary haunts in
the surrounding districts.   About the beginning of October,
while the mast still remains on the trees, they assemble in
flocks, and direct their course to the rich bottom lands.   At
this season they are observed in great numbers on the Ohio
and Mississippi.   The time of this irruption is known to the
Indians by the name of the *turkey month.*

The males, usually termed *gobblers,* associate in parties
numbering from ten to a hundred, and seek their food apart
from the females; whilst the latter either move about singly
with their young, then nearly two-thirds grown, or, in company with other females and their families, form troops, sometimes consisting of seventy or eighty individuals, all of whom
are intent on avoiding the old males, who, whenever opportunity offers, attack and destroy the young by repeated blows
on the skull.   All parties, however, travel in the same direction, and on foot, unless they are compelled to seek their
individual safety by flying from the hunter's dog, or their

march is impeded by a large river. When about to cross a
river, they select the highest eminences, that their flight may
be the more certain ; and here they sometimes remain for a
day or more, as if for the purpose of consultation, or to be duly
prepared for so hazardous a voyage. During this time the
males *gobble* obstreperously, and strut with extraordinary importance, as if they would animate their companions, and in-
spire them with the utmost degree of hardihood ; the females
and young also assume much of the pompous air of the males,
the former spreading their tails, and moving silently around.
At length the assembled multitude mount to the tops of the
highest trees, whence, at a signal note from a leader, the whole
together wing their way towards the opposite shore. All the
old and fat ones cross without difficulty, even when the river
exceeds a mile in width ; but the young, meagre, and weak,
frequently fall short of the desired landing, and are forced to
swim for their lives ; this they do dexterously enough, spreading their tails for a support, closing their wings to the body,
stretching the neck forwards, and striking out quickly and
forcibly with their legs. If, in thus endeavouring to regain
the land, they approach an elevated or inaccessible bank, their
exertions are remitted, they resign themselves to the stream
for a short time, in order to gain strength, and then, with one
violent effort, escape from the water. But in this attempt all
are not successful ; some of the weaker, as they cannot rise
sufficiently high in air to clear the bank, fall again and again
into the water, and thus miserably perish. Immediately after
these birds have succeeded in crossing a river, they for some
time ramble about without any apparent unanimity of purpose,
and a great many are destroyed by the hunters, although they
are then least valuable.

When the turkeys have arrived in their land of abundance,
they disperse in small flocks, composed of individuals of all
sexes and ages intermingled, who devour all the mast as they
advance ; this occurs about the middle of November. It
has been observed, that, after these long journeys, the turkeys

become so familiar as to venture on the plantations, and even approach so near the farmhouses as to enter the stables and corn-cribs in search of food; in this way they pass the autumn and part of the winter. During this season great numbers are killed by the inhabitants, who preserve them in a frozen state in order to transport them to a distant market.

Early in March they begin to pair; and, for a short time previous, the females separate from, and shun their mates, though the latter pertinaciously follow them, uttering their gobbling note. The sexes roost apart, but at no great distance, so that, when the female utters a call, every male within hearing responds, rolling note after note in the most rapid succession; not as when spreading the tail and strutting near the hen, but in a voice resembling that of the tame turkey when he hears any unusual or frequently repeated noise. Where the turkeys are numerous, the woods, from one end to the other, sometimes for hundreds of miles, resound with this remarkable voice of their wooing, uttered responsively from their roosting places. This is continued for about an hour; and, on the rising of the sun, they silently descend from their perches, and the males begin to strut, for the purpose of winning the admiration of their mates.

If the call be given from the ground, the males in the vicinity fly towards the individual, and, whether they perceive her or not, erect and spread their tails, throw the head backwards, distend the comb and wattles, strut pompously, and rustle their wings and body feathers, at the same moment ejecting a puff of air from the lungs. Whilst thus occupied, they occasionally halt to look out for the female, and then resume their strutting and puffing, moving with as much rapidity as the nature of their gait will admit. During this ceremonious approach, the males often encounter each other, and desperate battles ensue, when the conflict is only terminated by the flight or death of the vanquished.

This pugnacious disposition is not to be regarded as accidental, but as resulting from a wise and excellent law of nature, which always studies the good of the species, without regard to the individuals. Did not females prefer the most perfect of their species, and were not the favours of beauty most willingly dispensed to the victorious, feebleness and degeneracy would soon mark the animal creation ; but, in consequence of this general rule, the various races of animals are propagated by those individuals who are not only most to be admired for external appearance, but most to be valued for their intrinsic spirit and energy.

When the object of his pursuit is discovered, if the female be more than one year old, she also struts, and even gobbles, evincing much desire ; she turns proudly round the strutting male, and suddenly opening her wings, throws herself towards him, as if to terminate his procrastination, and, laying herself on the earth, receives his dilatory caresses. But should he meet a young hen, his strut becomes different, and his movements are violently rapid ; sometimes rising in air, he takes a short circular flight, and on alighting, drags his wings for a distance of eight or ten paces, running at full speed, occasionally approaching the timorous hen, and pressing her, until she yields to his solicitations. Thus are they mated for the season, though the male does not confine himself exclusively to one female, nor does he hesitate to bestow his attentions and endearments on several, whenever an opportunity offers.

One or more females, thus associated, follow their favourite, and roost in his immediate neighbourhood, if not on the same tree, until they begin to lay, when they change their mode of life, in order to save their eggs, which the male uniformly breaks, if in his power, that the female may not be withdrawn from the gratification of his desires. At this time the females shun the males during the greater part of the day ; the latter become clumsy and careless, meet each other peacefully, and so entirely cease to gobble, that the hens are obliged to court their advances, calling loudly and almost continually for them. The female may then be observed caressing the male, and imitating his peculiar gestures, in order to excite his amorousness.

The cocks, even when on the roost, sometimes strut and gobble, but more generally merely elevate the tail, and utter the *puff*, on which the tail and other feathers suddenly subside. On light or moonshining nights, near the termination of the breeding season, they repeat this action, at intervals of a few minutes, for several hours together, without rising from their perches.

The sexes then separate; the males, being much emaciated, cease entirely to gobble, retire and conceal themselves by prostrate trees, in secluded parts of the forest, or in the almost impenetrable privacy of a cane-break. Rather than leave their hiding-places, they suffer themselves to be approached within a short distance, when they seek safety in their speed of foot; at this season, however, they are of no value to the hunter, being meagre and covered with ticks. By thus retiring, using very little exercise, and feeding on peculiar grasses, they recover their flesh and strength, and when this object is attained, again congregate, and recommence their rambles.

About the middle of April, when the weather is dry, the female selects a proper place in which to deposit her eggs, secure from the encroachment of water, and, as far as possible, concealed from the watchful eye of the crow: this crafty bird espies the hen going to her nest, and having discovered the precious deposit, waits for the absence of the parent, and removes every one of the eggs from the spot, that he may devour them at leisure. The nest is placed on the ground, either on a dry ridge, in the fallen top of a dead leafy tree, under a thicket of sumach or briars, or by the side of a log; it is of a very simple structure, being composed of a few dried leaves. In this receptacle the eggs are deposited, sometimes to the number of twenty, but more usually from nine to fifteen; they are whitish, spotted with reddish brown, like those of the domestic bird. Their manner of building, number of eggs,

period of incubation, &c., appear to correspond throughout
the Union, as I have received exactly similar accounts from
the northern limits of the turkey range to the most southern regions of Florida, Louisiana, and the western wilds of Missouri.

The female always approaches her nest with great caution,
varying her course so as rarely to reach it twice by the same
route ; and, on leaving her charge, she is very careful to cover
the whole with dry leaves, with which she conceals it so artfully, as to make it extremely difficult, even for one who has
watched her movements, to indicate the exact spot ; hence
few nests are found, and these are generally discovered by
fortuitously starting the female from them, or by the appearance of broken shells, scattered around by some cunning lynx,
fox, or crow.   When laying or sitting, the turkey-hen is not
readily driven from her post by the approach of apparent
danger ; but, if an enemy appears, she crouches as low as
possible, and suffers it to pass.   A circumstance related by
Mr Audubon will show how much intelligence they display on
such occasions : having discovered a sitting hen, he remarked
that, by assuming a careless air, whistling, or talking to himself, he was permitted to pass within five or six feet of her ;
but, if he advanced cautiously, she would not suffer him to
come within twenty paces, but ran off twenty or thirty yards
with her tail expanded, when, assuming a stately gait, she
paused on every step, occasionally uttering a chuck.   They
seldom abandon their nests on account of being discovered by
man, but should a snake, or any other animal, suck one of the
eggs, the parent leaves them altogether.   If the eggs be removed, she again seeks the male, and recommences laying,
though otherwise she lays but one nest of eggs during the
season.   Several turkey-hens sometimes associate, perhaps for
mutual safety, deposit their eggs in the same nest, and rear
their broods together.   Mr Audubon once found three females
sitting on forty-two eggs.   In such cases the nest is constantly
guarded by one of the parties, so that no crow, raven, nor even
polecat, dares approach it.

The mother will not forsake her eggs, when near hatching; while life remains; she will suffer an enclosure to be made around and imprison her, rather than abandon her charge. Mr Audubon witnessed the hatching of a brood, while thus endeavouring to secure the young and mother. " I have laid flat," says he, " within a very few feet, and seen her gently rise from the eggs, look anxiously towards them, chuck with a sound peculiar to the mother on such an occasion, remove carefully each half-empty shell, and with her bill caress and dry the younglings, that already stand tottering and attempting to force their way out of the nest."

When the process of incubation is ended, and the mother is about to retire from the nest with her young brood, she shakes herself violently, picks and adjusts the feathers about the belly, and assumes a different aspect; her eyes are alternately inclined obliquely upwards and sidewise ; she stretches forth her neck in every direction, to discover birds of prey or other enemies; her wings are partially spread, and she softly clucks to keep her tender offspring close to her side. They proceed slowly ; and as the hatching generally occurs in the afternoon, they sometimes return to pass the first night in the nest. While very young, the mother leads them to elevated dry places, as if aware that humidity during the first few days of their life would be very dangerous to them, they having then no other protection than a delicate, soft, hairy down. In very rainy seasons wild turkeys are scarce, because, when completely wetted, the young rarely survive.

At the expiration of about two weeks, the young leave the ground, on which they had previously reposed at night under the female, and follow her to some low, large branch of a tree, where they nestle under the broadly curved wings of their vigilant and fostering parent. The time then approaches in which they seek the open ground or prairie land during the day in search of strawberries, and subsequently of dewberries, blackberries, and grasshoppers, thus securing plenty of food, and enjoying the influence of the genial sun. They frequently dust themselves in shallow cavities of the soil, or on anthills, in order to clean off the loose skin of their growing feathers, and rid themselves of ticks and other vermin.

The young turkeys now grow rapidly, and in the month of August, when several broods flock together, and are led by their mothers to the forest, they are stout, and quite able to secure themselves from the unexpected attacks of wolves, foxes, lynxes, and even cougars, by rising quickly from the ground, aided by their strong legs, and reaching with ease the upper limbs of the tallest tree. Amongst the numerous enemies of the wild turkey, the most dreaded are the large diurnal and nocturnal birds of prey, and the lynx (*Felis rufa*), who sucks their eggs, and is extremely expert at seizing both parent and young ; he follows them for some distance, in order to ascertain their course, and then, making a rapid circular movement, places himself in ambush before them, and waits until, by a single bound, he can fasten on his victim.

The following circumstance is thus related by Bartram :— "Having seen a flock of turkeys at some distance, I approached them with great caution, when, singling out a large cock, and being just on the point of firing, I observed that several young cocks were affrighted, and in their language warned the rest to be on their guard against an enemy, whom I plainly perceived was industriously making his subtle approaches towards them behind the fallen trunk of a tree about twenty yards from me. This cunning fellow-hunter was a large fat wild cat or lynx: he saw me, and at times seemed to watch my motions, as if determined to seize the delicious prey before me ; upon which I changed my object, and levelled my piece at him. At that instant, my companion, at a distance, also discharged his piece, the report of which alarmed the flock of turkeys, and my fellow-hunter, the cat, sprang over the log, and trotted off.

These birds are guardians of each other, and the first who sees a hawk or eagle gives a note of alarm, on which all within hearing lie close to the ground. As they usually roost in flocks,

perched on the naked branches of trees, they are easily discovered by the large owls, and, when attacked by these prowling birds, often escape by a somewhat remarkable manœuvre.
The owl sails around the spot to select his prey; but notwithstanding the almost inaudible action of his pinions, the quick
ear of one of the slumberers perceives the danger, which is
immediately announced to the whole party by a *chuck;* thus
alarmed, they rise on their legs, and watch the motions of the
owl, who, darting like an arrow, would inevitably secure the
individual at which he aimed, did not the latter suddenly drop
his head, squat, and spread his tail over his back; the owl then
glances over without inflicting any injury, at the very instant
that the turkey suffers himself to fall headlong towards the
earth, where he is secure from his dreaded enemy.

On hearing the slightest noise, wild turkeys conceal themselves in the grass or among shrubs, and thus frequently
escape the hunter, or the sharp-sighted birds of prey. The
sportsman is unable to find them during the day, unless he
has a dog trained for the purpose. It is necessary to shoot
them at a very short distance, since, when only wounded, they
quickly disappear, and, accelerating their motion by a sort of
half flight, run with so much speed, that the swiftest hunter
cannot overtake them. The traveller driving rapidly down
the declivity of one of the Alleghanies may sometimes see
several of them before him, that evince no urgent desire to
get out of the road; but on alighting, in hopes of shooting
them, he soon finds that all pursuit is vain.

In the spring, when the males are much emaciated by their
attendance on the females, it sometimes may happen that, in
cleared countries, they can be overtaken by a swift cur-dog,
when they will squat, and suffer themselves to be caught by
the dog, or hunter, who follows on horseback. But from the
knowledge we have gained of this bird, we do not hesitate to
affirm, that the manner of running down turkeys, like hares or
foxes, so much talked of, is a mere fable, as such a sport would
be attended with very trifling success. A turkey-hound will

sometimes lead his master several miles before he can a second time *flush* the same individual from his concealment; and even on a fleet horse, after following one for hours, it is often found impossible to *put it up.* During a fall of melting snow, turkeys will travel extraordinary distances, and are often pursued in vain by any description of hunters; they have then a long, straddling manner of running, very easy to themselves, but which few animals can equal. This disposition for running during rains or humid weather is common to all gallinaceous birds.

The males are frequently decoyed within gunshot in the breeding season by forcibly drawing the air through one of the wing-bones of the turkey, producing a sound very similar to the voice of the female; but the performer on this simple instrument must commit no error, for turkeys are quick of hearing, and, when frequently alarmed, are wary and cunning. Some of these will answer to the call without advancing a step, and thus defeat the speculations of the hunter, who must avoid making any movement, inasmuch as a single glance of a turkey may defeat his hopes of decoying them. By imitating the cry of the barred owl (*Strix nebulosa*), the hunter discovers many on their roosts, as they will reply by a gobble to every repetition of this sound, and can thus be approached with certainty about daylight, and easily killed.

Wild turkeys are very tenacious of their feeding grounds, as well as of the trees on which they have once roosted. Flocks have been known to resort to one spot for a succession of years, and to return after a distant migration in search of food. Their roosting-place is mostly on a point of land jutting into a river, where there are large trees. When they have collected at the signal of a repeated gobbling, they silently proceed towards their nocturnal abodes, and perch near each other: from the numbers sometimes congregated in one place, it would seem to be the common rendezvous of the whole neighbourhood. But no position, however secluded or difficult of access, can secure them from the attacks of the artful and vigilant hunter, who, when they are all quietly perched for the night, takes a stand previously chosen by day-light, and, when the rising moon enables him to take sure aim, shoots them down at leisure, and, by carefully singling out those on the lower branches first, he may secure nearly the whole flock, neither the presence of the hunter nor the report of his gun intimidating the turkeys, although the appearance of a single owl would be sufficient to alarm the whole troop : the dropping of their companions from their sides excites nothing but a buzzing noise, which seems more expressive of surprise than fright. This fancied security, or heedlessness of danger, while at roost, is characteristic of all the gallinaceous birds of North America.

The more common mode of taking turkeys is by means of *pens*, constructed with logs, covered in at top, and with a pas-sage in the earth under one side of it, just large enough to admit an individual when stooping. The ground chosen for this purpose is generally sloping, and the passage is cut on the lower side, widening outwards. These preparations being completed, Indian-corn is strewed for some distance around the pen, to entice the flock, which, picking up the grain, is gradually led towards the passage, and thence into the enclo-sure, where a sufficient quantity of corn is spread to occupy the leader until the greater part of the turkeys have entered. When they raise their heads and discover that they are prisoners, all their exertions to escape are directed upwards and against the sides of the pen, not having sagacity enough to stoop sufficiently low to pass out by the way they entered, and thus they become an easy prey, not only to the experienced hunter, but even to the boys on the frontier settlements.

In proportion to the abundance or scarcity of food, and its good or bad quality, they are small or large, meagre or fat, and of an excellent or indifferent flavour: in general, however, their flesh is more delicate, more succulent, and better tasted than that of the tame turkey : they are in the best order late in the autumn, or in the beginning of winter. The Indians value this food so highly when roasted, that they call it "the white man's dish," and present it to strangers as the best they can offer. It seems probable that in Mexico the wild turkey cannot obtain such substantial food as in the United States, since Hernandez informs us that their flesh is harder, and in all respects inferior to that of the domestic bird.

The Indians make much use of their tails as fans; the women weave their feathers with much art on a loose web made of the rind of the birch-tree, arranging them so as to keep the down on the inside, and exhibit the brilliant surface to the eye. A specimen of this cloth is in the Philadelphia Museum; it was found enveloping the body of an Indian female in the great saltpetre cave of Kentucky.

Among the benefits conferred by America on the rest of the world, the gift of this noble bird should occupy a distinguished place, as unquestionably one of the most useful of the feathered tribe, being capable of ministering largely to the sustenance and comfort of the human race. Though the turkey is surpassed in external beauty by the magnificent peacock, its flesh is greatly superior in excellence, standing almost unrivalled for delicacy of texture and agreeable sapidity. On this account it has been eagerly sought by almost all nations, and has been naturalised with astonishing rapidity throughout the world, almost universally constituting a favourite banquet-dish.

The turkey, belonging originally to the American continent, was necessarily unknown to the ancients, who, in this as in a thousand other instances, were deficient in our most common and essential articles of food. Readers unacquainted with the fact may well be surprised to learn, that although the introduction of this bird into Europe is comparatively modern, its origin has already been lost sight of, and that eminent naturalists of the last century, who lived so much nearer to the time of its first appearance, have expressed great uncertainty concerning its native country. Thus Belon, Aldrovandi, Gesner, Ray, &c., thought that it came originally from Africa and the East Indies, and endeavoured to recognise it in some of the domestic birds of the ancients. Belon and Aldrovandi supposed it to have been mentioned by ancient authors, but they mistook for it the *Numida meleagris* of Linné, which is actually an African bird, now almost naturalised in America, even in a wild state, so that it would be apparently more reasonable for America to regard that bird as indigenous, than that the old continent should lay claim to the turkey. In so soon losing sight of the origin of this bird, we see a strong exemplification of the ungrateful disposition of man, who can durably treasure up the memory of wrongs and injuries, but fails to recollect the greatest benefits he has received. It would be loss of time to combat the arguments advanced by authors, who have deceived themselves in attempting to deprive America of her just title to this bird, since they have been fully refuted by the eloquent Buffon ; but we may here introduce a sketch of its progress from America throughout Europe.

The first unquestionable description of the turkey was written by Oviedo in 1525, in the summary of his " History of the Indies." This bird was sent from Mexico to Spain early in the sixteenth century; from Spain it was introduced into England in 1524. Turkeys were taken to France in the reign of Francis the First, whence they spread into Germany, Italy, &c.; a few, however, had been carried to the latter country by the Spaniards some years previously. The first turkey eaten in France appears to have been served up at the wedding banquet of Charles the Ninth, in the year 1570. Since that period, they have been bred with so much care, that in England, as we read in ancient chronicles, their rapid increase rendered them attainable at country feasts, where they were a much esteemed dish as early as 1585. Europeans conveyed them to all their colonies, and thus were they gradually introduced into Asia, Africa, and even Oceanica.

The French distinguished them by the name of *Coq et Poule d'Inde* (cock and hen from India), because they were natives of the West Indies. Subsequently, for the sake of brevity,

they called them *Dindon*, an appellation which is yet retained.
The English name is still worse, as it conveys the false idea
that the turkey originated in Asia, owing to the ridiculous
habit, formerly prevalent, of calling every foreign object by
the name of Turk, Indian, &c.

Although the turkey is generally considered a stupid bird,
it is probable that his intellectual qualifications have not been
fairly appreciated, as he is susceptible of very lively emotions.
If any new and remarkable object attracts the attention of
the male, his whole appearance and demeanour undergo a
sudden and extraordinary change : relinquishing his peaceful
aspect, he boldly raises himself, his head and neck become
turgid, and the wattles, from an influx of blood, glow with
vivid red ; he bristles up the feathers of the neck and back,
his tail is vertically raised and expanded like a fan, and the
wing-feathers are extended until they touch the ground. Thus
transformed, he utters a low, humming sound, and advances
with a grave and haughty strut, occasionally accelerating his
steps, and, at the same time, rubbing the tips of the primary
feathers violently against the earth. During these manœuvres,
he now and then utters a harsh, interrupted, and dissonant
note, apparently expressive of the highest degree of rage :
this cry, sounding like *rook, oorook, oorook,* will be repeated
at the pleasure of any person who should whistle, or strike the
ear of the bird by any other acute or unusual sound. The
appearance of any red cloth is sure to awaken his anger, and
induce him to rush fearlessly on the disagreeable object, exert-
ing all his power to injure or destroy it.

In connection with the peculiar character of this bird, we
may advantageously quote the sentiments of the great Franklin, who expressed a regret that the turkey should not have
been preferred to the bald eagle as an emblem of the United
States. Certainly this eagle is a tyrannical and pusillanimous
bird, by no means an appropriate representative of a great and
magnanimous nation, as was the eagle chosen by the Romans.

" Others object to the bald eagle," says Franklin, in one of

his letters, " as looking too much like a dindon, or turkey.
For my own part, I wish the bald eagle had not been chosen
as the representative of our country; he is a bird of bad moral
character; he does not get his living honestly; you may have
seen him perched on some dead tree, where, too lazy to fish
for himself, he watches the labour of the fishing-hawk, and,
when that diligent bird has at length taken a fish, and is
bearing it to his nest for the support of his mate and young
ones, the bald eagle pursues him, and takes it from him.
With all this injustice, he is never in good case, but, like
those among men who live by sharping and robbing, he is
generally poor, and often very lousy.   Besides, he is a rank
coward; the little kingbird, not bigger than a sparrow, attacks
him boldly, and drives him out of the district.   He is, therefore, by no means, a proper emblem for the brave and honest
Cincinnati of America, who have driven all the *kingbirds* from
our country, though exactly fit for that order of knights which
the French call *Chevaliers d'industrie.*   I am, on this account,
not displeased that the figure is not known as a bald eagle,
but looks more like a turkey.   For, in truth, the turkey is, in
comparison, a much more respectable bird, and withal a true
original native of America.   Eagles have been found in all
countries, but the turkey was peculiar to ours.   He is, besides
(though a little vain and silly, 'tis true, but not the worse
emblem for that), a bird of courage, and would not hesitate to
attack a grenadier of the British Guards who should presume
to invade his farmyard with a red coat on."

But since the choleric temper and vanity of the tame turkey
are proverbial in various languages, in some of which its very
name is opprobrious, and often applied in derision to vain-
glorious and stupid people, we are better satisfied that its
effigy was not placed on the escutcheon of the United States.

Those who have not observed the turkey in its wild state,
have only seen its deteriorated progeny, which are greatly inferior in size and beauty.   So far from having gained by the
care of man, and the abundance of food accessible in its state

of domestication, this bird has degenerated, not only in Europe and Asia, but, what is certainly extraordinary, even in its native country. The domesticated turkey of America, accustomed as it is to roam in the woods and open fields almost without restraint, is in no respect superior to that of the European poultry-yard. I have, however, seen several very beautiful ones from Lancaster county, Pennsylvania, and Sussex county, New Jersey, that were said to be a cross-breed between the wild cock and tame hen. This crossing often occurs in countries where wild and tame turkeys are found; it is well known that they will readily approach each other; and such is the influence of slavery even upon the turkey, that the robust inhabitant of the forest will drive his degenerate kinsfolk from their own food and from their females, being generally welcomed by the latter and by their owners, who well know the advantages of such a connection. The produce of this commixture is much esteemed by epicures, uniting the luscious obesity of the one with the wild flavour of the other. A gentleman residing in Westchester county, New York, a few years since, procured a young female wild turkey, in order to make the experiment of crossing the breed; but, owing to some circumstance, it did not succeed, and in the ensuing spring this female disappeared. In the following autumn she returned, followed by a large brood; these were quite shy, but, by a little management, they were secured in a coop, and the mother allowed her liberty. She remained on the farm until the succeeding spring, when she again disappeared, and returned in autumn with another brood. This course she has repeated for several successive years.

Eggs of the wild turkey have been frequently taken from their nests, and hatched under the tame hen; the young preserve a portion of their uncivilised nature, and exhibit some knowledge of the difference between themselves and their foster-mother, roosting apart from the tame ones, and in other respects showing the force of hereditary disposition. The domesticated young, reared from the eggs of the wild turkey, are often employed as decoy birds to those in a state of nature. Mr William Bloom, of Clearfield, Pennsylvania, caught five or six wild turkeys, when quite chickens, and succeeded in rearing them. Although sufficiently tame to feed with his tame turkeys, and generally associate with them, yet they always retained some of their original propensities, roosting by themselves, and higher than the tame birds, generally on the top of some tree, or of the house. They were also more readily alarmed; on the approach of a dog, they would fly off, and seek safety in the nearest woods. On an occasion of this kind, one of them flew across the Susquehanna, and the owner was apprehensive of losing it; in order to recover it, he sent a boy with a tame turkey, which was released at the place where the fugitive had alighted. This plan was successful; they soon joined company, and the tame bird induced his companion to return home. Mr Bloom remarked, that the wild turkey will thrive more, and keep in better condition, than the tame, on the same quantity of food.

Besides the above-mentioned half-breed, some domesticated turkeys, of a very superior metallic tint, are sold in the Philadelphia and New York markets as wild ones. Many of these require a practised eye to distinguish their true character, but they are always rather less brilliant, and those I examined had a broad whitish band at the tip of the tail-coverts, and another at the tip of the tail itself, which instantly betrayed their origin, the wild ones being entirely destitute of the former, and the band on the tip of the tail being neither so wide nor so pure.

In the following description, we give the generic as well as the specific characters of the wild turkey, in order to make it complete.

The male wild turkey, when full grown, is nearly four feet in length, and more than five in extent. The bill is short and robust, measuring two inches and a half to the corner of the mouth; it is reddish, and horn colour at tip; the superior mandible is vaulted, declining at tip, and overhangs the inferior, being longer and wider; it is covered at base by a naked cere-like membrane, in which the nostrils are situated, they being half closed by a turgid membrane, and opening downwards; the inferior mandible slightly ascends towards the tip; the aperture of the ear is defended by a fascicle of small decomposed feathers; the tongue is fleshy and entire; the irides are dark brown ; the head, which is very small in proportion to the body, and half of the neck, are covered by a naked bluish skin, on which are a number of red wart-like elevations on the superior portion, and whitish ones on the inferior, interspersed with a few scattered black, bristly hairs, and small feathers, which are still less numerous on the neck; the naked skin extends farther downwards on the inferior surface of the neck, where it is flaccid and membraneous, forming an undulating appendage, on the lower part of which are cavernous elevations, or *wattles.* A wrinkled, fleshy, conic, extensible caruncle, hairy and pencillated at tip, arises from the bill at its junction with the forehead ; when the bird is quiescent, this process is not much more than an inch and a half long ; but when he is excited by love or rage, it becomes elongated, so as to cover the bill entirely, and depend two or three inches below it. The neck is of a moderate length and thickness, bearing on its inferior portion a pendant fascicle of black rigid hairs, about nine inches long. The body is thick, somewhat elongated, and covered with long truncated feathers; these are divided into very light fuliginous down at base, beyond which they are dusky; to this dusky portion succeeds a broad, effulgent, metallic band, changing now to copper colour or bronze gold, then to violet or purple, according to the incidents of light; and at tip is a terminal, narrow, velvet black band, which does not exist in the feathers of the neck and breast; the lower portion of the back, and the upper part of the rump, are much darker, with less brilliant golden violaceous reflections ; the feathers of the inferior part of the rump have several concealed, narrow, ferruginous, transverse lines, then a black band before the broad metallic space, which is effulgent coppery ; beyond

the terminal narrow black band is an unpolished bright bay
fringe.   The upper tail-coverts are of a bright bay colour, with
numerous narrow bars of bright shining greenish; all these
coverts are destitute of the metallic band, and the greater number have not the black subterminal one ; the vent and thighs
are plain brownish cinereous, intermixed with paler; the under
tail-coverts are blackish, glossed with coppery towards the tip,
and at tip are bright bay.

The wings are concave and rounded, hardly surpassing the
origin of the tail ; they have twenty-eight quill-feathers, of
which the first is shortest, and the fourth and fifth longest, the
second and ninth being nearly equal; the smaller and middling
wing-coverts are coloured like the feathers of the body ; the
greater coverts are copper violaceous, having a black band
near the whitish tip; their concealed web is blackish, sprinkled
with dull ferruginous : in old birds, the exterior web is much
worn by friction amongst the bushes, in consequence of which
those feathers exhibit a very singular, unwebbed, curved appearance, faithfully represented in the plate.   The spurious
wing, the primary coverts, and the primaries, are plain blackish
banded with white, which is interrupted by the shaft, and
sprinkled with blackish ; the secondaries have the white portion so large, that they may as well be described as white
banded with blackish, and are, moreover, tinged with ferruginous yellow ; this colour gradually encroaches on the white,
and then on the blackish, in proportion as the feathers approach the body, so that the tertials are almost entirely of that
colour, being only sprinkled with blackish, and having metallic
reflections on the inner web ; the anterior under wing-coverts
are brownish black, the posterior ones being grey ; the tail
measures more than a foot and a quarter, is rounded, and composed of eighteen wide feathers ; it is capable of being expanded and elevated, together with the superior tail-coverts,
so as to resemble a fan, when the bird parades, struts, or wheels.
The tail is ferruginous, mottled with black, and crossed by
numerous narrow undulated lines, of the same colour, which

become confused on the middle feathers ; near the tip is a broad
black band, then the feathers are again mottled for a short distance, and are widely tipped with ferruginous yellow.

The feet are robust and somewhat elongated ; the tarsus
measures more than six inches in length, being covered before
by large alternate pentagonal plates, and furnished, on the
inner posterior side, with a rather obtuse, robust, compressed
spur, nearly one inch long. The toes are three before, connected at base by a membrane, and one behind, touching the
ground only at tip, being articulated higher on the tarsus than
the others, and one-half shorter than the lateral toes, which are
equal ; the middle toe is more than four inches long, and the
posterior but little more than one inch ; they are all covered
by entire plates ; the sole is granulated : the colour of the feet
is red, the margins of the plates and scales, the membrane and
nails, being blackish ; the nails are oblong, wide, obtuse at tip,
rounded above, and perfectly plain beneath.

The female, or hen-turkey, is considerably smaller in size,
being three feet and a quarter long ; the bill and feet resemble
those of the male, but are proportionally smaller, the latter
being destitute of even a rudiment of spur ; the irides are like
those of the male ; the head and neck are not so naked as in
that sex, but are covered by small, decomposed feathers, of a
dirty greyish colour ; those of the back of the neck are tipt
with ferruginous, constituting a longitudinal vitta on that part;
the caruncle on the frontlet is rudimental, not susceptible of
being elongated ; the pectoral appendage is entirely wanting
in our specimen ; the general plumage is dusky grey, each
feather having a metallic band, less brilliant than that of the
male ; then a blackish band and a greyish terminal fringe ;
the black subterminal band is obsolete on the feathers of the
neck and of the whole inferior surface; those of the latter part,
with the feathers of the lower portion of the back, of the rump,
and the flanks, have their tips yellowish ferruginous, becoming
gradually brighter towards the tail. The vent and thighs are
dirty yellowish grey, without any reflections ; the under tail-

coverts are tipt and varied with rather deep ferruginous; the superior tail-coverts are like those of the male, but duller, and tipt with a broad, whitish ferruginous fringe. The wings are also duller, each covert being tipt with greyish; less white exists on the primaries, the bands being narrower, and the secondaries entirely destitute of them. The tail is similar in colour to that of the male. It is proper to remark, that the female which furnished the above description, and is figured in the plate, though certainly adult, had not attained to its full size and perfect beauty. It was procured in the month of March, on St John's river, Florida.

The young of both sexes resemble each other so closely, before the naked membrane acquires its tinge of red, as to be scarcely distinguishable; the females, however, when a few days old, are somewhat larger than the males, and have a weaker piping note; the males then begin to stand higher on their legs, which are stronger than those of the females, and soon exhibit the rudiments of spurs. On the approach of the first winter, the young males show a rudiment of the beard or fascicle of hairs on the breast, consisting of a mere tubercle, and attempt to strut and gobble; the second year, the hairy tuft is about three inches long; in the third, the turkey attains its full stature, although it certainly increases in size and beauty for several years longer. In a fine male specimen, evidently young, which I obtained in the Philadelphia market, the plumage is equally brilliant with that of the finest adult, although the frontal caruncle is only one inch in length, the pectoral appendage two inches, and the spur merely rudimental. The concealed portion of the plumage on the anterior part of the back is sprinkled with pale ferruginous, which disappears as the bird advances in age.

Females of four years old have their full size and colouring; they then possess the pectoral fascicle four or five inches long (which, according to Mr Audubon, they exhibit a little in the second year, if not barren), but this fascicle is much thinner than that of the male. The barren hens do not obtain this distinction until a very advanced age; and being preferable for the table, the hunters single them from the flock, and kill them in preference to the others. The female wild turkey is more frequently furnished with the hairy tuft than the tame one, and this appendage is gained earlier in life. The number of young hens without it has given rise to the incorrect assertion of a few writers, that the female is always destitute of it.

The weight of the hen generally averages about nine pounds avoirdupois. Mr Audubon has shot barren hens, in strawberry time, weighing thirteen pounds; and he has seen some few so fat as to burst open by falling from a tree, after being shot. The male turkeys differ more in bulk and weight: from the accounts I have received from various parts of the Union, fifteen or twenty pounds may be considered a fair statement of their medium weight; but birds of thirty pounds are not very rare, and I have ascertained the existence of some weighing forty. In relation to those surpassing the last-mentioned weight, according to the report of authors who do not speak from personal observation, I have not been able to find any, and am inclined to consider them as fabulous. Mr Audubon informs us he saw one in the Louisville market that weighed thirty-six pounds; the pectoral appendage of this bird measured more than a foot in length. Bartram describes a specimen of remarkable size and beauty, reared from an egg found in the forest, and hatched by a common hen. When this turkey stood erect, the head was three feet from the ground. The animal was stately and handsome, and did not seem insensible of the admiration he excited. Our plate, which is the first that has been given of the wild turkey, represents both sexes, reduced to one-third of their natural size; the male was selected from among many fine specimens, shot in the month of April, near Engineer Cantonment, on the Missouri. It weighed twenty-two pounds; but as the males are very thin at that season,* when in good order it must have weighed much more.

* The extraordinary leanness of this bird, at particular seasons of the year, has become proverbial in many Indian languages. An Omawhaw, who wishes to make known his abject poverty, says, " *Wah pawne zezecah ha go ba,*"—" I am as poor as a turkey in summer."

Though comparatively recent, the domestic state of the turkey has been productive of many varieties ; we need not, therefore, be surprised at the existence of numerous and remarkable differences in those animals which have been domesticated from time immemorial. The most striking aberration from the standard of the species, is certainly the tufted turkey, which is very rare, the crest being white in some specimens, and black in others. Tame turkeys sometimes occur of an immaculate black colour ; others are exclusively white ; some are speckled or variegated ; and all these varieties are continued by propagation under analogous circumstances. In the wild state, a white, or even a speckled turkey, is unknown ; and we may venture to say, that a plain black one has hardly ever occurred.

Moehring proposed the name of *Cynchramus* for this genus, as the term *Meleagris* was used by the ancients to indicate a different bird. All other naturalists have agreed with Linné, who, though fully aware of the fact, made use of the name we have adopted. But he included in the genus two allied species, which Gmelin very properly rejected, and placed in a separate genus, which he called *Penelope*, considering the turkey as *sui generis*. Latham again rendered the genus unnatural by restoring one of the objectionable Linnean species, perceiving that it was not properly placed in *Penelope ;* it is, in truth, a *Phasianus*. As now characterised, the present genus is exclusively American ; and by the discovery of a beautiful species closely allied to that of the United States, it now consists of two species. The ocellated turkey (*Meleagris ocelata*) inhabits Honduras, and may be distinguished from the common species by its smaller size, more brilliant plumage, and principally by having ocellated spots on the tail. It was first described by Cuvier, and has lately been figured in that magnificent periodical work, the " Planches Colorées " of Temminck and Laugier. A beautiful specimen has long been exhibited in the Charleston Museum.*

* Mr Audubon has recorded the following anecdote of a turkey, which he kept for some years in a tame state :—

" While at Henderson, on the Ohio, I had, among many other wild

M. Duponceau, so well known by his philological researches, has favoured us with the following table of names for the wild turkey in the different languages.

E. English pronunciation.——F. French.——S. Spanish.——G. German.

Algonkin	Mississay, E. . . .	Mackenzie.
Adayes	Owachuk, S. . . .	MS. Voc.
Atacapas	Skillig, S. . . .	MS. Voc.
Caddoes	Noe, E.	Dr Sibley.
Chetimachas . . .	Tsante hatineche hase, S.	MS. Voc.
Cherokee	Kainna; Oocoocoo, E. .	MS. Voc.
Chickasaws . . .	Fukit, E. . . .	MS. Voc.
Choctaws	Oopuh, E. . . .	MS. Voc.
Creeks	Pinewau, E. . . .	MS. Voc.
Delaware proper . .	Tschikenum, G. . .	Heckewelder and Zeisberger.
Delaware of New Jersey .	Tshikuuna, E. . .	MS. Voc.
Delaware of New Sweden .	Sickenem (Swedish) .	Luther's Catechism.
Huron	Ondetontak, F. . .	Père Sagard.
Wyandot (same people) .	Daigh-ton-tah, E. . .	Attwater in Archæol. Amer.
Illinois	Pireouah, F. . .	MS. Voc.
Knisteneaux . . .	Mes-sey-thew, E. . .	Mackenzie.
Miamis	Pilauoh	MS. Voc.
Nenticoke	Pahquun, E. . . .	MS. Voc.
Nottoway *	Kunum, E. . . .	MS. Voc.
Umawhaw (a branch of Sioux)	Ze-ze-kah, E. . . .	Say.
Onondagos (Iroquois) . .	Netachróchwa gatschínak, G.	Zeisberger's Dictionary, MS.
Osage { cock . .	Sukah tingah, E. . .	MS. Voc.
Osage { hen . .	Inchuga Sukah, E. . .	MS. Voc.
Ottos, or Wahtoktatah (Sioux)	Wa-ek-kung-ja, E. . .	Say.
Shawanese	Peléwa, G. . . .	Heckewelder.
Uchee †	Witch-pshah, E. . .	MS. Voc.
Unquachog (Long Island) .	Nahiam, E. . . .	MS. Voc.

birds, a fine male turkey, which had been reared from its earliest youth under my care, it having been caught by me when probably not more than two or three days old. It became so tame that it would follow any person who called it, and was the favourite of the little village. Yet it would never roost with the tame turkeys, but regularly betook itself, at night, to the roof of the house, where it remained until dawn. When two years old, it began to fly to the woods, where it remained for a considerable part of the day, to return to the enclosure as night approached. One morning I saw it fly off, at a very early hour, to the woods, and took no particular notice of that circumstance. Several days elapsed, but the bird did not return. I was going towards some lakes near Green River to shoot, when, having walked about five miles, I saw a fine large gobbler cross the path before me, moving leisurely along. Turkeys being then in prime condition for the table, I ordered my dog to chase it, and put it up. The animal went off with great

---

* Indians of Virginia, a branch of the Tuscaroras.

† *Uchees*, a nation of Florida Indians, speaking a curious language, full of particular sounds, not found in any other language ; they live among the Creeks

COOPER'S HAWK. (*Falco Cooperii.*)

PLATE X.—Fɪɢ. 1.

Philadelphia Museum, No. 403.—*My Collection.*

ASTUR COOPERII.—Bᴏɴᴀᴘᴀʀᴛᴇ? *

Falco Cooperii, *Bonap. Synop. App.* p. 433.—The Stanley Hawk, Falco Stanleii?
Aud. pl. 36, m. and f., *Orn. Biog.* i. p. 186.

Bᴜғғᴏɴ complained of the difficulty of writing a history of
birds, because he already knew eight hundred species, and
supposed that there might actually exist fifteen hundred, or
even, said he, venturing, as he thought, to the limit of probability, two thousand! What then would be his embarrassment at present, when nearly six thousand species are known,
and fresh discoveries are daily augmenting the number?

rapidity, and, as it approached the turkey, I saw with great surprise
that the latter paid little attention. Juno was on the point of seizing
it, when she suddenly stopped, and turned her head towards me. I
hastened to them, but you may easily conceive my surprise, when I saw
my own favourite bird, and discovered that it had recognised the dog,
and would not fly from it, although the sight of a strange dog would
have caused it to run off at once."

I have more than once known the domestic turkey-cock drive the
hen from her nest, sit upon the eggs until hatched, and perform all the
duties to the young incumbent on the female, and never during the
time allow her to approach. I once knew it take its place upon two
addled eggs, which a hen had long persevered upon, and was at last succeeded by the male, who kept his place for nearly a fortnight.—Eᴅ.

* We have the authority of the Prince of Musignano for considering
the Stanley hawk of Audubon identical with *Astur Cooperii;* and from
a comparison, as far as plates and descriptions will allow, they seem
at all events very closely allied. The bird is comparatively rare. Mr
Audubon observed them in Louisiana, and about the Falls of Niagara,
and describes them as very bold and intrepid; so much so, that one which
had seized a cock in a farm-yard, was repeatedly forced to the ground
before it could master its victim.

We would recommend for perusal the manners of the genus *Astur,*
as portrayed in this description. Any one who has witnessed the common species of Europe will at once perceive the truth of the delineation.—Eᴅ.

1. *Cooper's Hawk.* 2. *Palm Warbler.*
Falco Cooperii.   *Sylvia Palmarum.*

10.

The difficulties attending a general work on this subject are not, perhaps, experienced in an equal degree by one who confines himself to the history of a particular group, or of the species inhabiting a single district. Nevertheless, in a work like the present, which is not a monography limited to one genus or family, but embraces within its scope species belonging to all the different tribes, it is requisite, in order to explain their various relations and analogies, that the author should be more or less acquainted with the whole system of nature. To attempt, without the aid of methodical arrangement, a subject so vast, and apparently unlimited, would be hopeless. Hence the importance of a correct system of classification ; and the construction of one which shall exhibit, as far as practicable, the true affinities of objects, has exercised the attention of the most powerful minds that have been employed in the study of nature.

That division of the feathered class popularly called birds of prey has always been recognised as a separate and well-defined group. In the Linnean System they form the order Accipitres, and were, by that father of the science, distributed into three great natural divisions, which comprise nearly, if not quite, one-fifteenth part of all the known species of birds. The ulterior arrangement of one of these groups, the genus *Falco* of Linné, at present composed of between two and three hundred species, has much divided the opinions of naturalists. From the majestic eagle, the terror of the husbandman, to the feeblest hawk, preying on grasshoppers, it is undeniable that there exists in all these birds a great resemblance in some of the most prominent characteristics, which, being found to predominate in the fish-hawk, as well as the kite, and all other species of the falcon tribe, however dissimilar, indicate their separation as a peculiar family from all other birds. But that they are susceptible of division into smaller groups of inferior rank, no practical ornithologist will for a moment deny. Whether these minor groups shall be considered as trivial and secondary, or whether some of them ought not to be admitted as distinct and independent genera, is a question that has been much agitated, and respecting which ornithologists will probably for a long time continue to disagree. Equally great authorities might be cited in favour of either of these opinions, which, like many others of more importance that have divided mankind from the beginning of the world, may perhaps, after all, be considered as merely a dispute about words.

Admitting, however, as seems to be done by all parties, that this great genus may be subdivided with propriety, we look upon it as altogether a secondary question whether we shall call the minor groups genera, subgenera, or sections ; and we deem it of still less consequence, in a philosophical view, whether the names by which these groups are designated be taken from a learned or a vernacular language. It is our intention to pursue a middle course. We are convinced of the necessity of employing numerous subdivisions, not only in this, but also in its allied genus *Strix*. These, however, we cannot agree to admit as genera, preferring to call them subgenera, and giving them a name, but, when having occasion to mention a species belonging to any of them, to employ the name of the great genus.

The desire of avoiding too great a multiplication of groups, has caused some, even of the first ornithologists of our time, to employ sections that are not natural, and with false or inapplicable characters, and, as if they would compel nature to conform to their preconceived and narrow views, after having assigned decided limits to their groups, to force into them species not only widely different, but that do not even possess the artificial character proposed. We shall not imitate this irrational example. It shall rather be our object to compose natural groups, and, in obedience to this principle, whenever we meet with a group, or even a single species, clearly insulated, it shall at least be pointed out, not so much regarding the number of our subgenera, as the characters that unite the species of which they are respectively composed.

It is objected to the numerous subdivisions that have been proposed in our day, that they pass into and blend insensibly with each other. This is no doubt true; but is it not the same with regard to natural groups of every denomination? It is this fact which has induced us to consider them as subgenera, and not as distinct genera. We are told, however, by the advocates for numerous genera, that, in giving a name, we adopt a genus; but we do not see that this necessarily follows.

There are, we confess, other grounds on which we might be attacked with more advantage. We may, perhaps, be charged with inconsistency in refusing to admit, as the foundation of generic groups in the *Rapaces*, characters which are allowed, not only by ourselves, but by some of those who are most strenuously opposed to the multiplication of genera, to have quite sufficient importance for such distinction in other families. With what propriety, it might be asked, can we admit *Hydrobates* (*Fuligula*, Nob.), as distinct from *Anas*, and the various genera that have been dismembered from *Lanius*, at the same time that we reject, as genera, the different groups of hawks? To this we can only reply, that we are ourselves entirely convinced that all the subgenera adopted in our Synopsis among the *Falcones* of North America are quite as distinct from each other as *Coccyzus* and *Cuculus*, or *Corvus* and *Garrulus*. The latter genus we have admitted after Temminck, who is opposed to new genera among the hawks, though *Astur* and *Elanus* certainly require to be separated no less than the two genera that Temminck himself has established in the old genus *Vultur*.

No living naturalist (with the exception of those who, through a sort of pseudo-religious feeling, will only admit as genera groups indicated as such by Linné) has adhered longer than ourselves to large genera, at the same time that we could not deny the existence of subordinate natural groups. We will not pretend to deny that these are of equal rank with some recognised as genera in other families; and we can only say, that we consider it doubtful, in the present unsettled state of the science, what this rank ought to be. We therefore, in the instances above quoted, consider it of little importance whether these groups be considered as genera or subgenera.

But what is certainly of great importance, is to preserve uniformity in all such cases, to make co-ordinate divisions, and give corresponding titles to groups of equal value. This uniformity, however desirable, cannot, in the actual state of ornithology, be easily attained ; and we have decided, after much hesitation, to continue to employ subgenera. In doing this, we are moreover influenced by the great difficulty that is met with, in some cases, in determining the proper place of a species partaking of the characters of several groups, yet not in the least deserving to be isolated ; such as *Falco borealis*, which is almost as much an *Astur* as a *Buteo*, and has been placed by authors, according to their different views, in both these groups.

An extensive reform is evidently needed in the department of classification that relates to genera ; and we propose, with this view, to undertake at some future period a general work, when, erecting our system on a more philosophical basis, though we may restrict some, and enlarge other genera, we shall, in the instances to which we have alluded, as well as in a multitude of others, at least place them all on an equal footing.

Among the several groups into which the falcon tribe is divided, we come to one, composed of about sixty species, well marked, and, if kept within its proper bounds, very natural, to which authors have variously applied the name of *Accipiter*, *Sparvius*, and *Astur*, which last we have adopted.

Found in all parts of the globe, and destroying everywhere great numbers of birds and small quadrupeds, the hawks (by which English name we propose to distinguish this group more particularly) closely resemble each other in colour and changes of plumage, especially the North American and European species. They are eminently distinguished from all other falcons by their short wings, not reaching, by a considerable length, to the tip of their tail, which is even, or but very slightly rounded, and by their first quill-feather or primary, which is very short, while the fourth is constantly the longest. Their bill, suddenly curved from the base, is very strong and sharp; their head is narrowed before, with the eyes placed high, large, and fiery. Their feet are very long, and the toes especially, the middle one of which is much the longest, and all are armed with very strong sharp talons, well seconding the sanguinary nature of these fierce creatures; their outer toe is connected at base by a membrane to the middle one. The female is always one-third larger than the male, and the plumage of both is, in most species, dark above and white beneath—in the adult, barred with reddish or dusky. In the young bird the colour is lighter, the feathers skirted with ferruginous, and the white of the under parts streaked longitudinally with dusky, instead of being barred. The tail is uniform in colour with the back, with almost always a few broad bands of black, and sometimes of white, and a whitish tip.

The hawks (*Astures*) combine cunning with agility and strength. Sudden and impetuous in their movements, they make great havoc, especially among birds that keep in flocks, as pigeons, blackbirds, &c., and are the terror of the poultry yard. Fearless and sanguinary, they never feed, even when pressed by hunger, except on red and warm-blooded animals, whose quivering limbs they tear with savage delight. Birds they pluck very carefully, and quarter, before eating them, but swallow small quadrupeds entire, afterwards ejecting their skins rolled up into a ball. They always pursue and seize their prey upon the wing, not falling upon it from aloft, but, rapidly skimming the earth, make their insidious approaches sideways, and singling out their victim, dart upon it with fatal velocity. They never soar, like the kites and eagles, to the upper regions of the atmosphere; and it is only during the nuptial season that they are observed sailing in wide circles in the air. Their favourite haunts, during summer, are forests, building their nests on trees; in winter they spread over the

plains. Though generally observed alone, the male and his
companion are seldom far apart. During the youth of their
progeny, the parents keep them company, in order to teach
them to hunt their prey, and at such times they are observed
in families.

This group may be further subdivided into two sections,
to one of which the name of *Astur* has more strictly been
assigned, while the other has been distinguished by those of
Sparvius and *Accipiter*. The former, of which the goshawk of
Europe and North America (black-capped hawk of Wilson) is
the type, is characterised by its wings being somewhat longer,
body more robust and shorter, and much thicker tarsi. This
is the only species that inhabits the United States and Europe.

The second section, to which the present new species belongs, possessing all its characters in a pre-eminent degree,
equally with the hawk described by Wilson in its adult state
as *Falco Pennsylvanicus*, and in its youth as *Falco velox*, was
established on the sparrow-hawk of Europe, *Falco nisus*, but
the American species just mentioned are no less typical. The
hawks of this section are more elegantly shaped, being much
more slender; their wings are still shorter than in the other
section, reaching little beyond the origin of the tail, and their
tarsi slender and elongated, with a smooth and almost continuous covering.

Notwithstanding their smaller size and diminished strength,
their superior courage and audacity, and the quickness of their
movements, enable them to turn the flight of the largest birds,
and even sometimes, when in captivity together, to overcome
them. We have kept a sparrow-hawk (*Falco nisus*), which,
in the space of twenty-four hours that he was left unobserved,
killed three falcons which were confined with him.

The inextricable confusion reigning throughout the works
of authors who have not attended to the characters of the different groups of this genus, renders it next to impossible to decide, with any degree of certainty, whether our *Falco Cooperii*
has or has not been recorded. Though agreeing imper-

fectly with many, we have not been able, notwithstanding our most sedulous endeavours, to identify it with any. It is evidently a young bird, and we should not be surprised at its proving, when adult, a known species, perhaps one of the numerous species figured of late, and possibly *Le Grand Epervier de Cayenne* of Daudin, *Sparvius major*, Vieillot, stated to be one-third larger than the European sparrow-hawk. At all events, however, it is an acquisition to the ornithology of these States; and we have ventured to consider it as a new species, and to impose on it the name of a scientific friend, William Cooper of New York, to whose sound judgment, and liberality in communicating useful advice, the naturalists of this country will unite with us in bearing testimony, and to whom only the author, on the eve of his departure for Europe, would have been willing to intrust the ultimate revision and superintendence of this work.

The perfect accuracy with which Mr Lawson may be said to have outdone himself in the delineation of this bird, in all the details of its plumage, bill, and feet, will now at least have established the species in the most incontestable manner.

Our bird agrees very well with the falcon gentle, *Falco gentilis*, Linné; but as that species is referred to the young of the goshawk, we have preferred giving it a new name, to reviving one that might have created an erroneous supposition of identity. To the young goshawk, our hawk is, in fact, extremely similar in colour and markings, being chiefly distinguished from it by the characters of their respective sections, having the tarsi much more slender and elongated, and the wings still shorter; the tail is also considerably more rounded.

But it is to the sharp-shinned hawk (*Falco velox*) of Wilson, the *Falco Pennsylvanicus*, or *Falco fuscus* in its immature plumage, that our Cooper's hawk bears the most striking resemblance, and is in every particular most closely allied. Even comparing feather by feather, and spot by spot, they almost perfectly agree; but the much larger size of the present, it being more than twice the bulk, will always prevent their being confounded, even by the most superficial observer. Another good mark of discrimination may be found in the comparative length of the primaries; the second in *F. Cooperii* being subequal to the sixth, while in *F. velox* it is much shorter. The latter has also the fifth as long as the fourth; that, in our species, being equal to the third. The tail is also much more rounded, the outer feather being nearly an inch shorter than the middle one. In *F. velox* the tail is even, the outer feather being as long, or, if anything, longer than the middle. There is no other North American species for which it can be mistaken.

The bird represented in the plate, of which we have seen seven or eight specimens perfectly similar in size and plumage, was a male, killed in the latter part of September, near Bordentown, New Jersey. The stomach contained the remains of a sparrow. Another that we procured was shot on the 12th of December, while in the act of devouring on the ground a full-grown ruffed grouse which he had killed, though a larger and heavier bird than himself. Mr Cooper, the friend to whom we have dedicated this species, has recently favoured us with an accurate description of a specimen of a somewhat larger size, shot in the early part of November on the eastern part of Long Island.

The male Cooper's hawk is eighteen inches in length, and nearly thirty in extent. The bill is black, or rather blackish brown; the cere, greenish yellow; the angles of the mouth, yellow. The irides are bright yellow. The general colour above is chocolate brown, the feathers being whitish grey at base; on the head and neck above, they are blackish, margined with rufous, pure white towards the base, and greyish at the bottom, the white colour showing itself on the top and sides of the neck, and being much purer on the nucha. The back and rump are the same, but the feathers larger and lighter coloured, less margined with rufous, more widely greyish at base, and bearing each four regular spots of white in the middle of their length, which are not seen unless when the feathers are turned aside. The whole body beneath is white, each feather, including the lower wing-coverts and femorals, marked with a long, dusky medial stripe, broader and oblanceolate on the breast and flanks (some of the feathers of which have also a blackish band across the middle), the throat, and under wing-coverts; the long feathers of the flanks (or long axillary feathers) are white, banded with blackish; the vent and lower tail-coverts, pure white; the wings are nine inches long, and, when folded, hardly reach to the second bar of the tail from the base; the smaller wing-coverts and scapulars are like the back, the quills brown above (lighter on the shaft), and silvery grey beneath, regularly crossed by blackish bands, less conspicuous above; the space between the bands is white on the inner vanes at base; some of the secondaries and tertials are tipt and edged with rusty, and have more and more of white as they approach the body, so that those nearest may in fact be described as white banded with blackish. The first primary is very short, more so than the secondaries; the second is equal to the sixth, the third to the fifth, these two last mentioned being hardly shorter than the fourth, which, as in all *Astures*, is longest. The tail is full eight inches long, reaching five beyond the wings; its colour is ashy brown, much paler beneath, tipped with whitish, and crossed by four equidistant blackish bands, nearly one inch in breadth; the tail-coverts at their very base are whitish; the lateral feathers are lighter, and with some white on the inner webs. The legs and feet are yellow, slender, and elongated, but still do not reach, when extended, to the tip of the tail; the tarsus, feathered in front for a short space, is two and three-quarter inches long; as in other *Astures*, the middle toe is much the longest, and the inner, without the nail, is shorter than the outer, but taken with its much longer nail, is longer. The talons are black, and extremely sharp, the inner and the hind ones subequal, and much the largest, while the outer is the most delicate.

The female is larger, and measures two inches more in

length, but in plumage is perfectly similar to the male. As the male we have described and figured is evidently a young bird, it is very probable that the adult, after undergoing the changes usual in this group, obtains a much darker and more uniform plumage above, and is beneath lineated transversely with reddish. That in this supposed plumage the bird has not yet been found, is no reason to doubt its existence, as the species is comparatively rare. Even of the common *Falco fuscus,* though constantly receiving numerous specimens of the young, we have only been able to procure a single one in adult plumage during a period of four years.

We regret that this is all that is in our power to offer of the history of this species, which, as will be seen from the description, possesses in an eminent degree the characters of the group. From the circumstance of its being found here in autumn and winter, we are led to infer that it comes to us from the north.

PALM WARBLER. (*Sylvia palmarum.*)

PLATE X.—Fig. 2.

Motacilla palmarum, *Gmel. Syst.* i. p. 951, sp. 53, winter dress.—Sylvia palmarum, *Lath. Ind.* p. 544, sp. 136.—*Vieill. Ois. Am. Sept.* ii. p. 21, pl. 73 (and the other works of the same author), winter plumage.—*Nob. Add. Orn. U. S. in Jour. Ac. Ph.* v. p. 29.—*Id. Cat. Birds U. S. in Contr. Macl. Lyc. Ph.* i. p. 16, sp. 105.—*Id. Syn. Birds U. S.* sp. 105, in *Ann. Lyceum, N. Y.* ii. p. 78. —Motacilla ruficapilla, *Gmel. Syst.* i. p. 971, sp. 106, summer dress.—Sylvia ruficapilla, *Lath. Ind.* summer dress (not of *Vieill.*)—Ficedula Martinicana, *Briss. Av.* iii. p. 490, sp. 50, pl. 22, fig. 4, perfect plumage.—Le Bimbelé, ou Fausse Linotte, *Buff. Ois.* v. p. 330, winter dress.—Figuier à tete rousse, *Buff. Ois.* v. p. 306, summer dress.—Palm Warbler, *Lath. Syn.* iv. p. 489, sp. 131, winter dress.—Bloody-side Warbler, *Lath. Syn.* iv. p. 489, sp. 115 (not of *Penn.*), summer plumage.—*Philadelphia Museum.*

SYLVICOLA PALMARUM.—Jardine.

Sylvia palmarum, *Bonap. Synop.* p. 78.

THIS is one of those lively, transient visitants, which, coming in spring from warmer regions, pass through the middle States on their way to still colder and more northern countries, to breed. From the scarcity of the species, its passage has hitherto been unobserved ; and it is now, for the first time, introduced as a bird of the United States. Authors who have heretofore made mention of it, represent it as a permanent resident of St Domingo, and other islands of the West Indies, and even describe its nest and habits, as observed there.

In the United States, it is found during winter in Florida, where it is, at that season, one of the most common birds. In the month of November, they are very abundant in the neighbourhood of St Augustine, in East Florida, even in the town, and in other parts of the territory wherever the orange tree is cultivated, being rare elsewhere. They are found in great numbers in the orange groves near Charleston, South Carolina, at the same season, and have also been observed at Key West and the Tortugas, in the middle of February, and at Key Vacas in the middle of March. Their manners are sprightly, and a jerking of the tail, like the pewee, characterises them at first sight from a distance. The only note we have heard them utter is a simple chirp, very much like that of the black and yellow warbler, *Sylvia maculosa* (*Magnolia* of Wilson). They are fond of keeping among the thick foliage of the orange trees. A few are observed every year in spring on the borders of the Schuylkill, near Philadelphia, as well as in the central parts of New Jersey, on their passage to the north. They breed in Maine, and other parts of New England, where they are common during summer, and perhaps also in Canada, though probably not extending to the inhospitable climates of Hudson's Bay, whose natural productions are so well known.

The bird represented in the plate was shot near Bordentown, on the 17th of April, in the morning. It was a fine adult male, in the gayer plumage of the breeding season, in which it is now for the first time figured, and a description is subjoined.

Length, five inches and a quarter ; extent, more than eight inches ; bill, five-eighths of an inch long, very slender, straight, hardly notched, blackish, paler beneath ; feet, dusky grey, yellowish inside ; irides, dark brown, nearly black ; crown, bright chestnut bay ; bottom of the plumage lead colour all over, much darker beneath ; a well-defined superciliar line, and the rudiment of another, on the medial base of the upper mandible, rich yellow ; the same colour also encircles the eye ; streak through the eyes and cheeks, dusky olive, somewhat intermixed with dull chestnut ; upper parts olive green, each feather being dusky in the middle ; rump and upper tail-coverts yellow olive, all beneath bright yellow ; sides of the neck, breast, and flanks, with chestnut streaks ; superior wing-coverts blackish, margined and tipt with olive green, and somewhat tinged with chestnut ; inferior wing-coverts yellowish ; quills dusky, edged exteriorly with green, the outer one with white on the outer side, two exterior with a large white spot on the inner web at tip.

In the plumage here described, it has been mentioned by several authors, under the name of *Sylvia ruficapilla*, and by Latham is called the Bloody-side warbler. In that which we are about to describe, it was first made known by Buffon, who adopted the name of *Bimbelé*, given to it in the West Indies, and in this state it is figured by Vieillot as the *Sylvia palmarum*. The following description is drawn up from a specimen procured in Florida in winter.

Length, five inches ; bill, half an inch, slender, almost straight, and very slightly notched, blackish, paler beneath ; the feet are blackish ; irides, very dark brown. The general plumage above is olive brown, each feather being dusky along the middle ; the feathers of the head are dusky at base, as is the whole plumage ; then they are chestnut nearly to the tip (forming a concealed spot of that colour on the crown), where they are of the common colour, but somewhat darker ; the rump and superior tail-coverts are yellow olive ; a well-defined yellowish white line passes over the eye which is encircled with white ; the cheeks are dusky, as well as a streak through the eye ; the inferior parts are whitish, slightly tinged with yellowish, and with a few blackish streaks on each side of the throat, and on the breast and flanks ; the belly is immaculate, and more richly tinged with yellow, the inferior tail-coverts being pure yellow ; the wing-coverts are of the colour of the feathers of the back, the blackish centre being more extended and deeper ; the wings have no bands ; the quill-feathers are blackish, edged externally with pale yellow olive, becoming whitish towards the tip ; the five outer ones are subequal ; the tail is even, its feathers are somewhat pointed, edged externally with yellow olive, internally with whitish, the outer one also externally whitish ; the two outer ones with a large pure white spot on their inner vane at tip ; the third and fourth, each side, with an inner white terminal margin.

In this plumage, this bird resembles so nearly *Sylvia coronata* in its most humble dress, that it is distinguishable only on a close examination. However, the bill is longer and more slender, the crown-spot chestnut instead of yellow, the feathers being destitute of the white which is observable in the other by separating the feathers; the rump is olive yellow, not pure yellow, and that colour extending on the tail-coverts, which it does not in *Sylvia coronata.* The under parts tinged with yellow, and especially the pure yellow tail-coverts, which are pure white in *S. coronata,* will sufficiently distinguish them.

It is a remarkable circumstance that there is no obvious difference to be observed between the plumage of the sexes, notwithstanding the statements of authors to the contrary. This is the case, however, in *S. coronata,* and in almost all the warblers that change periodically from a dull to a bright plumage ; and, in fact, in most birds in which this change takes place.

According to Buffon and Vieillot, this bird is a permanent resident in the West Indies, where, as they state, the name is sometimes applied to it of *Fausse linotte.* We, however, can perceive scarcely any resemblance, except in its dull state of plumage, to a similar state of the red-poll finch. The name of *Bimbelé,* by which it is known among the negroes of those countries, is derived from the recollection of an African bird, to which, probably, the resemblance is not more evident. Unfortunately, this propensity of limited minds to refer new objects, however distinct, to those with which they are acquainted, seems to have prevailed throughout the world, and is found exemplified nowhere more absurdly than in the Anglo-American names of plants and animals.

The food of this little warbler consists chiefly of fruits and small seeds. Its song is limited to five or six notes ; but though neither brilliant nor varied, it is highly agreeable, the tones being full, soft, and mellow. While other birds of its kind build in thickets and humble situations, this proud little creature is said always to select the very lofty tree from which it takes its name, the palmist (a species of palm), and to place its nest in the top, in the sort of hive formed at the base or insertion of the peduncle which sustains the clusters of fruit.

Such are the facts we have gathered from authors ; but as the singular description of the nest coincides exactly with the manner of building of the *Tanagra Dominica,* and as, moreover, the palm warbler appears not to be known in its gayer vesture in the West Indies, we cannot easily believe that it breeds elsewhere than where we have stated—that is, in the temperate, and even colder regions of America—and that what has been mistaken for its nest in reality belongs to the above-named, or some other bird.

The first accounts of this species were given, as we have already stated, by Buffon, and from him subsequent writers appear to have copied what they relate of it. The bird which he described must have been a very young specimen, as its colours are very dull, much more so than the one figured and described by Vieillot, who supposes, though erroneously, Buffon's specimen to have been a female. Even Vieillot's, which is certainly our species in its winter dress, is much duller in colour than those we received from Florida ; and these again are far less brilliant than the bird in our plate, represented as it appears for a few days in the spring in Pennsylvania and New Jersey, and is found throughout summer in Maine ; thus exhibiting the several gradations of change which the plumage undergoes.

Naturalists cannot be too circumspect in receiving reports, even from the most respectable sources,—their own senses affording the only authentic testimony to be relied on. From information derived from Mr T. Peale, who had no opportunity for making comparisons, we erroneously stated in the first volume of this work that *Sylvia celata* (Say) was one of the most common birds in Florida during winter, keeping among the orange trees, &c. All this statement had reference to the present species ; and as soon as the specimens brought by Mr Peale as *Sylvia celata* were shown to us, the error was immediately perceived. We therefore hasten to correct this mistake, which would be otherwise of more consequence, inasmuch as no one else could, for a long time, detect it. This species resembles, it is true, *S. celata* (whose range must remain limited to the Rocky Mountains), and perhaps still more *S. rubricapilla* (Wilson), but it is not of the same subgenus, *Dacnis,* and it may readily be known by the white spots of the tail-feathers.

When the genus *Sylvia,* containing upwards of two hundred and fifty species, shall have been properly studied, it will be found practicable to divide it into several more sections, subgenera, and even perhaps genera. This bird, along with many other North American species, will constitute a highly natural group, very distinct from the true *Sylvia,* of which *S. atricapilla* may be considered as the type. We presume that it is the group we have in view to which Mr Swainson has given the name of *Sylvicola,* in his "Synopsis of Mexican Birds." Our species is erroneously placed by Buffon among his *Demi-fins,* corresponding to our *Dacnis* and Wilson's *Worm-eaters.*

WHITE-TAILED HAWK. (*Falco dispar.*)

PLATE XI.—FIG. 1.

Falco dispar, *Temm. et Laug. Pl. Col.* 319, young female.—*Nob. App. to Synopsis of N. A. Birds in Ann. Lyc. New York,* p. 435.—Milvus (now Elanoides) leucurus, *Vieill.* (Alcon blanco, D'Azara), *Nouv. Dict. d'Hist. Nat.* xx. p. 556. —Falco melanopterus, *Nob. Jour. Ac. Ph.* v. p. 28.—*Id. Cat. Birds U. S.* sp. 16, *in Contr. Macl. Lyc.* i. p. 11.—*Id. Synopsis of N. A. Birds,* sp. 16, *in Con. Lyc. N. Y.*—Le Faucon Blanc, *Sonnini's d'Azara,* iii. p. 96, sp. 36.—*My Collection.*

ELANUS DISPAR.—Jardine.

Falco dispar, *Bonap. Synop. App.* p. 435.

THIS beautiful hawk, which we recently discovered to be an inhabitant of North America, is so strikingly similar to the black-winged hawk (*Falco melanopterus* *) of the old continent, that we have hitherto considered them as identical, contrary to the opinion of Vieillot, whose authority, it is true, could in this case be of little weight, as he had not seen the species, but, like many others, had merely given it a name ; his sole knowledge of it being derived from the work of D'Azara. We have now yielded only to the decision of Temminck (who has lately introduced the young into his *Planches Colorées*), but not without much reluctance, especially as that distinguished ornithologist has evidently not been at the trouble of comparing the two species, otherwise he would certainly not have omitted noticing their affinities and differential characters, since, in the history of species so closely allied as these

* *Falco melanopterus,* Daud. *Orn.* ii. p. 152, sp. 124 ; Lath. *Ind. Suppl.* p. 6, sp. 16. *Falco Sonninensis,* Lath. *Ind. Suppl.* p. 12, sp. 38. *Elanus cæsius,* Savigny, *Ois. d'Egyp.* p. 98, pl. 2, f. 2 ; Vieill. *Nouv. Dict. d'Hist. Nat.* viii. p. 240 (now *Elanoides*). *Elanus melanopterus,* Leach, *Zool. Misc.* iii. p. 4, pl. 122 ; Vigors, *Descr. Austral. Birds,* in Tr. Lin. Soc. xv. p. 185. *Le Blac,* Le Vaillant, *Ois. d'Afr.* i. p. 147, pl. 36 male, 37 young male. *Black-winged falcon,* Lath. *Syn. Suppl.* ii. p. 28, sp. 23. *Sonnini's falcon,* Lath. *Syn. Suppl.* ii. p. 52, sp. 59.

The inspection of original drawings, in a collection that Mr Gray, of the British Museum, was kind enough to show me lately in London, has enabled me to add to these already numerous synonyms, *Falco axillaris,* Lath. *Ind. Suppl.* (*Circus axillaris,* Vieill. !), from New Holland.

Drawn from Nature by A.Rider

Drawn by R.W. Lawson

1. White tailed Hawk.   2. Female Coerulean Warbler.
   Falco Dispar.   Sylvia Azurea.

11.

two, the differential characters are of more importance and utility than the most laboured descriptions.

This comparison we have carefully instituted between our American specimens and others from Africa and Java. They agreed perfectly, especially with that from Java, in every, the minutest character, even feather by feather, much better than birds of prey of the same species and from the same country do generally. They are even more alike than different specimens from the old continent of the black-winged itself, since that species is said to vary considerably in the black markings, which extend more or less on the wings in different individuals. Nevertheless, a constant, though trivial, differential character, added to the difference of locality, has induced us to follow Temminck's course, in which we should never have ventured to take the lead. This character consists in the tail being in *Falco dispar* constantly irregular, while in *F. melanopterus* it is even ; or, to explain it more clearly, the outer tail-feather is rather the longest in the African, and more than half an inch shorter than the next in the American species. This essential character is much more conspicuous in Temminck's plate than in ours, owing to the tail being spread. In the black-winged, also, the lower wing-coverts are destitute of the black patch so conspicuous in the American bird ; a female from Java has, however, a slight indication of it, but no trace of it is observable in our African males.

By admitting this to be a distinct species from the black-winged hawk, we reject one more of those supposed instances, always rare, and daily diminishing upon more critical observation, of a common habitation of the same bird in the warm parts of both continents, without an extensive range also to the north. A steady and long-protracted exertion of its powerful wings would have been requisite to enable it to pass the vast and trackless sea which lies between the western coast of Africa, the native country of the black-winged hawk, and the eastern shores of South America. Yet, were the species identical, this adventurous journey must have been performed.

For, even admitting several centres of creation, we cannot believe that Nature,* who, notwithstanding her luxuriant abundance, evidently accomplishes all her ends with the greatest economy of means, has ever placed, aboriginally, in different parts of the globe, individuals of the same species, but has always given to each the power of extending its range according to volition, in any direction where it should find climate, food, or other circumstances most appropriate.

The white-tailed hawk is one of those anomalous species which connect the generally received divisions of the great genus *Falco.* It participates in the form and habits of the kites (*Milvus*), while in its other relations it approaches the true falcons (*Falco*), and at the same time presents traits peculiar to itself. Savigny has, therefore, very properly considered its near relative, the black-winged, as the type of a peculiar group, which he elevates to the rank of a genus, but which we for the present shall adopt as a subgenus only. Subsequent observations have confirmed Le Vaillant's opinion that the swallow-tailed hawk (*Falco furcatus*) is closely related to it ; and associated with a few other recently discovered species, they have been considered as a distinct group, under Savigny's name of *Elanus.* Vieillot adopted the group as a genus, but, for what reason we know not, has since changed the name to *Elanoides.* The hawks of this group are readily distinguished from all others by the superior length of the second primary of their elongated wings ; by their bill, rounded above, curved from the base, and not toothed ; their hirsute cere, thick, short, and wholly reticulated tarsi, half-feathered before ; toes entirely separated, and powerful nails. The head is flattened above, the gape wide, and the eyes large, deep sunk, and with the orbits greatly projecting above. The colours are also similar in the different

---

* The word *nature* being taken in so many different acceptations, we think proper to state, that, with Ranzani, we mean by it " the aggregate of all created beings, and of the laws imposed on them by the Supreme Creator."

species, being white, or pale (bluish white, &c.), with more or less of black. The comparatively even tail of the two allied species of which we are treating eminently distinguishes them from the others of the subgenus, which have the tail exceedingly forked. They are remarkable also for another characteristic, that of having the nails rounded beneath, and not canaliculate, a circumstance that occurs besides only in the subgenus *Pandion.** This character, which we formerly attributed to all the *Elani*, and which we believe we first observed not to exist in the fork-tailed species, has induced Mr Vigors, the English ornithologist, to separate the latter as a new genus, under the name of *Nauclerus*.

The female white-tailed hawk is sixteen and a half inches long, and three feet five and a half inches in extent. The bill is black, and measures from the corners of the mouth one inch and a half; the sides of the mouth, posterior portion of the lower mandible, and cere, bright yellow orange; bristles on the cere white, as well as those first on the lores; those nearest the eye, black; irides, brownish red; eyelids, white; cilia, long and black; orbits, black, wider before the eye; front line over the orbits, sides of the head, neck, and body, and whole inferior surface of the bird, together with the thighs, pure white; head, pearl grey, becoming gradually darker from the pure white front towards the neck and back, which are entirely bluish ash, as well as the rump, scapulars, secondaries, and greater wing-coverts; smaller and middle wing-coverts, deep glossy black; spurious wing, blackish; lining of the wing and inferior coverts, pure white, the latter with a wide black patch; primaries on both surfaces, slate colour, the shafts black, and, the first excepted, margined exteriorly, and slightly at tip with dusky, and interiorly with whitish; the margin of the inner web is of a remarkably close texture, with a very soft surface; the first primary is a little shorter than the third, the second longest; the two outer

* In *Pandion*, however, it is the middle nail that is rounded; in this species, it is the lateral and posterior only.

ones are slightly serrated on their outer web. When closed, the wings reach within less than an inch of the tip of the tail. The tail is seven inches long, slightly emarginated, and with the outer feather more than half an inch shorter than the adjoining one ; the middle feathers are very pale bluish slate, all the others pure white ; shafts above, black towards the tip, and beneath white ; that of the exterior tail-feather white, tipt with dusky above towards the base ; feet, bright yellow orange ; tarsus, one inch and a half long, feathered in front half its length, the remainder covered with small reticulated scales ; toes separated to the base ; nails, large, black, very acute, and, with the exception of the middle one, perfectly rounded beneath ; the middle one is very sharp on the inner side.

The male is of a smaller size ; the upper surface, instead of being bluish slate, is more of a dirty greyish, slightly tinged with ferruginous ; the tail is less purely white. These sexual differences are the more worthy of note, as they are the reverse of what is exhibited in other hawks. It is, however, possible that they are not to be found in very old males.

The young of both sexes, but especially the young males, are somewhat darker, and are strongly tinged with ferruginous, principally on the head, neck, and wings ; the breast being entirely of that colour. A specimen of the African species in this state is figured by Le Vaillant, whose plates in general are tolerably accurate ; but how great is the disappointment of the ornithologist to find the tarsi represented as covered distinctly with plates, as in other hawks ! We cannot let pass this opportunity of exhorting engravers, draughtsmen, and all artists employed on works of natural history, never to depend on what they are accustomed to see, but in all cases to copy faithfully what they have under their eyes ; otherwise, taking for granted what they ought not, they will inevitably fall into these gross errors. Even the accurate Wilson himself, or rather perhaps his engraver, has committed the same error in representing the feet of the swallow-tailed hawk. Of what

consequence, it will perhaps be said, is the form of the scales
covering the foot of a hawk? But these afford precisely one
of the best representative characters of groups, and it will,
therefore, not be thought unnecessary to caution artists in this
and similar cases.

The young, as described by Temminck, is in a more advanced stage of plumage; the front, forepart of the neck,
thighs, flanks, and under tail-coverts, are of pure white; the
breast and belly are of the same colour, but are marked with
reddish spots and brown lines; the occiput, nucha, back, and
scapulars are brownish, mixed with whitish, and more or less
tinged with cinereous; all these feathers having wide margins
of whitish and reddish; the upper tail-coverts are black, with
reddish margins; the inferior, marbled with black and white;
the quills are bluish, terminated with white; the tail is of a
greyish white, with black shafts; all the feathers have dark
cinereous towards the point, and are tipt with white.

This species is an inhabitant of a great portion of the American continent, as the *Alcon blanco* of Paraguay, so well described by D'Azara, is undoubtedly the same bird. Vieillot
undertook to classify it from D'Azara's description, applying
to it the name of *Milvus leucurus;* but, after more attentive
consideration, he perceived that it was not a *Milvus,* but an
Elanus. He consequently removed it to that genus which he
called *Elanoides,* at the same time asserting, that, with the
swallow-tailed hawk, it ought to constitute a different section
from the black-winged hawk; from which, upon actual comparison, it is with difficulty shown to be even specifically
distinct! Such are the absurdities into which authors are
betrayed, through the highly reprehensible practice, to which
some are addicted, of attempting to classify and name animals
they have never seen, from the descriptions or mere indications of travellers. Though, by such means, they may sometimes gain the credit of introducing a new species, and thus
deprive future observers, who may risk their fortunes, or even
their lives, in pursuit of imperfectly known animals, of their

best reward, they cannot fail to incur the merited reprobation of all honourable and fair-dealing naturalists.

Though this bird ranges so widely over the American continent, it is everywhere a rare species, and in the United States appears to be confined to the southern extremity. The specimen figured in the plate, of the natural size, was shot in December, in the neighbourhood of St Augustine, East Florida, at the residence of my near relation, Colonel Achilles Murat, whose kind hospitality afforded to Mr Titian Peale every facility for the prosecution of his scientific researches. It was observed by Mr Peale, about the dawn of day, sitting on the dead branch of an old live oak, attentively watching the borders of an adjacent salt marsh, which abounded with *Arvicola hispidus*, and the different species of sparrow which make their residence in the southern parts of the Union. It was very shy, and, on his approach, it flew in easy circles at a moderate elevation ; and such was its vigilance, that the greater part of a day was spent in attempting to get within gunshot. At length, the cover of interposing bushes enabled him to effect his purpose. It was a beautiful female, in perfect adult plumage. This sex, in the perfect state, is now for the first time represented, Temminck's plate representing the young female only ; and even the figures of the African analogue, in Le Vaillant's works, exhibit only the male in the young and adult states. As usual in the tribe of predaceous birds, the female is much larger than the male, and is therefore entitled to precedence.

Though this species is so rare, its near relative, the black-winged hawk, appears, on the contrary, to be very numerous. In Africa, where it was first discovered, and which is probably its native country, it is rather a common species, and has a very extensive range. Le Vaillant frequently observed it on the eastern coast of that little known continent, from Duyven-Hoek to Caffraria, where, however, it is less common. The same traveller found it to inhabit also in the interior, in the Cambdebo, and on the shores of the Swart-kop and Sunday rivers. It is very common in Congo, and numerous also in Barbary, Egypt, and far distant Syria. The researches of Ruppel in the interior of north-eastern Africa, already so productive, and from which so much more may be expected, have furnished specimens of this species, of which we owe two to the kindness of Dr Creitzschmaer, the learned and zealous director of the museum of the free city of Frankfort—an institution which has risen up with such wonderful rapidity. We are also informed that it is an inhabitant of India, which is rendered probable by a specimen from Java in my collection. It is found in New Holland, being numerous in the autumn of New South Wales, where it is migratory, and preys chiefly on field-mice, but is seldom known to attack birds. It is there observed at times to hover in the air, as if stationary and motionless. Though occasionally met with on the African coast of the Mediterranean, not a solitary individual has ever been known to visit the opposite shores of Italy, Spain, or Turkey, nor has it been met with in any other part of Europe.

When at rest, it is generally seen perched on high bushes, where the pure white of the lower parts of its body renders it very conspicuous at a distance. It utters a sharp piercing cry, which is often repeated, especially when on the wing, though Mr Peale assures us that our individual uttered no cry. Like its closely related species, it does not attack small birds, except for the purpose of driving them from its favourite food, which consists of hemipterous insects, chiefly of the *Gryllus* and *Mantis* genera, as well as other insects, and some reptiles. In the stomach of our specimen, however, Mr Peale found, besides the usual food, fragments of an *Arvicola hispidus,* and one or two feathers, apparently of a sparrow : but it is not a cowardly bird, as might be suspected from its affinity to the kites, and from its insignificant prey, since it successfully attacks crows, shrikes, and even the more timid birds of its own genus, compelling them to quit its favourite haunts, which it guards with a vigilant eye. They build in the bifurcation of trees. The nest is broad and shallow, lined internally with moss and feathers. The female is stated to lay four or five eggs ; the nestlings at first are covered with down, of a reddish-grey colour.

The African species is said to diffuse a musky odour, which is retained even after the skin is prepared for the museum ; but we are inclined to believe that it is in the latter state only that it possesses this quality. Mr Peale did not observe any such odour in the bird he shot, but being obliged, for want of better food, to make his dinner of it in the woods, found it not unpalatable.

FEMALE CŒRULEAN WARBLER. *(Sylvia azurea.)*

PLATE XI.—Fig. 2.

Wilson's American Ornithology, Cœrulean Warbler, Sylvia cœrulea, vol. ii. 141, pl. 17, fig. 5, for the male.—Sylvia azurea, *Stephens, Cont. Shaw's Zool.* x. p. 653.—*Nob. Obs. Jour. Ac. Nat. Sc. Ph.* iv. p. 193, male.—Sylvia bifasciata, *Say, in Long's Expedition to the Rocky Mountains,* i. p. 170, male.—*Philadelphia Museum,* No. 7309, male ; 7310, female.

SYLVICOLA CŒRULEA.—Swainson.

Male, vol. i. p. 283.

The merit of having discovered this bird is entirely due to the Peale family, whose exertions have contributed so largely to extend the limits of natural history. The male, which he has accurately described and figured, was made known to Wilson by the late venerable Charles Wilson Peale, who alone, and unaided, accomplished an enterprise, in the formation of the Philadelphia Museum, that could hardly have been exceeded under the fostering hand of the most powerful Government. To the no less zealous researches of Mr Titian Peale the discovery of the female is recently owing, who, moreover, evinced his sagacity by determining its affinities, and pointing out its true place in the system. Although it preserves the principal characters of the male, yet the difference is sufficiently marked to deserve an especial notice in this work.

The specimen here represented was procured on the banks of the Schuylkill, near Mantua village, on the 1st of August 1825. It was very active, skipping about on the branches of an oak, attentively searching the leaves, and crevices of the bark, and at intervals taking its food on the wing, in the manner of the flycatchers. It warbled in an undertone, not very unlike that of the blue-grey flycatcher of Wilson (*Sylvia cœrulea*, L.), a circumstance that would lead to the supposition of its being a male in summer dress ; but on dissection it proved to be a female.

The female azure warbler is four and three-quarter inches long, and eight and a quarter in extent.* Bill, blackish above, pale bluish beneath ; feet, light blue ; irides, very dark brown ; head and neck above, and back, rich silky green, brighter on the head, and passing gradually into dull bluish on the rump ; line from the bill over the eye, whitish, above which is the indication of a blue-black line widening behind ; a dusky streak passes through the eye ; cheeks, dusky greenish ; beneath, entirely whitish, strongly tinged with yellow on the chin ; sides of the neck, breast, flanks, and vent, streaked with dark bluish ; the base of the whole plumage is bluish white ; inferior tail-coverts, pure white ; wings and tail, very similar to those of the male, though much less brilliant ; smaller wing-coverts, bluish, tipt with green ; middling and large wing-coverts, blackish, widely tipt with white, constituting two very apparent bands across the wings, the white slightly tinged with yellowish at tip ; spurious wing, blackish ; quill-feathers, blackish, edged externally with green, internally and at tip with whitish, the three nearest the body more widely so ; the inferior wing-coverts, white ; tail, hardly rounded, feathers, dusky slate, slightly tinged with bluish externally, and lined with pure white internally, each with a white spot towards the tip on the inner web. This spot is larger on the outer feathers,

---

* The dimensions given by Wilson of the male must be rather below the standard, as they are inferior to those of the female ; whereas all the specimens we examined were larger, as usual.

and decreases gradually until it becomes inconspicuous on the two middle ones.

The description of the male need not here be repeated, having been given already with sufficient accuracy by Wilson, to whose work the reader is referred. On a comparison of the description and figures, he will find that the chief difference between the sexes consists in the female being green instead of blue, in her wanting the black streaks, and in being tinged with yellow beneath.

We have to regret our inability to add much to Wilson's short and imperfect account of the species. It is by no means more common at this time than it was when he wrote, which may account for the difficulty of ascertaining the period of its migrations, and for the circumstance of our having never met with the nest, and our want of acquaintance with its habits. We can only add to its history, that it is found in the trans-Mississippian territory, for the *Sylvia bifasciata* of Say, accurately described in Long's first expedition, is no other than the male. We have examined the specimen shot at Engineer Cantonment.

Although the undisputed merit of first making known this species belongs to Wilson, yet the scientific name that he applied to it cannot be retained, inasmuch as it is preoccupied by the blue-grey warbler, a Linnean species, which Wilson placed in *Muscicapa*, but which we consider a *Sylvia*, notwithstanding that it does in some degree aberrate from the typical species of that genus.* Under such circumstances, we cannot hesitate in adopting the name substituted by Mr Stephens, the continuator of Shaw's compilation.

* See my "Observations on the Nomenclature of Wilson's Ornithology."

Drawn from Nature by A. Rider.

Engraved by W.H.Lizars.

Blue Hawk or Hen Harrier.

Falco Cyaneus.

12.

BLUE HAWK, OR HEN-HARRIER. (*Falco cyaneus.*)

PLATE XII.

Wilson's American Ornithology, vol. vi. p. 67, pl. 51, fig. 1, for the young (under the name of Marsh Hawk, Falco uliginosus).—Falco cyaneus, *Linn. Syst.* i. p. 126, sp. 10.—*Gmel. Syst.* i. p. 276, sp. 10.—*Iter Poseg.* p. 27, adolescent male.— *Lath. Ind. Orn.* i. p. 39, sp. 94.—*Montagu, in Trans. Lin. Soc.* ix. p. 182.— *Meyer, Tasch. Deutschl. Vog.* i. p. 145.—*Temm. Man. Orn.* i. p. 72.—*Ranz. El. Zool.* iii. pl. 7, p. 137, sp. 28.—*Brehm, Lehrb. Eur. Vog.* i. p. 59.—*Selby, Ill. Br. Orn.* i. p. 26, pl. 10, fig. 1, male ; fig. 2, female.—*Savi. Orn. Tosc.* i. p. 63.—*Nob. Cat. and Syn. Birds U. S.* sp. 22.—Falco pygargus, *Linn. Syst.* i. p. 126, sp. 11.—*Gmel. Syst.* i. p. 277, sp. 11, female and young.— Falco Hudsonius, *Linn. Syst.* i. p. 128, sp. 19.—*Gmel. Syst.* i. p. 277, sp. 19, young, American.—Falco Bohemicus, *Gmel. Syst.* i. p. 276, sp. 107.—*Lath. Ind.* p. 38, sp. 93, adult male.—Falco albicans, *Gmel. Syst.* i. p. 276, sp. 102.— *Lath. Ind.* p. 38, sp. 93, adult male.—Falco griseus, *Gmel. Syst.* i. p. 275, sp. 100.—*Lath. Ind.* p. 37, sp. 86.—*Gerard, Tabl. Elem.* p. 37, adolescent male.—Falco montanus, var. B, *Gmel. Syst.* i. p. 278, sp. 106.—*Lath. Ind.* p. 48, sp. 116.—Falco cinereus, *It. Poseg.* p. 27, adolescent male.—Falco albicollis, *Lath. Ind.* p. 36, sp. 81, adult South American male.—Falco Buffonii, *Gmel. Syst.* i. p. 277, sp. 103, female and young, American.—Falco uliginosus, *Gmel. Syst.* i. p. 278.—*Lath. Ind.* p. 40, sp. 95.—*Sabine, Zool. App. to Frankl. Exp.* p. 671, young American.—Falco rubiginosus, *It. Poseg.* p. 29.— *Lath. Ind.* p. 27, sp. 56, young.—Falco ranivorus, *Daud. Orn.* ii. p. 170.— *Lath. Ind. Suppl.* p. 7, young.—Falco europogistus, *Daud. Orn.* ii. p. 110, adolescent male.—Circus europogistus, *Vieill. Ois. Am. Sept.* i. p. 36, pl. 8, adolescent male.—Circus Hudsonius, *Vieill. l. c.* i. p. 36, pl. 9, young.— Circus uliginosus, *Vieill. l. c.* i. p. 37, female and young. —Circus variegatus, *Vieill. l. c.* i. p. 37, male changing.—Circus gallinarius, *Vieill. Nouv. Dict. d'Hist. Nat.* iv. p. 459.—Circus cyaneus, *Id.* xxxi. p. 410.—Circus cyaneus, *Boie.*—Circus ranivorus, *Vieill. Nouv. Dict. d'Hist. Nat.* iv. p. 456, young African.—Falco strigiceps, *Nills. Orn. Suec.* i. p. 21.—Falco torquatus, *Briss. Orn.* i. 345, sp. 7 ; *Id.* 8vo, p. 100, male and female ; *Brunn.* sp. 14.— Falco montanus cinereus, *Briss. Orn.* i. p. 355, sp. 9, var. A ; *Id.* 8vo, p. 112, adolescent male.—Accipiter freti Hudsonis, *Briss. Orn.* vi. *App.* p. 18, sp. 47.—Lanarius cinereus, *Briss. Orn.* i. p. 365, sp. 17 ; *Id.* 8vo, p. 106.— Lanarius albicans, *Briss.* i. p. 367, sp. 18.—Subbuteo, *Gesner, Av.* p. 48.— Pygargus accipiter, *Ray, Syn.* p. 17, sp. 5.— *Will. Orn.* p. 40, pl. 7.—Falco plumbeus, cauda tesselata, *Klein, Av.* p. 52, sp. 22.—Lanarius, *Aldr. Orn.* i. pl. 381, 382, adult male.—Lanarius cinereus, sive Falco cinereo-albus, *Frisch,* pl. 79, 80, adult male.—Falco montanus secundus, *Aldr.—Will.* pl. 9, adult male.—Albanella, *Storia degli Ucc.* i. pl. 35, adult male.—Falco pygargo, *Id.* i. pl. 31, female.—Autre Oiseau St Martin, *Belon, Hist. Ois.* p. 104.—L'Oiseau St Martin, *Buff. Ois.* i. p. 212 ; *Id. Pl. Enl.* 459, adult male.—*Gerardin, Tabl. Elem. Orn.* i. p. 43.—La Soubuse, *Buff. Ois.* i. p. 215, pl. 9 ; *Id. Pl. Enl.* 443, young female ; 480, young male.—*Gerardin, Tabl. Elem. Orn.* i. p. 37, female and young.—Le Grenouillard, *Le Vaill. Ois. Afrique,* i. p. 63, pl. 23, young.—Kore oder Halbweyhe, *Bechst. Tasch.*

Deutsch. p. 25, sp. 20.—*Meyer and Wolf, Ois. d'Allem.* liv. 27, pl. 5, adult
male ; pl. 6, female.—*Naumann, Vog. Deutsch.* ed. 2, i. pl. 39, fig. 1, adult
male ; fig. 2, adult female ; pl. 38, fig. 2, young male.—Mause Habicht,
Missilauche, *Meyer, Boehm. Abh.* vi. p. 313, adult male.—Blue Hawk, *Edw.*
v. p. 33, pl. 225, adult male.—Marsh Hawk, *Edw.* p. 173, pl. 291.—*Penn.*
Arct. Zool. sp. 105.—*Lath. Syn.* i. p. 90, sp. 75, var. A, female and young.—
Ash-coloured Mountain Falcon, *Lath. Syn.* i. p. 94, sp. 78, var. A, adolescent
male.—Hen-harrier, *Edw.* pl. 225, very old male.—*Will. (Angl.)* p. 172.—*Alb.*
ii. pl. 5.—*Hayes, Brit. Birds,* pl. 1.—*Lewin, Brit. Birds,* i. p. 18.—*Penn.*
Brit. Zool. i. sp. 58, p. 28.—*Lath. Syn.* i. p. 88, sp. 74 ; *Id. Suppl.* p. 22,
adult male.—Ring-tail Hawk, *Edw.* iii. pl. 107.—*Penn. Arct. Zool.* sp. 106,
female and young.—Ring-tail, *Will. (Ang.)* p. 72.—*Alb.* iii. pl. 3.—*Hayes,*
Brit. Birds, pl. 2.—*Lewin, Brit. Birds,* i. pl. 18, female ; *Id.* pl. 2, fig. 4,
the egg.—*Penn. Brit. Zool.* sp. 59.—*Lath. Syn.* i. p. 89, sp. 75 ; *Id. Suppl.*
p. 22, female and young.—White-rumped Bay Falcon, *Lath. Syn.* p. 54, sp.
34, var. B, young.—Hudson's Bay Ring-tail, *Lath. Syn.* i. p. 91, sp. 76,
young.—White Lanner, *Lath. Syn.* i. p. 87, sp. 73, adult male.—Grey
Falcon, *Penn. Brit. Zool.* i. sp. 49.—*Lewin, Brit. Birds,* i. pl. 15.—*Lath.*
Syn. i. p. 82, sp. 67, adolescent male.—New York Falcon, *Penn. Arct. Zool.*
ii. p. 209, adolescent male.—Ranivorous Falcon, *Lath. Syn. Suppl.* female
and young.—White-necked Falcon, *Lath. Syn. Suppl.* p. 30, sp. 101, adult
male, South American.—Cayenne Ring-tail, *Lath. Syn.* i. p. 91, sp. 76, var.
A, young,—Falco glaucus, the Sharp-winged Hawk, of a pale sky-blue colour,
the tip of the wings black, *Bartr. Trav.* p. 290, adult male.—Falco sub-
cœruleus, the Sharp-winged Hawk, of a dark or dusky blue colour, *Bartr.*
Trav. p. 290, adolescent male.—Falco ranivorus, the Marsh Hawk, *Bartr.*
Trav. p. 290, young.—*Philadelphia Museum.*—*My Collection.*

CIRCUS CYANEUS.—Bechstein.*

Falco (subgen. Circus) cyaneus, *Bonap. Synop.* p. 32.—Buteo (Circus) cyaneus ?
var. Americanus.—Hen-harrier, *North. Zool.* ii. p. 55.

As will be perceived upon a slight inspection of our long and
elaborate list of synonyms, this well-known species is found

---

* There still appears to be a difference of opinion regarding the
identity of the European *Circus cyaneus,* or hen-harrier, and the North
American *Falco uliginosus* of Wilson.

Wilson was of opinion that they were the same, but only judged
from descriptions, being unable to obtain specimens from Britain for
comparison. The Prince of Musignano thinks they are the same, and
repeats the asseveration in his latest correspondence, but still perceives
some differences of habit and in the changes of plumage that would
weigh far if something more decided could be established. Mr Swainson
and Dr Richardson describe it with a query, under the name of
C. Americanus, and give numerous measurements of specimens, which
neither agree with each other, or generally with those of Britain.
Though I cannot at present fix on characters, I strongly suspect that

in almost every part of the globe; and not only does it seem to have been considered everywhere distinct, but nearly every

North America will at least possess one species distinct from that of Europe, and that the real European one, from its close alliance, is yet confounded with it ;—there is no bird where I have found so much variation in the dimensions. I have always observed the American birds larger than those of this country, and the tarsi stronger and proportionally longer. In America, the species seems remarkably abundant, and certainly differs slightly in habit, &c. ; but it is well known that animals, as well as birds, will accommodate themselves to a difference of circumstances.

The group to which this species belongs is intimately connected on both sides by, as it were, intermediate species—on the one by *Astur*, on the other by *Buteo*. Those true to the type are, however, at once known, and may be named as that now under discussion, the *C. histrionicus*, *Montagui*, &c. It seems distributed over the world. The colours and changes of plumage in all are very similar ; both sexes are clothed with the same livery until after the second moult, and are so similar as hardly to be distinguished, except by the difference of size. In the males, the change is to shades of grey ; in the females, to a lighter tinge of the reddish or darker browns, which appear to be the prevalent colours of the whole. The feathers of the rump assume a different colour, generally pure white, and show a prominent bar or band during flight. Their form is long and slender, and of no great power ; the wings are of considerable capacity, exhibiting the form adapted for a buoyant rather than rapid flight ; the tail ample. When sitting on the ground or on a rock, for they very seldom perch on trees, the attitude is very erect, like that of the sparrow-hawks ; but the most remarkable feature is the owl-like disk which surrounds the face, and is, in fact, nearly similar to that in the long-tailed hawk-owls.

The habits of those in Great Britain differ considerably according to the district they inhabit. In a country possessing a considerable proportion of plain and mountain, where I have had the greatest opportunities of attending to them, they always retire, at the commencement of the breeding season, to the wildest hills, and during this time not one individual will be found in the low country. For several days previous to commencing their nest, the male and female are seen soaring about, as if in search of, or examining, a proper situation, are very noisy, and toy and cuff each other in the air. When the place is fixed, and the nest completed, the female is left alone, and, when hatching, will not suffer the male to visit the nest, but on his approach rises and drives him with screams to a distance ! The nest is made very frequently in a heath-bush by the edge of some ravine, and is composed of sticks,

different appearance which it assumes during its progress
through the various and extraordinary changes that its plumage

with a very slender lining. It is sometimes also formed on one of those
places called *scars*, or where there has been a rush on the side of a steep
hill after a mountain thunder-shower ; here little or no nest is made,
and the eggs are merely laid on the bare earth, which has been scraped
hollow. In a flat or level country, some common is generally chosen,
and the nest is found in a whin or other scrubby bush, sometimes a
little way from the ground, as has been remarked in the descriptions of
the American birds. The young are well supplied with food, I believe,
by both parents, though I have only seen the female in attendance ;
and I have found in and near the nest the common small lizard, stone-
chats, and young grouse.

When the young are perfectly grown, they, with the old birds, leave
the high country, and return to their old haunts, hunting with regularity the fields of grain, and now commit great havoc among the young
game. At night they seem to have general roosting-places, either among
whins or long heath, and always on some open spot upon the ground.
On a moor of considerable extent, I have seen seven in the space of one
acre. They began to approach the sleeping ground about sunset, and
before going to the roost, hunted the whole moor, crossing each other,
often three or four in view at a time, gliding along in the same manner
as that described by Dr Richardson of the *C. Americanus*. Half an
hour may be spent in this way. When they approach the roost, they
skim three or four times over it, to see that there is no interruption,
and then at once drop into the spot. These places are easily found in
the day ; and the birds may be caught by placing a common rat-trap,
or they may be shot in a moonlight night. In both ways I have procured many specimens.

When kept in confinement, they generally roosted on the ground, in
a corner of the cage, three or four huddled together ; once or twice I
have found them perching ; during the day, they rested mostly on the
ground ; and only when alarmed rose to the cross-bars. I have never
seen them perch in a wild state.

Their flight is accurately described by our author ; and when hunting in this country, it is performed in the same manner, flying low
over the ground, beating the brushwood or rough cover, and along the
hedges. They never take their prey on the wing ; but when pursuing,
make a slight dash, and follow it to the place chosen for refuge. I
once shot an old female which had driven a covey of partridges into a
thick hedge, and was so intent upon watching her prey, that she allowed
me to approach openly from a distance of nearly half a mile. They
are often met with about the sea-coast ; and I have seen one repeatedly

undergoes according to sex and age, has in each country given rise to a nominal species. At the same time, however, that names were thus inconsiderately multiplied for one bird, two really distinct were always confounded together. Analogous in their changes, similar in form and plumage, it was reserved for the acute and ingenious Montagu to point out the difference, and establish the two species by permanent characters. The new one was called by him *Falco cineraceus,* and is known by the English name of ash-coloured harrier. It is figured and accurately described in all its states of plumage by Vieillot in his "Galerie des Oiseaux," where he has dedicated it to its discoverer, calling it *Circus Montagui ;* thus fully apologising for having in his article *Buzzard* of the "New Dictionary of Natural History," declared it to be a state of the other. How far, however, it may be considered a compliment to change the name given to a species by its discoverer, in order to apply even his own to it, we are at a loss to imagine.

The principal distinctive characters of the two species are to be found in the relative length of the wings and tail, and in the proportional lengths of the primaries. In the ash-coloured harrier, the sixth primary is shorter than the first, the second is much longer than the fifth, and the third is the longest ; the wings, when closed, reach to the tip of the tail. In the hen-harrier, the first primary is shorter than the sixth, the second subequal to the fifth, and the third equal to the fourth, the longest ; the wings closed, not reaching, by more than two inches, to the tip of the tail, which is also but slightly rounded in the latter, while in the ash-coloured it is cuneiform. Other

come to the stake-nets on the Solway Firth, and eat the dead fish that were left there. In hunting, they pursue a regular beat or tract for many days together. I have repeatedly watched a bird for miles, day after day, follow nearly the same line, only diverging on the appearance of prey ; and so nearly at one time do they pass the different ranges of their course, that I have placed myself in cover about the time they were expected, often with success : if they returned at all, there was never more than a quarter of an hour of variance from their usual time.—ED.

minor differences are besides observable in the respective sexes and states of both; but as those we have indicated are the only ones that permanently exist, and may be found at all times, we shall not dwell on the others, especially as Montagu's species appears not to inhabit America. We think proper to observe, however, that the adult male of *Falco cineraceus* has the primaries wholly black beneath, while that of the *F. cyaneus* has them black only from the middle to the point; and that the tail-feathers, pure white in the latter, are in the former spotted beneath. The female in our species is larger than the corresponding sex of the other, though the males in both are nearly of equal size; and the collar that surrounds the face is strongly marked in ours, whereas it is but little apparent in the other. The *F. cineraceus* has two white spots near the eyes, which are not in the *F. cyaneus*. The young of the former is beneath rusty, without spots. Thus slight but constant differences are seen to represent a species, while the most striking discrepancies in colour, size, and (not in this, but in other instances) even of form, prove mere variations of sex or age! We cannot wonder at the two real species having always been confounded amidst the chaotic indications of the present.

Even Wilson was not free from the error which had prevailed for so long a period in scientific Europe, that the ring-tail and hen-harrier were two species. Though he did not publish a figure of the present in the adult plumage of the male, he was well acquainted with it as an inhabitant of the southern States; for there can be no doubt that it is the much desired blue hawk which he was so anxious to procure—the only land bird he intended to add to his "Ornithology," or at least the only one he left registered in his posthumous list. It was chiefly because he was not aware of this fact, and thought that no blue hawk existed in America corresponding to the European hen-harrier, that Mr Sabine, in the Appendix to Franklin's "Expedition," above quoted, persisted in declaring that the marsh-hawk was a distinct species peculiar to

America, of which he supposed the Hudson's Bay ring-tail to be the young. The differences which he detected on comparing it with the European ring-tail must have been owing to a different state of plumage of his specimen of this ultra-changeable species. If, however, he had not mentioned the colours merely, as bringing it nearer to the ash-coloured falcon of Montagu, we might be inclined to believe that the specimen he examined was indeed a young bird of that species, which, though as yet unobserved, may, after all, possibly be found in North America. At all events, Wilson's, and the numerous American specimens that have passed under our examination, were all young hen-harriers.

After having stated that the error of considering the hen-harrier and ring-tail as different species had prevailed for years in Europe, it is but just to mention, that Aldrovandi, Brisson, Ray, and others of the older authors, were perfectly in accordance with Nature on this point. It was perhaps with Linné, or at least with Buffon, Gmelin, Pennant, and Latham himself, who afterwards corrected it, that the error originated. Latham, confident of his own observations and those of Pennant, who had found males of the species said to be the female of the *Falco cyaneus* (hen-harrier), and not reflecting that these males might be the young, exclaims, "Authors have never blundered more than in making this bird (the ring-tail) the same species with the last mentioned (hen-harrier) ;" an opinion that he was afterwards obliged to recant. In physical science we cannot be too cautious in rejecting facts, nor too careful in distinguishing, in an author's statement, what has passed under his own eyes, however extraordinary it may seem, from the inference he draws from it. Thus, to apply the principle in this instance, Latham might have reconciled the fact of males and females being found in the plumage of the ring-tail, with the others, that no females were ever found under the dress of the hen-harrier, and that some ring-tails would gradually change into hen-harriers.

Whether or not the marsh-hawk of America was the same

with the ring-tail of Europe, Wilson would not take upon himself to pronounce, as he has left to his bird the distinctive name of *Falco uliginosus ;* though he positively states that, in his opinion, they are but one species, and even rejects as false, and not existing, the only character on which the specific distinction was based, that of the American having " strong, thick, and short legs," instead of having them long and slender. For want of opportunity, however, of actually comparing specimens from both continents, he could choose no other course than the one he has followed ; and so great appears to have been the deference of ornithologists for this extraordinary man, that, while they have unhesitatingly quoted as synonymous with the European hen-harrier the African specimens described by Le Vaillant, and even the various nominal species created or adopted by Vieillot as North American, the *Falco uliginosus* of former authors has been respected, probably, as the marsh-hawk of Wilson !   But the latter is not, more than the others, entitled to be admitted as distinct, being merely the present in its youthful dress.

The hen-harrier belongs to the subgenus *Circus,* which in English we shall call harrier, the name of buzzard being appropriated to the *Buteones.*   Though perfectly well marked in the typical species, such as this, the group to which our bird belongs passes insensibly into others, but especially into that called *Buteo,* some even of the North American species being intermediate between them.   Whenever the groups of falcons shall be elevated to the rank of genera, it will perhaps be found expedient to unite *Circus* and *Buteo,* as they do not differ much more from each other than our two sections of hawks—those with long and slender legs, and those with short stout legs, *Astur* and *Sparvius* of authors, the line of demarcation being quite as difficult to be drawn.

The harriers are distinguished in their tribe by their weak, much compressed bill, destitute of a tooth or sharp process, but with a strongly marked lobe ; their short and bristly cere; their long, slender, and scutellated tarsi ; their slender toes, of which the outer are connected at base by a membrane ; their nails, subequal, weak, channelled beneath, much incurved, and extremely sharp. A very remarkable characteristic is exhibited in their long wings, subequal to the tail, which is large, and even, or slightly rounded at tip: their first quill is very short, always shorter than the fifth, and the third or fourth is the longest. Their slender body and elegant shape chiefly distinguish them from their allies the buzzards. They may be further subdivided into those in which the female at least is possessed of that curious facial ring of scaly or stiff feathers, so remarkable in the owls, and those entirely destitute of it. One species only is found in the United States which belongs to the first section, and cannot be confounded with any other than that from which we have thought proper to distinguish it at the beginning of this article. In this section, the female differs essentially from the male, the young being similar to her in colour. The latter change wonderfully as they advance in age, to which circumstance is owing the wanton multiplication that has been made of the species. In those which compose the second section, the changes are most extraordinary, since, while the adult male is of a very uniform light colour, approaching to white, the female and young are very dark, and much spotted and banded: they are also much more conspicuously distinguished by the rigid facial ring.

These birds are bold, and somewhat distinguished for their agility, especially when compared with the buzzards ; and in gracefulness of flight they are hardly inferior to the true falcons. They do not chase well on the wing, and fly usually at no great height, making frequent circuitous sweeps, rarely flapping their wings, and strike their prey upon the ground. Their food consists of mice, and the young of other quadrupeds, reptiles, fishes, young birds, especially of those that build on the ground, or even adult water-birds, seizing them by surprise, and do not disdain insects ; for which habits they are ranked among the ignoble birds of prey. Unlike most other large birds of their family, they quarter their victims previously to swallowing them, an operation which they always perform on the ground.   Morasses and level districts are their favourite haunts, being generally observed sailing low along the surface, or, in the neighbourhood of waters, migrating when they are frozen.   They build in marshy places, among high grass, bushes, or in the low forks or branches of trees; the female laying four or five round eggs, entirely white or whitish, without spots.   During the nuptial season, the males are observed to soar to a considerable height, and remain suspended in the air for a length of time.

The male hen-harrier is eighteen inches long, and forty-one in extent ; the bill is blackish horn colour ; the cere greenish yellow, almost hidden by the bristles projecting from the base of the bill ; the irides are yellow.   The head, neck, upper part of the breast, back, scapulars, upper wing-coverts, and middle tail-feathers, pale bluish grey, somewhat darker on the scapulars; the upper coverts, being pure white, constitute what is called a white rump, though that part is of the colour of the back, but a shade lighter ; breast, belly, flanks, thighs, under wing-coverts, and under tail-coverts, pure white, without any spot or streak.   The wings measure nearly fourteen inches, and, when closed, reach only two-thirds the length of the tail, which is eight and a half inches long, extending by more than two inches beyond them ; the primaries, of which the first is shorter than the sixth, the second and fifth sub-equal, and the third and fourth longest, are blackish, paler on the edges, and white at their origin, which is more conspicuous on their inferior surface ; the secondaries have more of the white, being chiefly bluish grey on the outer web only, and at the point, which is considerably darker.   The tail is but very slightly rounded.   All the tail-feathers have white shafts, and are pure white beneath; the middle ones are bluish grey, the lateral almost purely white; somewhat greyish on the outer vane, and obsoletely barred with blackish grey on the inner. The feet are bright yellow, and the claws black; the tarsus is three inches long, and feathered in front for an inch.

The female is larger, being between twenty and twenty-one inches long, and between forty-four and forty-seven in extent ; the tarsi, wings, and tail, proportionally longer, but strictly corresponding with those of the male. The general colour above is chocolate-brown, more or less varied with yellowish rufous ; the space round the orbits is whitish, and the auriculars are brown ; the small stiff feathers forming the well-marked collar or ruff are whitish rusty, blackish brown along the shaft ; the feathers of the head and neck are of a darker brown, conspicuously margined with yellowish rusty ; on the nucha, for a large space, the plumage is white at the base, as well as on the sides of the feathers, so that a little of that colour appears, even without separating them ; those of the back and rump are hardly, if at all, skirted with yellowish rusty, but the scapulars and wing-coverts have each four regular large round spots of that colour, of which those farthest from the base lie generally uncovered ; the upper tail-coverts are pure white, often, but not always, with a few rusty spots, constituting the so-called white rump, which is a constant mark of the species in all its states of plumage. The throat, breast, belly, vent, and femorals, pale yellowish rusty, streaked lengthwise with large acuminate brown spots, darker and larger on the breast, and especially the under wing-coverts, obsolete on the lower parts of the body, which are not spotted. The quills are dark brown, whitish on the inner vane, and transversely banded with blackish ; the bands are much more conspicuous on the inferior surface, where the ground colour is greyish white. The tail is of a bright yellowish rusty, the two middle tail-feathers dark cinereous ; all are pure white at the origin, and regularly crossed with four or five broad blackish bands ; their tips are more whitish, and the inferior surface of a greyish white, like that of the quills, but very slightly tinged with rusty, the blackish bands appearing to great advantage, except on the outer feathers, where they are obsolete, being less defined even above.

The young male is almost perfectly similar in appearance to

the adult female (which is not the case in the ash-coloured harrier), being, however, more varied with rusty, and easily distinguished by its smaller size. It is in this state that Wilson has taken the species, his very accurate description being that of a young female. The male retains this plumage until he is two years old, after which he gradually assumes the grey plumage peculiar to the adult: of course they exhibit almost as many gradations as specimens, according to their more or less advanced age. The ash and white appear varied or mingled with rusty; the wings, and especially the tail, exhibiting more or less indications of the bands of the young plumage. The male, when he may be called already adult, varies by still exhibiting the remains of bands on the tail, more or less marked or obliterated by the yellowish edges of the feathers of the back and wings, and especially by retaining on the hind head a space tinged with rusty, with blackish spots. This space is more or less indicated in the greater part both of the American and European specimens I have examined. Finally, they are known by retaining traces of the yellowish of the inferior surface, in larger or smaller spots, chiefly on the belly, flanks, and under tail-coverts.

For the greater embellishment of the plate, we have chosen to represent one of these very nearly but not quite adult males, in preference to a perfectly mature bird, which may easily be figured to the mind by destroying every trace of spot or bar. It is, moreover, in this dress that the adult is met with in the middle and northern States, where it is very rare, and we have never seen a specimen quite mature, though the young are tolerably common; as if the parents sent their children on a tour to finish their education, then to return and marry, and remain contentedly at home. The specimen here figured was shot on Long Island, and was preserved in Scudder's Museum, New York.

Its total length is eighteen inches; breadth, forty-one; the bill, bluish black; cere, irides, and feet, yellow; claws, black. The plumage above is bluish ashy, much darker on the scapulars, and, with the feather-shafts, blackish; beneath, white, slightly cream-coloured on the breast; the belly, flanks, and lower tail-coverts with small arrow-shaped spots of yellowish rusty; the long axillary feathers are crossed with several such spots, taking the appearance of bands; the upper tail-coverts are pure white; the primaries, dusky blackish at the point, edged with paler, and somewhat hoary on the outer vane; at base, white internally and beneath. The tail is altogether of a paler ash than the body, tipped with whitish, and with a broad blackish subterminal band; all the tail-feathers are pure white at their origin under the coverts, the lateral being sub-banded with blackish and white on their inner vanes, and the outer on the greater part of the outer web also; the shafts are varied with black and white.

The hen-harrier's favourite haunts are rich and extensive plains and low grounds. Though preferring open and champaign countries, and seeming to have an antipathy to forests, which it always shuns, it does not, like the ash-coloured harrier, confine itself to marshes, but is also seen in dry countries, if level. We are informed by Wilson, that it is much esteemed by the Southern planters, for the services it renders in preventing the depredations of the rice-birds upon their crops. Cautious and vigilant, it is not only by the facial disk that this bird approaches the owls, but also by a habit of chasing in the morning and evening, at twilight, and occasionally at night, when the moon shines. Falconers reckon it among the ignoble hawks. Cruel, though cowardly, it searches everywhere for victims, but selects them only among weak and helpless objects. It preys on moles, mice, young birds, and is very destructive to game; and does not spare fishes, snakes, insects, or even worms. Its flight is always low, but notwithstanding, rapid, smooth, and buoyant. It is commonly observed sailing over marshes, or perched on trees near them, whence it pounces suddenly upon its prey. When it has thus struck at an object, if it reappears quickly from the grass or reeds, it is a proof that it has missed its aim; for, if otherwise, its prey is devoured on the spot.

It breeds in open wastes, frequently in thick furze coverts, among reeds, marshy bushes, the low branches of trees, but generally on the ground. The nest is built of sticks, reeds, straw, leaves, and similar materials heaped together, and is lined with feathers, hair, or other soft substances; it contains from three to six, but generally four or five, pale bluish-white eggs, large and round at each end; the young are born covered with white down, to which succeed small feathers of a rust colour, varied with brown and black. If any one approaches the nest during the period of rearing the young, the parents evince the greatest alarm, hovering around, and expressing their anxiety by repeating the syllables, *geg, geg, gag ;* or *ge, ge, ne, ge, ge.* Crows manifest a particular hostility to this species, and destroy numbers of their nests.

The hen-harrier is widely spread over both continents, perhaps more than any other land-bird, though it is nowhere remarkably numerous. In the northern countries of America it is a migratory species, extending its wanderings from Florida to Hudson's Bay. It is not known to breed in the northern, or even in the middle States, where the adults are but rarely seen. In the southern parts of the Union, and especially in Florida, they are rather common, in all their varieties of plumage. The species is also found in the West Indies, Cayenne, and probably has an extensive range in South America. It is found throughout Britain, Germany, Italy, the north of Africa, and the northern portion of Asia. It is very common in France and the Netherlands, is found in Russia and Sweden, but does not inhabit the north of Norway, being by no means an arctic bird. It is again met with in the southern parts of Africa, near the Cape of Good Hope, and is not uncommon all along the eastern coast of that continent. In Switzerland, and other mountainous countries, it is of very rare occurrence.

Drawn from Nature by A Ryder.   *Engraved by W.H.Lizars.*

1. Steller's Jay. 2. Lapland Longspur. 3. Female.
Garrulus Stelleri.   Emberiza Lapponica.

13.

STELLER'S JAY. (*Garrulus Stelleri.*)

PLATE XIII.—FIG. 1.

Corvus Stelleri, *Gmel. Syst.* i. p. 370, sp. 27.—*Lath. Ind.* p. 158, sp. 20.—*Nob. Suppl. Syn. Birds U. S.* sp. 63, bis, *in Zool. Journ. Lond.* v. p. 2 ; *Id. in App. Gen. N. A. Birds in Ann. Lyc. N. Y.* p. 438.—Garrulus coronatus ? Swainson, *Syn. Birds Mex.* sp. 67, *in Phil. Mag.* N. S. i. p. 437, old bird ? —Garrulus Stelleri, *Vieill. Nouv. Dict. d'Hist. Nat.* xii. p. 481.—Geai de Steller, *Daud. Orn.* ii. p. 248.—Steller's Crow, *Penn. Arct. Zool.* sp. 139.— *Lath. Syn.* i. p. 387, sp. 21 ; *Id. 2d Suppl.* viii. p. 111, sp. 8 ; *Id. Gen. Hist.* iii. p. 56, sp. 58.—Collection of Mr Leadbeater, in London.

GARRULUS STELLERI.—Vieillot.*

Garrulus Stellaris, *North. Zool.* ii. p. 294.—Pica Stelleri, *Wagl. Syst. Av.* Pica, No. 10.

To the enlightened liberality and zeal for science of that distinguished collector, Mr Leadbeater, of London, we, and the American public, are now indebted for the appearance of the first figure ever given of this handsome jay. Trusting his precious specimens twice to the mercy of the waves, he confided to us this, together with several other still more rare and valuable North American birds, which no consideration would have induced him to part with entirely, to have them drawn, engraved, and published, on this side of the Atlantic. It is the frequent exercise of similar disinterestedness in the promotion of scientific objects that has procured for Mr Leadbeater the distinction with which he is daily honoured by learned bodies and individuals.

The Steller's jay is one of those obsolete species alluded to in the preface to this volume. It is mentioned by Pallas as

---

* This species, though very similar, is distinct from the *Garrulus coronatus*, Swainson, which it is not impossible may yet be added as a straggler to the northern continent. We may here mention the splendid Columbia jay, the *Pica Bullockii* of Wagler, which Mr Audubon has figured. It is a native of Mexico and California, and a specimen was procured by Mr Audubon from the Columbia river. It may be considered only as a straggler, and very rare. This, and one or two others, the Indian *Garrulus erythrophynchus*, are remarkable for the length of their tail. The body of the American bird is not so large as that of the common jay, but the total length is thirty-one inches.—ED.

having been shot by Steller, when Behring's crew landed
upon the coast of America.   It was first described by Latham,
from a specimen in Sir Joseph Banks's collection, from Nootka
Sound, and on his authority has been admitted into all subsequent compilations.   The species is indeed too well characterised to be doubted, and appears, moreover, to have been
known to Temminck, as it is cited by him as a true jay in
his " Analysis of a General System."   Nevertheless, adhering
strictly to our plan of not admitting into the ornithology of the
United States any but such as we had personally examined,
we did not include this species either in our Catalogue or
Synopsis of the birds of this country ; and it is but recently
that Mr Leadbeater's specimen has enabled us to add it to
our list.

   In elevating our subgenus *Garrulus* to the rank of a genus,
we merely conform to the dictates of Nature, in this instance
coinciding with Temminck, whose intention it is, as he informs us, to include in it the jays and magpies, leaving the
name of *Corvus* for those species which are distinguished by
their black plumage and short and even tails.   These birds
are on every account well worthy of this distinction, and we
cheerfully adopt an arrangement which we deem consonant
with nature ; but we cannot agree to the change of termination (*Garrula*) which he has attempted to introduce, under
the pretence that his genus is more extensive than the genus
Garrulus of former authors.   That genus was, in fact, formed
by Brisson, and afterwards by Linné united with *Corvus*.
This latter genus of Linné certainly contained within itself
the constituents of several very natural genera ; but the additions made to it by Gmelin and Latham rendered it an utter
chaos, where every new species with a stout bill took its place,
in defiance of the genuine characters.   Under such circumstances the task of the ornithologist who professed to be guided
by philosophical principles was, doubtless, not merely to subdivide, but to make an entire reformation,   Illiger, with his
usual judgment, perceived the evil, and attempted its remedy ;

but his genus was still too extensive, and, besides, was not natural, as it included the wax-wings, a very distinct genus, that had always been forced into others. The only advantage it possessed over that of Latham was, that all the species it comprised exhibited its artificial characters. As restricted by Brisson, Vieillot, and lately adopted by Temminck, by whom it was previously much limited, it is perfectly natural ; though we cannot help remarking that some even of the eighteen species enumerated by the latter in his article on the generalities of the crows, in the *Planches Colorées*, may again be separated, such as *Corvus Columbianus*, Wils., which ought, perhaps, to constitute a genus by itself. Vieillot, and other recent writers on ornithology, have long since adopted the genus *Garrulus* as distinct even from *Pica*, though we prefer retaining the latter merely as a subgenus of *Garrulus*, since it is absolutely impossible to draw the line of separation between them without resorting to minute and complicated distinctions.

The jays and magpies, in fact, require to be distinguished from the crows, as a genus, on account of their form, colour, habits, and even their osseous structure. Their upper mandible, somewhat inflected at tip, and the navicular shape of the lower, afford obvious characteristic marks. Their wings, too, are rather short, and do not reach by a considerable space to the tip of the tail, which is long, and more or less rounded, sometimes greatly wedge-shaped. On the contrary, the crows have long wings, reaching almost or quite to the extremity of the tail, which is short, and even at tip. The identity in the shape of the wings and tail, and even the colours of their plumage, which agree in all the species and in different climates, render the crows a very natural and well-marked group. The black plumage and offensive odour, which cause them to be viewed everywhere with disgust, and even somewhat of superstitious dread, are far from being characteristics of the neat and elegant jays.

The true *Corvi* are distinguished by the following traits :—

Bill, very stout; feet, very strong; general form, robust; flight, highly sustained, straight or circular, as if performing evolutions in the air. They live, travel, and breed in large bands; affect wide plains and cultivated grounds, only retiring to the adjacent forests to roost, and are always seen on high and naked trees, but never on thickets, shrubs, or bushes. Their voice is deep and hoarse. They are more or less fond of cattle, some species preying on the vermin that infest them. Though devouring all kinds of food, yet their propensity is decidedly carnivorous. Their black, unvaried colours are remarkably opposed to the bright and cheerful vesture of the jays, whose plumage is of a much looser texture, the feathers being longer and much more downy.

The jays are again more particularly distinguished from the magpies by their head-feathers being long and silky, and always erectile (especially when the bird is excited or angry), even when they are not decidedly crested, as is the case in many species. Their colours are also gayer, and more brilliant, with more or less of blue. The species of both these sections are garrulous, noisy, and inquisitive. Together with the crows, they are eminently distinguished by their stout, cultrate bill, generally covered at base with setaceous, incumbent, porrect feathers, hiding the nostrils. The female is similar to the male in appearance, and the young differ but little, and only during the first year, from the adult. They are very shy, suspicious, possessed of an acute sense of smelling, and evince great sagacity in avoiding snares. They are omnivorous in the fullest extent of the word, feeding on grains, insects, berries, and even flesh and eggs. When they have caught a small bird, which they can only do when feeble and sickly, or ensnared, they place it under their feet, and with their bill tear it to pieces, swallowing each piece separately. Nevertheless, they give the preference to grains or fruits. The northern species are wary and provident, collecting stores of food for the winter. They are very petulant; their motions quick and abrupt, and their sensations lively. When alarmed,

by the appearance of a dog, fox, or other living or dead object,
they rally together by a peculiar note, as if they would impose
upon it by their numbers and disagreeable noise. When on
the ground, they display great activity; or if on trees, they
are continually leaping about from branch to branch, and
hardly ever alight on dead or naked ones. They are generally met with in forests, seldom in open plains; their favourite
resort is among the closest and thickest woods. Less suspicious and cunning than the crows, or even the magpies, they
may be decoyed into snares and taken in great numbers, especially by imitating the voice of one of their own species in
difficulties, or by forcing a captive individual to cry. They
live in families or by pairs the greater portion of the year;
and though considerable numbers may be seen travelling at
once, they always keep at intervals from each other, and never
in close flocks like the crows. They are easily tamed, and are
susceptible of attachment; learn readily to articulate words
and imitate the cries of different animals. They have a
troublesome propensity to purloin and conceal small objects not
useful to themselves; and as jewels and precious metals are
peculiarly apt to attract their notice, they have been the cause,
when kept as pets, of serious mischief. Every one is familiar
with the story of the thieving magpie, become so celebrated by
the music of Rossini, and which is founded on fact.

The jays breed in woods, forests, orchards, preferring old
and very shaded trees, placing their nests in the centre against
the body, or at the bifurcation of large limbs. The nest is
built without art, and is formed of twigs and roots, whose
capillary fibres serve as a lining inside; the eggs are from four
to six. The old ones keep the food for their young in the
œsophagus, whence they can bring it up when wanted. The
young are born naked, and remain for a long period in the
nest, being still fed for some time by the parents after they
are full fledged.

Unlike the melancholy crows, which step gravely, lifting
one foot after the other, the jays and magpies move about

nimbly by hopping, and are constantly in motion while on the ground. Their flight is, moreover, neither protracted nor elevated, but merely from tree to tree, and from branch to branch, shooting straight forward at once when wishing to go any distance, now and then flapping their wings, and hovering as they descend, when about to alight. It is quite the reverse with the crows ; and all these characters are of the greatest importance in the establishment of natural groups.

While the true *Corvi*, by their stout and almost hooked bill, and the carnivorous habits of some species, exhibit on the one hand the gradual passage from the vultures, and on the other, by the slender-billed species, the transition to the crow, blackbirds, and troopials ; the affinities of the jays present nice gradations to the genera already dismembered from *Corvus*, such as *Nucifraga, Pyrrhocorax, Bombycilla,* and at the same time form other links with *Lanius,* and even with *Turdus* and *Acridotheres.*

There is one remarkable analogy of the jays which we cannot pass over in silence. It is, however, singular, and hitherto unsuspected, with the titmouse (*Parus*). Form, habits, even the peculiar looseness of texture of the plumage, all are similar in these genera, hitherto estimated so widely different. This resemblance extends even to colour in some species : it might even be asked, what else, in fact, is the Canada jay than a large titmouse ; and what the crested titmouse but a small jay ? The blue colour of the typical jays predominates, moreover, in other *Pari* ; and the *P. caudatus* of Europe has also the long, cuneiform tail of some, no less than *P. bicolor* their crest.

The genus *Garrulus* has an extensive geographical range, being found in all latitudes and longitudes. It is composed of about thirty species, nearly half of which may more properly be called jays : of the latter, there are but two in Europe ; and though we have doubled the number given by Wilson, we think that others will yet be discovered in the wild western tracts of this continent. There exist imperfect accounts of two or three species inhabiting the countries near the Rocky Mountains, one of which is probably that here described, and others may prove to be some of the newly discovered Mexican species, one of which, the *Garrula gubernatrix* of Temminck, is so proudly beautiful.

The Steller's jay is more than twelve inches long. The bill measures one inch and a half, is entire, and totally black ; the bristly feathers over the nostrils are also wholly black. The feathers of the head are greatly elongated, forming a large crest, more than two and a half inches long, and, with the whole head and neck, entirely deep brownish black, greyish on the throat ; the feathers on each side of the front are slightly tipped with bright and light azure, thus forming a dozen or more of small dots on that part ; on the neck, the brown becomes lighter, and extends down on the back, occupying the scapulars as well as the inner wing-coverts ; on the middle of the back the brown becomes somewhat tinged with bluish, and blends gradually into a fine bright blue colour, covering the rump and the upper tail-coverts ; all the inferior parts from the neck, at the lower part of which the dusky colour passes into blue, are blue, somewhat tinged with grey, which is the general colour of the base of the plumage. The wings are nearly six inches in length ; the fourth, fifth, and sixth primaries being subequal and longest. All the outer wing-coverts and the secondaries are blue, faintly crossed with obsolete blackish lines ; the under wing-coverts are dusky ; the primaries are dark dusky, and, with the exception of the outer ones, at tip are edged or tinged with blue ; on the inner vane, the secondaries are blackish, but on the outer, they are deep glossy blue. The tail is five inches and a half long, and but slightly rounded ; it is of a deep glossy azure blue, more brilliant on the outer vanes of the feathers, the inner being slightly tinged with dusky ; an indication of obliterated transverse blackish lines may be perceived in certain lights on almost all the tail-feathers in our specimen, and we have no doubt that on others they are more marked ; the shafts both of the quills and tail-feathers are black. The tarsus is an inch and three-quarters long; the femorals, blackish, slightly mixed with bluish at the joint; the feet and nails are entirely black.

This description is taken from the individual represented in the plate, which was killed near the Oregon or Columbia river. Another specimen, from Mexico, also in Mr Lead-beater's collection, exhibited greater brilliancy of plumage, being principally distinguished, as nearly as our recollection serves, by the black colour of the anterior parts being less extended, and by having more of silvery bluish (indicated in our bird) on the front, extending to the throat and eyebrows, and somewhat round the head. This, without any hesitation, we considered as a more perfect specimen, a mere variety of age, and would have had our figure made from it; but having been informed that an English ornithologist (his name and that of the species were not mentioned, or, if they were, we have forgotten them) considered it as a new Mexican species, we have preferred, notwithstanding our conviction, strictly copying the less brilliant specimen procured in the United States territory to the more beautiful one from Mexico. The appearance of *Garrulus coronatus* of Mr Swainson in the Synopsis before quoted reminded us of the circumstance, and we have therefore quoted it with doubt. Our two birds agree perfectly in markings and dimensions. Of the habits of the Steller's jay, little or nothing is known. It inhabits the western territory of the United States beyond the Rocky Mountains, extending along the western coasts of North America, at least from California to Nootka Sound; is common on the Oregon, and found also in Mexico on the tableland, and in Central America.

It is a curious fact in ornithological geography, that of the four jays now admitted into the fauna of the United States, while the common blue jay, the only eastern representative of the genus, spreads widely throughout the continent, the three others should be confined in their range each to a particular section of country. Thus, the Canada jay is the northern, the Florida jay is the southern, and the present the western representative of the genus. It is probable that another species at least, our *Garrulus ultramarinus*, from Mexico, will soon be admitted as the central jay. To the latter bird, Mr Swainson, who had probably not seen my paper describing it (published more than two years ago in the *Journal of the Academy of Natural Sciences*), gives the name of *G. sordidus ;* at least, judging from his short phrase, and the dimensions and locality, they are the same.

LAPLAND LONGSPUR. (*Emberiza Lapponica.*)

PLATE XIII.—Fig. 1, Male ; Fig. 2, Female.

Fringilla Lapponica, *Linn. Syst.* i. p. 317, sp. 1.—*Faun. Suec.* sp. 235.—*Gmel. Syst.* i. p. 900, sp. 1.—*Retz. Faun. Suec.* p. 242, sp. 119.—*Forst. Ph. Tr.* lxii. p. 404.—*Fabr. Faun. Grœnl.* p. 119, sp. 8.—*Lath. Ind.* p. 440, sp. 18.—*Ubers.* i. p. 289, sp. 18.—Fringilla montana, *Briss. Orn.* iii. p. 160, sp. 38.—*Klein, Av.* p. 92, sp. 10.—Fringilla calcarata,'*Pallas, It.* p. 710, sp. 20, t. E ; *Id.* in 4to, French transl. iii. pl. 1.—*Meyer and Wolf, Tasch. Deutschl.* i. p. 176, sp. 13.—Emberiza Lapponica, *Nilsson, Orn. Suec.* i. p. 157, sp. 76.—*Ranz. El. Zool.* vi. p. 24.—Emberiza calcarata, *Temm. Man. Orn.* i. p. 322.—*Brehm, Lebhr. Eur. Vog.* i.'p. 221.—*Richardson, App. to Parry's 2d Voy.* p. 345.— Passerina Lapponica, *Vieill. Nouv. Dict. d'Hist. Nat.* xxv. p. 12.—Plectro- phanes calcaratus, *Meyer, Tasch.* iii. p. 176, sp. 13.—Plectrophanes Lapponica, *Selby, in Trans. Linn. Soc.* xv. p. 156, pl. 1, young.—Montifringilla congener, *Aldrov. Orn.* ii. p. 821, pl. 823.—Le Grand Montain, *Buff. Ois.* iv. p. 134.— Le Pinson de Montagne, *Gerardin, Tabl. Elem. d'Orn.* i. p. 186.—Lerchen Finck, *Bechst. Naturg. Deutsch.* iii. p. 246, sp. 16.—*Naum. Nachtr.* iii. p. 25, pl. 20, B., female ; plate 40, male in autumn.—Greater Brambling, *Alb.* iii. p. 59, pl. 63.—Lapland Finch, *Penn. Arct.* ii. sp. 259.—*Lath. Syn.* iv. p. 263, sp. 14.—*Ubers.* iii. p. 256, sp. 14.—*My Collection.*

PLECTROPHANES LAPPONICA.—Selby.*

Emberiza (subgen. Plectrophanes) Lapponica, *Bonap. Synop. App.* p. 440.— *North. Zool.* ii. p. 248.

This species, long since known to inhabit the desolate arctic regions of both continents, is now for the first time introduced into the fauna of the United States, having been omitted both in our Synopsis and Catalogue. It is entitled to be

---

* See vol. i. p. 325.

ranked among the birds of this country from the fact that a few stragglers out of the numerous bands which descend in winter to comparatively warm latitudes show themselves almost every year in the higher unsettled parts of Maine, Michigan, and the north-western territory.   Even larger flocks are known not unfrequently to enter the territory of the Union, where, contrary to what is generally supposed, they are observed to alight on trees, as well as on the ground, notwithstanding their long and straight hind nail.   We think it highly probable that some individuals, especially in their youth, visit in cold winters the mountainous districts of the middle States, as they are well known in Europe to wander or stray to the more temperate climates of Germany, France, England, and especially Switzerland; in all which countries, however, the old birds are never seen.   It is not extraordinary that they should never have been observed in the Atlantic States, as they are nowhere found in maritime countries.

No figure of the adult male in perfect plumage has before now, we believe, been given; and no representation at all is to be met with in the more generally accessible books or collections of plates.   Mr Selby has lately published a figure of the young in the *Linnean Transactions* and it will also, we presume, appear in his splendid work, which yields to none but Naumann's, Wolf's, and Wilson's, in point of accuracy and character.   That recorded by him appears to be the first instance of an individual having been found in Britain.   The species is common in the hilly districts of eastern Europe, but is chiefly confined within the polar circle, though found abundantly in all the northern mountainous districts of Europe and Asia, particularly Siberia and Lapland.   It is sometimes known to descend in autumn and winter, and, though very rarely, in spring, either singly and astray, or in immense clouds, into the north and middle of Germany. Great numbers were seen in the neighbourhood of Frankfort on the Main in the middle of November 1821.   In France,

they are restricted to the loftiest and most inaccessible mountains, where they are very rare ; so much so, that in those of
the Vosges, Gerardin only met with a single specimen after
six years' researches, though more frequent in the mountains
of Dauphiné. They are common during summer in arctic
America, and are found at Hudson's Bay in winter, not
appearing before November : near the Severn river they
haunt the cedar trees, upon whose berries they feed exclusively.
These birds live in large flocks, and are of so social a disposition, that when separated from their own species, or when
in small parties, they always join company with the common
lark of Europe, or in America, with some of the different
snow-birds. They feed chiefly on seeds, especially of the
dwarf willows growing in frozen and mountainous countries,
but occasionally also on leaves, grass, and insects. They
breed on small hillocks, in open marshy fields ; the nest is
loosely constructed with moss and grasses, lined with a few
feathers. The female lays five or six oblong eggs, yellowish
rusty, somewhat clouded with brown. The Lapland longspur,
like the larks, never sings but suspended aloft in the air,
at which time it utters a few agreeable and melodious notes.

As may be seen by the synonyms at the head of this article,
this bird has been condemned by nomenclators to fluctuate
between different genera. But between *Fringilla* and *Emberiza*
it is not difficult to decide, as it possesses all the characters
of the latter in an eminent degree, even more so than its near
relative the snow-bunting, which has never been misplaced.
It has even the palatine knob of *Emberiza*, and much more
distinctly marked than in the snow-bunting (*Emberiza nivalis*).
It has been erroneously placed in *Fringilla* merely on account
of its bill being somewhat wider and more conic.

Meyer has lately proposed, for the two just mentioned nearly
allied species, a new genus under the name *Plectrophanes*
(corresponding to the English name we have used). This we
have adopted as a subgenus, and are almost inclined to admit
as an independent genus, being well characterised both by form

and habits. The two species of *Plectrophanes* to which we
apply the name of longspur, together with the buntings, are
well distinguished from the finches by their upper mandible,
contracted and narrower than the lower, their palatine tubercle,
&c. From the typical *Emberizæ* they differ remarkably by the
length and straightness of their hind nail, and the form of
their wings, which, owing to the first and second primaries
being longest, are acute. In the true buntings, the first quill
is shorter than the second and third, which are longest. This
species, in all its changeable dresses, may at once be known by
its straight and very long hind nail, which is twice as long as
the toe. The bill is also stronger and longer than in the other
species.

The longspurs are strictly arctic birds, only descending in
the most severe and snowy winters to less rigorous climates,
and never to the temperate zone, except on the mountains.
Hence they may, with the greatest propriety, be called snow
birds. They frequent open countries, plains, and desert regions,
never inhabiting forests. They run swiftly, advancing by successive steps like the lark (which they resemble in habits,
as well as in the form of their hind nail), and not by hopping,
like the buntings. The conformation of their wings also
gives them superior powers of flight to their allied genera, the
buntings and finches. Their moult appears to be double, and,
notwithstanding Temminck's and my own statement to the
contrary, they differ much in their summer and winter plumage. Owing to this, the species have been thoughtlessly multiplied : there are, in reality, but two, the present, and snow-
bunting of Wilson.

The male Lapland longspur, in full breeding dress, is nearly
seven inches long, and twelve and a quarter in extent ; the bill
is nearly half an inch long, yellow, blackish at the point ; the
irides are hazel and the feet dusky ; the head is thickly furnished with feathers ; the forepart of the neck, throat, and the
breast are glossy black ; the hind head is of a fine reddish
rusty ; a white line arises from the base of the bill to the eye,

behind which it becomes wider, descending on the sides of the neck somewhat round the breast; the belly and vent are white; the flanks posteriorly with long blackish streaks; the back and scapulars are brownish black, the feathers being skirted with rusty; the smaller wing-coverts are blackish, margined with white; the greater coverts margined with rufous, and white at tip, forming two white bands across the wings; the primaries are blackish, edged with white; secondaries emarginated at tip, dusky, edged with rusty; the wings, when closed, reach to three-fourths the tail; the tail is two and a half inches in length, rather forked, and of a blackish colour; the outer feather on each side with a white cuneiform spot, and the outer web almost entirely white; the second with a white cuneiform spot only. The hind nail is almost an inch long.

The adult female is somewhat smaller than the male. In spring, she has the top of the head, the shoulders, back, and wing-coverts, brownish black, the feathers being edged with rusty; the sides of the head, blackish, intermixed with rusty; over the eyebrows a whitish line, as in the male, tinged with rusty; the nucha and rump are brownish rusty, with small black spots; the throat is white, encircled with brown; remaining inferior parts, white; wings and tail as in the other sex.

· The male in autumn and winter has the bill brownish yellow; irides and feet, brownish. Head, black, varied with small spots of rusty; auriculars, partly encircled with black feathers; throat, yellowish white, finely streaked with deep black. Fore neck and breast, black, mixed with greyish white; the line passing through the eye down the breast, yellowish white, becoming darker on the breast; lower surface from the breast, white, spotted on the flanks. Wings, deep blackish chestnut, crossed by two white lines; primaries on the inside at tip, margined with white. Tail, forked, brownish black, all the feathers margined with rusty, the two outer with a white cuneiform spot at tip.

The dress of the female in autumn and winter is as follows : Head, and neck above, shoulders, and back, greyish rusty, with blackish spots, the rusty predominating on the neck and rump ; the superciliar line, whitish rusty, uniting with a white streak from the angle of the bill ; throat, white each side, with a brownish line ; upper part of the breast, greyish, spotted with black ; inferior parts, white ; the flanks with longitudinal blackish marks.

The young of both sexes, during the first year, are of a yellowish brown above, tinged with greyish, streaked and spotted with blackish, the shafts of the feathers being of that colour ; the cheeks and auriculars are brownish, the latter mixed with black ; a small blackish spot, that spreads as the bird advances in age, is already visible near the opening of the ears ; above the eye is a broad streak of pale brownish ; the throat is yellowish white, slightly streaked with brown, and with a blackish line on each side coming from the corner of the lower mandible ; the lower portion of the neck and breast is of a dingy reddish white, more intense, and thickly spotted with blackish brown on the breast and flanks ; the belly and vent are almost pure whitish. The wing-coverts and second-aries are blackish brown, margined with dark rusty, and tipped with white ; the primaries are dusky brown, paler at the edge. The tail-feathers are dusky, and also margined with deep rusty ; the outer bearing a reddish white conic spot, which is merely longitudinal, and narrow on the next. The bill is entirely of a dirty yellowish brown ; the feet are dusky brown ; the hind nail, though still longer than its toe, is much shorter, and not quite so straight. The figures represent an old male and a young female.

1.Florida Jay. 2.Northern Three-toed Woodpecker. 3 Young Red-headed Woodpecker.

Garrulus Floridanus. Picus Tridactylus.   *Picus Erythrocephalus*

14.

FLORIDA JAY. (*Garrulus Floridanus.*)

PLATE XIV.—Fig. 1.

Garrulus cyaneus, *Vieill. Nouv. Dict. d'Hist. Nat.* xii. p. 476. —Garrulus cœru-
lescens, *Vieill. Nouv. Dict. d'Hist. Nat.* xii. p. 480.—Garrulus cœrulescens,
Ord, in Jour. Ac. Nat. Sc. Philad. i. p. 346.—Corvus Floridanus, *Nob. Syn.*
Am. Birds, sp. 64, *in Ann. Lyc. N. Y.—Id. Cat. Birds U. S.* sp. 64, *in*
Contr. Macl. Lyc. Phil.—Corvus Floridanus, Pica glandaria minor, the Little
Jay of Florida, *Bart. Trav.* p. 290.—Pica glandaria cœrulea non cristata, *Bart.*
Trav. p. 172.—Le Geay azurin, and Le Geay gris-bleu, *Vieill. Nouv. Dict.*
l. c.—Philadelphia Museum, No. 1378, male ; 1379, female.—*My Collection.*

GARRULUS FLORIDANUS.—Bonaparte.

A single glance at the plate on which this fine bird is represented, and at that of the preceding, or Steller's jay, will
suffice, better than the longest description, to show the error
committed by Latham in quoting, in his recent work, "General
History of Birds," the name of this species among the synonyms
of that dedicated to Steller. In fact, the large crest of that
species (of which the present is altogether destitute), and its
black head, the light brown back, and bluish collar of this ;—
but it is needless to carry the comparison between them any
further ; they are too dissimilar to suffer it, and we shall re-
serve pointing out differences until required by closely related
species, of which more striking examples will not long be
wanting.

Mistakes of this kind are perhaps unavoidable in a compilation of such extent as the work we have mentioned, and if they
proceeded from a laudable desire of excluding nominal species,
evinced throughout, we should refrain from censure ; but when,
on the contrary, we find in the same work such repeated instances of an inconsiderate multiplication of species, they cannot be too severely condemned.

Vieillot, in the case of this bird, has fallen into the contrary,
and much more common error, of making two species out of
it—one from personal observation, and the other by compilation. This mistake has already been corrected by Mr Ord,

in a valuable paper which he drew up on his return from Florida, where he enjoyed the advantage of studying this species in its native haunts.

"When we first entered East Florida," says Mr Ord, "which was in the beginning of February, we saw none of these birds; and the first that we noticed were in the vicinity of St Augustine, on the 13th of the above-mentioned month. We afterwards observed them daily in the thickets near the mouth of the St Juan. Hence we conjectured that the species is partially migratory. Their voice is not so agreeable as that of the *Garrulus cristatus*, or crested blue jay of the United States; they are quarrelsome, active, and noisy, and construct their nests in thickets. Their eggs I have not seen." "The blue jay, which is so conspicuous an ornament to the groves and forests of the United States, is also common in Florida. This beautiful and sprightly bird we observed daily, in company with the mocking-bird and the cardinal grosbeak, around the rude habitations of the disheartened inhabitants, as if willing to console them amid those privations which the frequent Indian wars, and the various revolutions which their province has experienced, have compelled them to bear." The Florida jay, however, is a resident in that country, or only removes from section to section. It is not confined to Florida, where it was first noticed by Bartram, being found also in Louisiana, and in the west extends northward to Kentucky, but along the Atlantic not so far. In East Florida it is more abundant, being found at all seasons in low thick covers, clumps, or bushes. They are most easily discovered in the morning about sunrise, on the tops of young live oaks, in the close thickets of which they are found in numbers. Their notes are greatly varied, and in sound have much resemblance to those of the thrush and the blue jay, partaking a little of both. Later in the day it is more difficult to find them, as they are more silent, and not so much on the tree-tops as among the bushes, which are too thickly interwoven with briers and saw-palmettos, to be traversed; and unless the birds are killed on the spot,

which they seldom are when struck with fine shot, it is next
to impossible to come at them in such situations. This species,
like its relatives, is omnivorous, but being inferior in strength,
does not attack large animals. The stomachs of our specimens
contained small fragments of shells, sand, and half-digested
seeds.

The blue jays, though also found in the same localities, are
not so numerous ; they keep more in the woods, and their note
is louder.

The Florida jay is eleven and a half inches long, and nearly
fourteen in extent ; the bill is one inch and a quarter long,
hardly notched, and of a black colour, lighter at tip ; the incumbent setaceous feathers of the base are greyish blue, mixed
with a few blackish bristles ; the irides are hazel brown ; the
head and neck above and on the sides, together with the
wings and tail, are bright azure ; the front, and a line over
the eye, bluish white ; the lores and cheeks of a duller blue,
somewhat mixed with black ; the back is yellowish brown,
somewhat mixed with blue on the rump, the upper tail-coverts
being bright azure ; the inner vanes and tips of the quills are
dusky, their shafts, as well as those of the tail-feathers, being
black. All the lower parts are of a dirty pale yellowish grey,
more intense on the belly, and paler on the throat, which is
faintly streaked with cinereous, owing to the base of the plumage appearing from underneath, its feathers having blackish,
bristly shafts, some of them without webs. From the cheeks
and sides of the neck, the blue colour passes down along the
breast, and forms a somewhat obscure collar ; the under wing
and under tail coverts are strongly tinged with blue, which
colour is also slightly apparent on the femorals ; the inferior
surface of the wings and tail is dark silvery grey ; the base of
the plumage is plumbeous ash, blackish on the head ; the wings
are four and a half inches long, and reach, when closed, hardly
beyond the coverts of the tail, which is five and a half inches
long, extending beyond the wings three and a half ; the spurious feather is extremely short ; the first primary (often mis-

taken for the second), is as short as the secondaries ; the five succeeding are subequal, the third and fourth being rather the longest. The tail is somewhat wedge-shaped, the outer feather being half an inch shorter than the next, and one inch and a half shorter than the middle one. The tarsus is an inch and a quarter long, and black, as well as the toes and nails.

The female is perfectly similar to the male, being but a trifle less in size, and quite as brilliant in plumage.

Two years since it fell to our lot to describe and apply the name of ultramarine jay (*Garrulus ultramarinus*) to a species found in Mexico, closely resembling this, and to which Mr Swainson, in his " Synopsis of Mexican Birds," has lately given the name of *Garrulus sordidus*, his specimen being probably a young one. The principal distinctive characters may be found in its larger dimensions, but especially in the shape of its tail, which is perfectly even, and not in the least cuneiform, as it generally is in the jays. The back, though it is also somewhat intermixed with dusky, is much more blue than in our species, and indeed the whole azure colour is somewhat more brilliant and silky ; the bluish collar is wanting, and the under wing, but especially the under tail coverts, are much less tinged with blue. The wings, moreover, are proportionally larger.

NORTHERN THREE-TOED WOODPECKER.
(*Picus tridactylus*)

PLATE XIV.—Fig. 2.

Picus tridactylus, *Linn. Syst.* i. p. 177, sp. 21.—*Gmel. Syst.* i. p. 439, sp. 21.—*Faun. Suec.* sp. 103.—*Act. Holm.* 1740, p. 222.—*Phil. Trans.* lxii. p. 388.—*Scop. Ann.* i. sp. 56.—*Georgi, Reise,* p. 165.—*Borowsk, Nat.* ii. p. 138. sp. 8.—*Lath. Ind.* p. 243, sp. 56.—*Meyer and Wolf, Tasch. Deutsch. Vog.* i. p. 125, sp. 8.—*Temm. Man. Orn.* i. p. 401, young.—*Brehm, Lehr. Eur. Vog.* i. p. 142.—*Ranz. Elem. Orn.* ii. p. 184, sp. 9, tab. 7, fig. 4.—Picus hirsutus, *Vieill. Ois. Am. Sept.* ii. p. 68, pl. 124, adult male.—Picoides, *Lacepede.*—Dendrocopos tridactylus, *Koch, Baierische Zool.*—Tridactylia hirsuta, *Stephens in Shaw's Zool.* ix. p. 219.—Picus tridactylus anomalus, *Mus. Petr.* 368.—Picchio a tredita, *Stor. degli Ucc.* ii. pl. 180.—Pic. tridactyle ou Picoide, *Temm. l. c.*—Dreizehiger Specht, *Bechst. Nat. Deutschl.* ii. p. 1044.—*Naum·*

Vog. Nachtr. pl. 41, fig. 81.—*Meyer and Wolf, Ois. d'Allem. Cah.* 26, pl. 4, male ; pl. 6, female—Northern Three-toed Woodpecker, *Edwards,* pl. 114, male.—Three-toed Woodpecker, *Penn. Arct. Zool.* sp. 168.—*Lath. Syn.* ii. p. 600, sp. 51 ; *Id. Suppl.* p. 112.—*Philadelphia Museum,* male.—*My Collection,* male, female, and young.

APTERNUS ARCTICUS.—Swainson.

Picus (Apternus) Arcticus, *Sw. North. Zool.* ii. p. 313.

THIS species is one of those which, from their habitation being in the extreme north, have a wide range round the globe. It is, in fact, met with throughout northern Asia and Europe, from Kamtschatka to the most eastern coasts of the old continent ; and in America, is very common at Hudson's Bay, Severn river, Fort William on Lake Superior, and throughout the north-west in hilly and wooded tracts. In the United States it is only a rare and occasional winter visitant, never having been received by us, except from the northern territory of the State of Maine. The species, contrary to what is observed of most other arctic birds, does not appear to extend so far south, comparatively, as in Europe, though it is not improbable that on this continent it may also inhabit some unexplored mountainous districts, resembling the wild regions where only it is found in Europe. In both continents, the species affects deep forests among mountains, the hilly countries of northern Asia and Europe, and the very lofty chains of central Europe, whose elevation compensates for their more southern latitude. It is exceedingly common in Siberia, is abundant in Norway, Lapland, and Dalecarlia, among the gorges of Switzerland and the Tyrol, especially in forests of pines. It is not uncommon in the Canton of Berne, in the forest near Interlaken, though very rare in Germany and the more temperate parts of Europe. It is well known to breed even in Switzerland, and deposits, in holes formed in pine trees, four or five eggs of a brilliant whiteness ; its voice and habits are precisely the same as those of the spotted woodpeckers. Its food consists of insects and their larvæ and eggs, and sometimes seeds and berries. It is easily decoyed by imitating its voice.

This species is eminently distinguished among the North American and European woodpeckers by having only three toes, the inner hind toe being wanting; besides which it has other striking peculiarities, its bill being remarkably broad and flattened, and its tarsi covered with feathers half their length; the tongue is, moreover, not cylindrical, but flat and serrated at the point, which conformation we have, however, observed in the three European spotted woodpeckers, and in the American *Picus varius, villosus, pubescens,* and *querulus.* In all these species the tongue is flat, with the margins projecting each side, and serrated backwards, plain above, convex beneath, and acute at the tip.*

Linné, Brisson, and other anterior writers, confounded this northern bird with a tropical species, the southern three-toed woodpecker, *Picus undulatus* of Vieillot, which inhabits Guiana, and, though very rarely, Central America, but never so far north as the United States. It is the southern species of which Brisson has given us the description, while Linné described the present. It is nevertheless probable that he had the other in view when he observes, that in European specimens the crown was yellow, and in the American red, though, as he states, from Hudson's Bay. The latter mistake was corrected by Latham, who, however, continued to consider the

---

* Mr Swainson has thought the three-toed woodpeckers of sufficient importance to form a subgenus; and I rather think that he will be right in his views. These birds were included by Koch in his genus *Dendrocopus,* of which they possess the general form and colour, but differ chiefly in the structure of the foot. I believe more species will be discovered in the south parts of America; and Mr Swainson, although he does not enter minutely into the distinctions, considers that there are two confounded under the northern three-toed woodpecker. The present bird he denominates *Apternus Arcticus,* and retains *tridactylus* for the three-toed woodpecker of Pennant and Edwards, the *Picus tridactylus* of Forster. The chief differences are in the greater size of the former, the difference of marking, and the relative proportion of the wings. The Northern Expedition observed the first only on the eastern declivities of the Rocky Mountains, where the common species was also procured. This investigation may be worth the while of those persons who have the opportunity.—Ed.

southern as no more than a variety, in which he was mistaken, since they are widely distinct; but as he had no opportunity of seeing specimens, he is not to be censured, especially as he directed the attention of naturalists to the subject. The merit of firmly establishing the two species is, we believe, due to Vieillot. Besides several other traits, the northern bird is always to be distinguished in every state of plumage from its southern analogue by that curious character whence Vieillot took his highly characteristic name (*Picus hirsutus, Pic à pieds vétus*), the feathered tarsi, a peculiarity which this alone possesses to the same extent. The plumage is a uniform black above in the adult, with the top of the head yellow in the male, while the southern, whose tarsi are naked, is black undulated with white, the male having the sinciput red. It is worthy of remark that the three-toed group found in arctic and in tropical America should have no representative in the intermediate countries.

Although these are the only three-toed woodpeckers noted as such in the books, several others are known to exist, some of which, long since discovered, have through inadvertence, or want of proper discrimination, been placed among the four-toed species. The three-toed woodpeckers have been formed into a separate genus, a distinction to which they might indeed be considered entitled if they all possessed the other characters of the present; but, besides that this character appears to be insulated, and of secondary importance (since all forms of the bill known among the four-toed species are met with among the three-toed, which ought, therefore, to make as many groups as there are forms, instead of a single one), the naturalist is perplexed by the anomalous species that inhabit India, of which one has only a stump destitute of nail, and another merely a very small nail without the toe; and, as if Nature took delight in such slow and gradual transitions, two others, furnished with both toe and nail, have the toe exceedingly short, and the nail extremely small! This serves to demonstrate that *Picus*, like other natural groups, admits of subdivision.

These, however, ought not to be separations ; and the genus has been left comparatively untouched by the great innovators of our day, who have only established three genera from it. The first of these, *Colaptes*, of which *P. auratus* of North America may be considered the type, comprises the species that have four toes and slightly curved bills, forming the passage to *Cuculus ;* another, for which the name of *Picus* is retained, includes the four-toed species with straight bills, and the third for the three-toed species indiscriminately. The only foreign three-toed species in our collection, the beautiful *Picus Bengalensis* of authors (*Picus tiga* of Horsfield), widely spread through tropical Asia and the adjacent islands, and, though long since known, always ranked as four-toed, has the bill precisely similar to the four-toed species, being even re- markably compressed, and very sharp on the ridge.

The male northern three-toed woodpecker is ten inches long, and sixteen in extent ; the bill measures one inch and a quarter, is of a blackish lead colour, bluish white at the base of the lower mandible ; it is very broad at base, cuneiform and obtuse at tip, and much depressed throughout, the ridge being very much flattened : both mandibles are perfectly straight, the upper pentagonal, the lower obtusely trigonal ; the tongue is somewhat shorter than that of other species of the genus ; the bristly feathers at the base of the bill are very thick and long, a provision which Nature has made for most arctic birds ; in this they measure half an inch, and are blackish, white at base, somewhat mixed with reddish white ; the irides are bluish black ; the whole head and neck above and on the sides, back, rump, scapulars, smaller wing and tail coverts, constituting the whole upper surface of the bird, of an uniform, deep, glossy black, changing somewhat to green and purple, according to the incidence of light ; the feathers of the front are tipt with white, producing elegant dots of that colour (which per- haps disappear with age) ; the crown of the head is ornamented with a beautiful oblong spot one inch in length, and more than half an inch broad, of a bright silky golden yellow, faintly tinged with orange, and the feathers in this place very fine, and somewhat rigid ; they are black at their base, and marked with white at the limits of the two colours; the base of the plumage elsewhere is uniformly plumbeous ash : each side, from the corner of the mouth, arises a broad white line, form-ing a white space before the eye, prolonged on the neck ; beneath this there is a black one, which, passing from the base of the lower mandible, joins the mass of black of the body ; a tuft of setaceous white feathers advances far upon the bill beneath ; the throat, breast, middle of the belly, and tips of the under tail-coverts are pure white ; the sides of the breast, flanks broadly, and base of the tail-coverts, and even of some of the belly feathers, are thickly waved with lines of black and white, as well as the femoral and short tarsal feathers: in very old birds, as the one represented in the plate, these parts are considerably less undulated, being of a much purer white ; the wings are five inches long, reaching two-thirds the length of the tail ; the spurious feather is exceedingly short, the first primary hardly longer than the seventh ; and the four following subequal and longest ; the smaller wing-coverts, as mentioned, glossy black ; all the other upper coverts, as well as the quills, are of a dull black, the primaries being somewhat duller ; these are regularly marked on both webs with square white spots, larger on the inner webs, and as they approach the base ; the secondaries are merely spotted on the inner vane, the spots taking the appear-ance of bands ; the tips of all the quills are unspotted, the lower wing-coverts are waved with black and white, similar to the flanks ; the tail is four inches long, of the shape usual in the woodpeckers, and composed of twelve feathers, of which the four middle, longest, and very robust and acute, are plain deep black, the next on each side is also very acute, and black at base, cream white at the point, obliquely and irregularly tipt with black ; the two next to these are cream white to the tip, banded with black on the inner vane at base, the more exterior being much purer white and somewhat rounded ; the exterior of all is very short and rounded, and banded throughout with black and pure white : the tarsus is seven-eighths of an inch long, feathered in front for nearly half its length, and, with the toes and nails, dark plumbeous ; the nails are much curved and acute, the hind one being the largest.

The above is a minute description of our finest male specimen, with which all those we have examined coincide more or less. By comparing, however, this description with the detailed ones found in some works, we must conclude that the species is subject to variations in size and plumage, which, according to the erroneous impression given by authors, could not be satisfactorily accounted for by difference of sex, age, or locality : thus in some specimens the cervix is described white, or partly whitish, instead of being wholly black ; the back is also said to be waved with white, which is indeed the case, and with the cervix also, but only in young birds. There is a circumstance, however, that could not be explained by supposing a difference of age, for while some specimens are seen with no appearance of white or yellow on the crown, but having that part, as well as the body, rich shining black, others, with a good deal of lemon yellow on that part, are of a duller black, much varied with white. As in other doubtful and intricate cases, these obscurities are dissipated by a close inspection and unprejudiced observation of nature, and we feel much gratification in being enabled to unveil to ornithologists the mystery of these diversities of plumage in this species, by merely pointing out the sexual differences, as well as those originating in the gradual change from youth to maturity in both sexes ; which, when understood, will not be found more extraordinary than in other species.

The adult female has never been recognised by any author, nor, hitherto, even by ourselves, having been misled by others in taking the young for her ; and this we have only discovered by inspecting a great many specimens. She is precisely similar to the male, even in the minutest particulars, excepting the absence of yellow on the head, this part being of a rich and glossy black.

The young of both sexes are of a dull blackish; the setaceous feathers of the nostrils are greyish, somewhat tinged with rusty; all the feathers of the crown are tipt with white, constituting thick dots on that part, to which they give a silvery appearance; the cheek-bands are obscure and much narrower; the cervix is more or less varied with white, and the feathers of the back being banded with white, gives to that part a waved appearance; the under parts are more thickly waved with black: six, instead of four, of the middle tail-feathers are almost wholly black, the outer of the six having only two or three whitish spots on the outer web. The remaining parts, with due allowance, are similar to the adult.

The young male gradually assumes the yellow, which is at first but little extended, and of a pale lemon colour, through which are yet for some time seen the white dots attributed to the female. She indeed has them very conspicuous in youth, as they are not confounded with any yellow, but loses them entirely as she advances to the adult state.

YOUNG RED-HEADED WOODPECKER.
(*Picus erythrocephalus.*)

PLATE XIV.—Fig. 3.

See *Wilson's American Ornithology*, vol. i. p. 142, pl. 9, fig. 1, for the adult.—Picus erythrocephalus, *Linn. Syst.* i. p. 174, sp. 7.—*Mus. Adolph. Frid.* ii. p. 21.—*Briss. Orn.* iv. p. 52, sp. 19, pl. 3, fig. 1 ; *Id.* 8vo, ii. p. 50.—*Gmel. Syst.* i. p. 429, sp. 7.—*Borowsk. Nat.* ii. p. 136, sp. 4.—*Lath. Ind.* p. 227, sp. 9, adult.—*Vieill. Ois. Am. Sept.* ii. p. 60, pl. 112, adult ; pl. 113, young. —Picus obscurus, *Gmel. Syst.* i. p. 429.—*Lath. Ind.* p. 228, sp. 11, young.— Picus capite toto rubro, the Red-headed Woodpecker, *Catesby, Car.* i. pl. 20, adult.—Picus capite colloque rubris, *Klein, Av.* p. 28, sp. 12, adult.—Picus capite toto rubro, *Kalm, It.* iii. pl. 43, adult.—Picchio di testa rossa, *Storia degli Ucc.* pl. 170, adult.—Pic noir à domino rouge, *Buff. Ois.* vii. p. 55, adult. —Pic de Virginie, *Buff. Pl. enl.* 117, adult.—Pic tricolor, *Vieill. l. c.* adult and young.—Red-headed Woodpecker, *Penn. Arct. Zool.* sp. 160.—*Kalm, Trav. (Angl.)* ii. p. 86.—*Lath. Synop.* ii. p. 561, adult.—White-rumped Woodpecker, *Lath. Syn.* ii. p. 563, sp. 10, young.·

MELANERPES ERYTHROCEPHALUS.—Swainson.

See vol. i. p. 146, male.

The state in which the common red-headed woodpecker is here represented has given rise to a nominal species; and it is, in fact, so difficult to recognise for that bird, that we have thought proper, after the example of Vieillot, to give an exact figure of it. We feel no diffidence in affirming, that in this, through the exertions of Messrs Rider and Lawson, we have fully succeeded; and it will perhaps be allowed to be the best representation of a bird ever engraved. We have nothing to add to Wilson's excellent account of the manners of this very common species, and, therefore, shall limit ourselves to the description of the young as represented.

The young red-headed woodpecker is nine and a half inches long and seventeen inches in extent. The bill is short and robust, being but one-eighth more than an inch in length; the upper mandible has the ridge slightly curved; the bill is horn colour, whitish at base beneath; the setaceous feathers covering the nostrils are very short, and not thick, rufous grey, tipt with black; the whole head, neck, and upper parts of the breast (which are red in the adult), are blackish, each feather broadly edged with whitish, giving the throat the appearance of being whitish, streaked with blackish; the auriculars are plain dusky black; from the breast beneath all is dingy white, the feathers of the breast and lower tail-coverts having dusky shafts; the back and scapulars are black, the feathers being margined with whitish grey; the rump and upper tail-coverts pure white; the wings are five inches and a half long; the spurious feather very short, the first primary subequal to the fifth, the second to the fourth, the third being longest; the smaller wing-coverts are uniform with the back; the larger are of a deeper black, and tipt with pure white; the spurious wing is wholly deep black; the under wing-coverts are pure white, blackish along the margin of the wing ; the primaries are plain black, tipt and edged externally with whitish ; the secondaries are white, shafted with black, and with an acuminate, broad, subterminal band, which, running from one to the other, takes a zigzag appearance ; the tail is four inches long, and, like those of all the woodpeckers we have examined, composed of twelve feathers, of which the outer on each side is extremely short and inconspicuous, and pure white, with a black shaft. All the others, which are very acute, longer, and more acuminate, and stiffer as they approach the centre, are black, and, except the two middle ones, slightly whitish each side of the shaft at tip, the outer being also of that colour on its outer margin. The feet are dark plumbeous, the tarsus being seven-eighths of an inch long, and feathered for a short space in front.

The young of both sexes are, no less than the adult, perfectly alike ; as they advance in age, the margins of the feathers disappear, and the black becomes deep and glossy, and all the colours much purer ; the scarlet of the head comes on very gradually, so that specimens are found with merely a reddish tinge, and generally with a few dots on the hind neck ; it is one of these specimens, with a few streaks of red, that we have selected for the sake of ornamenting the plate, as well as to exemplify the manner in which the change takes place. No such mark appears at first.

In the adult, the whole head, neck, and breast, are bright and deep scarlet, with the feathers black at base ; the back, scapulars, and smaller wing-coverts, are rich glossy black ; the rump, upper tail-coverts, and from the breast beneath, white, the bottom of the plumage being plumbeous, and the tail-coverts with blackish shafts ; the wings and tail are black ; the lower wing-coverts pure white, with the margin of the wing deep black ; the secondaries are white, shafted to near the tip with black ; the last of the primaries being also white at tip, and on the greater part of the base of the outer vane ; the small lanceolate outer feather is white, black on the shaft and base of the inner vane ; the two next only being tipt with white, the outer of which is also white on the exterior margin.

EVENING GROSBEAK.   (*Fringilla vespertina.*)

PLATE XV.—Fig. 1.

Fringilla vespertina, *Cooper, in Ann. Lyc. New York,* i. p. 220.—*Nob. Cat. Bird
  U. S. sp.* 188, *in Contr. Macl. Lyc. Phila.* i. p. 21.—*Id. Syn. Birds, U. S.
  sp.* 188, *in Ann. Lyc. N. Y.* ii. p. 113.—*Id. Suppl. in Zool. Journ. London,*
  iv. p. 2.—*Cabinet of the Lyceum of Nat. Hist. of New York.*—Mr Leadbeater's Collection in London.

COCCOTHRAUSTES? VESPERTINA.—Cooper.*

Coccothraustes vespertina, *North. Zool.* ii. p. 269.

FEW birds could form a more interesting acquisition to the fauna of any country than this really fine grosbeak. Beautiful in plumage, peculiar in its habits, important to systematical writers, it combines advantages of every kind. It was named and first described by Mr Cooper, and little has since been discovered of its history to be added to the information he has collected and given us in the journal above quoted. The species appears to have an extensive range in the northern and north-western parts of this continent, being met with from the extremity of the Michigan territory to the Rocky Mountains, within the same parallels. It is common about the head of Lake Superior, at Fond du Lac, and near the Athabasca Lake. A few were observed by Mr Schoolcraft, during the first week of April 1823, about Sault Sainte Marie, Michigan territory, where they remained but a short time,

---

* In another note we remarked, that this genus was supplanted in America by Mr Swainson's genus *Guiraca.* At that time we overlooked the evening grosbeak, which approaches nearer to our idea of the genus *Coccothraustes* than some of those which have lately been placed within it. And it will remain at present a question whether the present form be retained under that title, and the aberrant species separated, or the reverse. We do not consider that this bird can range with *Guiraca.*— Ed.

Drawn from Nature by A. Rider.

Engraved by W. H. Lizars.

1. Evening Grosbeak. 2. Female Rose breasted Grosbeak. 3. Female White-winged Crossbill. 4. Female Indigo Finch.
Fringilla Vespertina.   Fringilla Ludoviciana.   Loxia Leucoptera.   Fringilla Cyanea.

and have not appeared since ; and by Major Delafield, in the month of August of the same year, near the Savannah river, north-west from Lake Superior. They appear to retire during the day to the deep swamps of that lonely region, which are covered with a thick growth of various trees of the coniferous order, and only leave them in small parties at the approach of night. Their note is strange and peculiar ; and it is only at twilight that they are heard crying in a singular strain. This mournful sound, uttered at such an unusual hour, strikes the traveller's ear, but the bird itself is seldom seen ; though, probably from its unacquaintance with man, it is so remarkably tame and fearless as almost to suffer itself to be caught with the hand.

The specimen of the evening grosbeak presented to the Lyceum of New York by Mr Schoolcraft, from which Mr Cooper established the species, was thought, until lately, the only one in possession of civilised man ; but we have since examined two others shot early in the spring on the Athabasca Lake, near the Rocky Mountains, and preserved among the endless treasures of Mr Leadbeater of London. From the more perfect of these, our plate, already engraved from Mr Cooper's specimen, has been faithfully coloured ; and the subjoined description is carefully drawn up from a perfect specimen now before us, which Mr Leadbeater, with the most obliging liberality, has confided to our charge.

Although we consider the grosbeaks (*Coccothraustes*) as only a subgenus of our great genus *Fringilla*, they may with equal propriety constitute one by themselves, as the insensible degrees by which intermediate species pass from one form into another (which determined us in considering them as a subgenus, and not a genus) are equally observable between other groups, though admitted as genera. *Coccothraustes* is as much entitled to be distinguished generically from *Fringilla*, as *Turdus* from *Sylvia ;* and at all events, its claim is fully as good, and perhaps better, than its near relation, *Pyrrhula*. In the present work, however, we have preferred retaining things as we found them, until we can apply ourselves to the work of a general reform, as announced in a previous article of this work. Though we regard the grosbeaks as a subgenus, others, going to the opposite extreme, have erected them into a separate family, composed of several genera. The evening grosbeak is, however, so precisely similar in form to the haw-finch-type of the group, as to defy the attempts of the most determined innovators to separate them. Its bill is as broad, as high, quite as strong and turgid, with both mandibles equal, the upper depressed and rounded above, and the commissure straight. It conforms even, in a slight degree, in the rhomboidal shape of the ends of the secondaries—a character so conspicuous in its analogue, to which, in the distribution and transitions of its tints, though very different, it also bears a resemblance. It is, however, of the four North American species of its group, the only one so strictly allied, for even the cardinal grosbeak, the most nearly related of these species, on account of its short, rounded wings and other minor traits, might be separated, though fortunately it has not as yet, to our knowledge; the others have been already.

The evening grosbeak is eight and a half inches long; its bill is of a greenish yellow, brighter on the margins, seven-eighths of an inch long, five-eighths broad, the same in height; the capistrum and lora are black; the front is widely bright yellow, prolonged in a broad stripe over the eye to the ears; the hind crown is black, intermixed with yellow, visible only on separating the feathers, but leading to the suspicion that at some period the yellow extends perhaps all over the crown; the sides and inferior parts of the head, the whole neck, above and beneath, together with the interscapulars and breast, are of a dark olive brown, becoming lighter by degrees; the scapulars are yellow, slightly tinged with greenish; the back, rump, with the whole lateral and inferior surface, including the under wing and under tail coverts, yellow, purer on the rump, and somewhat tinged with olive brown on the belly. Although these colours are all very pure, they are not definitely separated, but pass very insensibly into each other ; thus the black of the crown passes into the dark brown of the neck, which, becoming lighter by degrees, is blended with the yellow of the back. The same thing takes place beneath, where the olive brown of the breast passes by the nicest gradations into the yellow of the posterior parts ; the whole base of the plumage is pale bluish plumbeous, white before the tips of the feathers ; the femorals are black, skirted with yellow ; the wings are four and a half inches long ; the smaller, middling, and exterior larger wing-coverts, are deep black, as well as the spurious wing ; those nearest the body are white, black at the origin only ; the quills are deep black, the three outer being subequal and longest, attenuated on their outer web at the point, and inconspicuously tipt with whitish ; the secondaries are marked with white on their inner web, that colour extending more and more as they approach the body, the four or five nearest being entirely pure white, like their immediate coverts, and slightly and inconspicuously edged with yellow externally ; the tail is two and a half inches long, slightly forked, and, as well as its long superior coverts, very deep black ; the outer feather on each side has, on the inner vane, towards the tip, a large, roundish, white spot, which seems disposed to become obliterated, as it is much more marked on one than on that of the other side which corresponds to it, and does not exist in all specimens ; a similar spot is perceptible on the second tail-feather, where it is, however, nearly obliterated ; the feet are flesh-colour, the nails blackish, the tarsus measuring three-quarters of an inch.

No difference of any consequence is observable between the sexes, though it might be said that the female is a little less in size, and rather duller in plumage.

FEMALE ROSE-BREASTED GROSBEAK. (*Fringilla Ludoviciana.*)

PLATE XV.—FIG. 2.

See *Rose-breasted Grosbeak Loxia rosea* (*Ludoviciana*) *Wils. Am. Orn.* ii. p. 135, pl. 17, fig. 1, for the male.—Loxia Ludoviciana, *Linn. Syst.* i. p. 306, sp. 38.— *Gmel. Syst.* i. p. 862, sp. 38.—*Lath. Ind.* p. 379, sp. 25.—Fringilla punicea, *Gmel. Syst.* i. p. 921, sp. 81.—*Lath. Ind.* p. 444, sp. 34, adult male.—Loxia maculata, *Gmel. Syst.* i. p. 861, sp. 87.—*Lath. Ind.* p. 379, sp. 26, young.— Loxia obscura, *Gmel.* i. p. 862, sp. 88.—*Lath. Ind.* p. 379, sp. 27, female.— Coccothraustes Ludoviciana, *Briss. Orn.* iii. p. 247, sp. 14, pl. 12, fig. 2 ; *Id.* Svo, i. p. 378.—Coccothraustes rubricollis, *Vieill. Gal. Ois.* i. part ii. p. 67, pl. 58 (very bad), and *Dict.*—Pyrrhula Ludoviciana, *Sabine, Zool. App. to Frankl. Exp.* p. 675.—Fringilla Ludoviciana, *Nob. Obs. Nom. Wils. Orn.* sp. 80.—*Id. Cat. Birds U. S.* sp. 189.—*Id. Syn. Birds U. S.* sp. 189.—Guiraca Ludoviciana, *Swainson, Syn. Mex. Birds*, sp. 76, in *Phil. Mag. N. S.* i. p. 438.—Le Rose-gorge, *Buff. Ois.* iii. p. 460.—Gros-bec de la Louisiane, *Buff. Pl. enl.* 153, fig. 2, male.—Moineau à poitrine et ventre pourprés, *Sonn. Buff.* xlviii. p. 240.—Red-breasted Grosbeak, *Penn. Arct. Zool.* sp. 212.—*Lath. Syn.* iii. p. 126, sp. 24.—Red-breasted-Finch, *Penn. Arct. Zool.* sp. 275.— *Lath. Syn.* iii. p. 272, sp. 30, adult male.—Dusky Grosbeak, *Penn. Arct. Zool.* sp. 216.—*Lath. Syn.* iii. p. 127, sp. 26, female.—Spotted Grosbeak, *Penn. Arct. Zool.* sp. 213.—*Lath. Syn.* iii. p. 126, sp. 25, young.—*Philadelphia Museum*, No. 5806, male ; 5807, female.

GUIRACA LUDOVICIANA.—Swainson.

See vol. i. p. 277.

THOUGH several figures have been published of the very showy male rose-breasted grosbeak, the humble plumage of the female and young has never been represented. It would, however, have better served the purposes of science if the preference had been given to the latter, though less calculated to attract the eye, inasmuch as striking colours are far less liable to be misunderstood or confounded in the description of species than dull and blended tints. It will be seen by the synonymy that nominal species have in fact been introduced into the systems. But if it be less extraordinary that the female and young should have been formed into species, it is certainly unaccountable that the male itself should have been twice described in the same works, once as a finch, and once as a grosbeak. This oversight originated with Pennant, and later compilers have faithfully copied it, though so easy to rectify.

The female rose-breasted grosbeak is eight inches long, and twelve and a half inches in extent. The bill has not the form either of the typical grosbeaks or of the bullfinches, but is intermediate between them, though more compressed than either. It is three-quarters of an inch long, and much higher than broad ; instead of being pure white, as that of the male, it is dusky horn-colour above, and whitish beneath and on the margins; the irides are hazel brown ; the crown is of a blackish-brown, each feather being skirted with lighter olive brown, and faintly spotted with white on the centre ; from the nostrils a broad band passes over the eye, margining the crown to the neck ; a brown streak passes through the eye, and the inferior orbit is white; more of the brown arises from the angle of the mouth, spreading on the auriculars ; on the upper part of the neck above, the feathers are whitish, edged with pale flaxen, and with a broad, oblong, medial, blackish brown spot at tip; on the remaining part of the neck and inter-scapulars, this blackish spot is wider, so that the feathers are properly of that colour, broadly edged with pale flaxen ; the back and rump, and the upper tail-coverts, are of a lighter brown, with but a few merely indicated and lighter spots ; the whole inferior surface of the bird is white, but not very pure ; the sides of the throat are dotted with dark brown, the dots occupying the tips of the feathers ; the breast and flanks are somewhat tinged with flaxen (more dingy on the latter), and each feather being blackish along the middle at tip, those parts appear streaked with that colour; the middle of the throat, the belly, and under tail-coverts are unspotted ; the base of the plumage is everywhere plumbeous ; the wings are rounded, less than four inches long, entirely dusky brown, somewhat darker on the spurious wing, all the feathers, both quills and coverts, being lighter on their edges ; the exterior webs of the middle and larger wing-coverts are whitish at tip, constituting two white bands across the wings ; the primaries are whitish at the origin beneath the spurious wing; the secondaries are inconspicuously whitish externally at tip, that nearest the body

having a very conspicuous whitish spot; the lower wing-
coverts are of a bright buff; and as they are red in the male,
afford an excellent essential character for the species; the tail
is three inches long, nearly even, and of a paler dusky brown;
the two outer feathers are slightly edged internally with
whitish, but without the least trace of the large spot so conspicuous in the male, and which is always more or less apparent in the young of that sex; the feet are dusky, the
tarsus measuring seven-eighths of an inch.

The young male is at first very similar to the female, and is,
even in extreme youth, paler and somewhat more spotted; but
a little of the beautiful rose colour, of which the mother is
quite destitute, soon begins to make its appearance, principally
in small dots on the throat: this colour spreads gradually, and
the wings and tail, and soon after the head, blacken, of course
presenting as they advance in age a great variety of combinations.

For the description of the beautiful adult male we shall
refer to Wilson, whose description is good, and the figure
accurate; but not having stated any particulars about the
habits of the species, we shall subjoin the little that is known
of them. Though long since recorded to be an inhabitant
of Louisiana, whence it was first received in Europe, recent
observations, and the opinion of Wilson, had rendered this
doubtful, and it was believed to be altogether an arctic bird,
averse to the warm climates of the southern States, and hardly
ever appearing even in the more temperate. Its recent discovery in Mexico is, therefore, a very interesting and no less
remarkable fact; and we may safely conclude that this bird
migrates extensively according to season, spending the summer
in the north, or in the mountains, and breeding there, and
in winter retiring southward, or descending into the plains;
being, however, by no means numerous in any known district,
or at any season, though perhaps more frequent on the borders
of Lake Ontario. Its favourite abode is large forests, where

it affects the densest and most gloomy retreats.   The nest is placed among the thick foliage of trees, and is constructed of twigs outside, and lined with fine grasses within ; the female lays four or five white eggs, spotted with brown.   This may also be called an " evening grosbeak," for it also sings during the solemn stillness of night, uttering a clear, mellow, and harmonious note.

We have placed this species in our subgenus *Coccothraustes.* It is probably because he laboured under the mistake that all the grosbeaks removed from *Loxia* had been placed in *Pyrrhula* by Temminck, that Mr Sabine has made it a bullfinch ; and in truth the bill very much resembles those of that genus, so that the species is intermediate between the two.   Mr Swainson places it, together with the blue grosbeak, *Fringilla (Coccothraustes) cœrulea,* in a new genus which he calls *Guiraca,* but without as yet characterising it.   These species have, it is true, a bill somewhat different from that of the typical *Coccothraustes* (as may be seen by comparing this with the evening grosbeak), being much less thick and turgid, and higher than broad ; the upper mandible being larger than the lower, and covering its margins entirely, compressed on the sides, making the ridge very distinct (not rounded above), and curved from the base, but at tip especially ; the margins of both are angular.   The representation of the bill in Wilson's plate of the male is remarkably exact.

FEMALE WHITE-WINGED CROSSBILL.   (*Loxia leucoptera.*)

PLATE XV.—Fig. 3.

See *Wilson's American Ornithology,* iv. p. 48, pl. 31, for the young male.—Loxia leucoptera, *Gmel. Syst.* i. p. 844, sp. 12.—*Vieill. Gal. Ois.* i. p. 56, pl. 52, young male.—*Nob. Obs.* sp. 84.—*Id. Cat. and Syn. Birds U. S.* sp. 195.— Loxia falcirostra, *Lath. Ind.* p. 371, sp. 2.—Le Bec-croisé leucoptère, *Sonn. Buff.* xlvii. p. 65.—*Vieill. Nouv. Dict. Hist. Nat.* 2d ed. iii. p. 339.—White-winged Crossbill, *Lath. Syn.* iii. p. 108, sp. 2 ; *Id. Suppl.* p. 148.—*Dixon, Voy.* t. 20, p. 358, female.—*Penn. Arct. Zool.* ii. sp. 208.—*My Collection*, male, female, young, and middle-aged.

LOXIA LEUCOPTERA.—Gmelin ?

See vol. ii. p. 42.

The white-winged crossbill, first made known by Latham in his celebrated "Synopsis," was subsequently introduced on his authority into all the huge compilations of the last century. Wilson gave us the first figure of it, which is that of the male, and promised a representation of the female, together with " such additional facts relative to its manners as he might be able to ascertain." It is to fulfil Wilson's engagement that we now give a correct figure of the other sex of this species, which we are also enabled to describe minutely, in all its different states of plumage. This has never before been done, though Vieillot, since Wilson's time, has compiled some account of its habits, described the female, and recently published a bad enough figure of the male in his "Galerie des Oiseaux."

The English name was bestowed by its discoverer, the scientific was imposed on it by the compiler Gmelin, who, like the daw in the fable, though with much better success, appropriated to himself the borrowed plumes of others, making Latham's new species his own, by being the first to give them scientific names, which the discoverer himself was afterwards obliged to adopt in his "Index Ornithologicus." In the present instance, however, he took the liberty of altering Gmelin's name, most probably with the view of giving one analogous to that of *Loxia curvirostra*, and indicative of the remarkable form of the bill. That character having since been employed as generic, the propriety of Latham's change has ceased to exist, and, in fact, the advantage is altogether on the side of Gmelin. We have therefore respected the right of priority, even in the case of an usurper.

The female white-winged crossbill is five inches and three-quarters long, and nearly nine in extent; the bill is more than five-eighths long, of a dark horn-colour, paler on the edges : as is the case in the whole genus, it is very much compressed

throughout, but especially at the point, where the edges almost
unite into one : both mandibles are curved (the lower one upwards) from the base, the ends crossing each other ; the upper
has its ridge distinct, and usually crosses to the left in both
sexes, and not, as Wilson appears to intimate, generally in
one sex only; the lower mandible is considerably shorter; the
tongue is short, cartilaginous, and entire ; the irides are of
a very dark hazel ; the small setaceous feathers covering the
nostrils, which is one of the characteristics of the genus, are
whitish grey ; the bottom of the plumage is everywhere slate
colour ; the head, and all the upper parts, down to the rump,
are of a greyish green, strongly tinged with olive, each feather
being marked with black in the centre, giving the plumage a
streaked appearance, as represented in the plate ; the rump is
pure pale lemon yellow ; the upper tail-coverts are blackish,
margined with whitish olive ; the front, and a broad line over
and round the eye and bill, are slightly distinguished from the
general colour of the head by the want of olivaceous, being
greyish white, and as the feathers are very small, appear
minutely dotted with black : the curved blackish spot, more
apparent in the colours of the male, is slightly indicated on the
sides of the head ; the sides of the head and neck, the throat,
and the breast, are of a greyish white, also streaked with
blackish, and somewhat tinged with yellowish on the sides of
the breast ; the flanks become of a dingy yellowish grey, and
have large, dull, blackish blotches ; the belly and vent are of
a much purer whitish, and the streaks are on that part long,
narrow, and well defined ; the under tail-coverts are blackish,
with broad white margins ; the wings are three inches and a
half long, reaching, when closed, to the last of the tail-coverts;
the first three primaries are subequal and longest, the fourth
being but little shorter, and much longer than the succeeding;
the general colour of the wing is black, the smaller coverts each
margined with olive ; the middle and longer coverts broadly
tipt with white, forming a double band across the wings, so
conspicuous as to afford the most obvious distinguishing cha-

racter of the species; all the quills are slightly edged with paler, the tertials being also tipt with white; the under wing-coverts are of a dark silvery, as well as the whole inferior surface of the wing; the tail measures two and a half inches, being as usual composed of twelve feathers; it is black, and deeply emarginate, the feathers acute, and slightly edged with paler; the feet are short, rather robust, and blackish; the tarsus five-eighths of an inch in length, somewhat sharp behind, with its covering entire before; the toes are divided to the base, very short, the middle one considerably the longest, but much less than half an inch long, the lateral one subequal (all these being remarkable characters of the genus); the hind toe long, and stoutest; the nails strong, much curved, and sharp, the hind one the longest, and twice as large as the lateral.

The male described by Latham, Wilson, and Vieillot as in full plumage, but which, with Temminck, we have good reasons for believing to be between one and two years old, differs from the female in being a trifle larger, and of a crimson red where she is olive grey; the base of the plumage is also considerably darker, approaching to black on the head, which colour predominates in several parts of the plumage, round the eye, on the front, in a broad line curving and widening from the eye, each side of the neck, and appearing distinctly on the back, where it generally forms a kind of band descending from the base of the wing: the rump is of a beautiful rose-red; the black of the wings and tail is deeper; the white, pure, and more extended; the lining of the quills, and especially of the tail-feathers, more conspicuous; the belly is of a pure whitish, much less streaked, &c.

The bird which, from analogy, we take for the adult male, though we have no positive evidence for deciding whether it is in the passage to or from the preceding, differs only in having a light buff orange tinge where the other has crimson: it agrees with it in all its minute markings, the patch on the sides of the head is better defined, and the wings and tail are of a still deeper black, the edges of the quills and tail-feathers being very conspicuous, and almost pure white. All these facts conspire to favour our opinion. In this state, the bird is rare, as might be expected, and has not before been noticed by any naturalist: we have not represented it, only that we might not multiply figures of the same species.

The very young male before assuming the red, at the age of one year, exactly resembles the female, being only more greyish, and less tinged with olive, and having the rump greenish yellow, instead of yellow.

The four above-described states of plumage are selected from a number of specimens shot on the same day and out of the same flock. The changes of these birds must still rank among the unexplained phenomena of natural history. An illustration might be attempted, by supposing a double moult to take place in the birds of this genus; but besides that we ought to be cautious in admitting an hypothesis like this, not founded on observation, it would be entirely untenable in the present instance, from the fact that all the variations of plumage are found at the same period of the year; thus proving that age, and of course sex, but not season, produce these changes; and we must provisionally admit that, contrary to what takes place in all other birds, these (the crossbills), together with the pine bullfinches, lose, instead of acquiring, brilliancy of colours as they advance in age.

This species inhabits, during summer, the remotest regions of North America, and it is therefore extraordinary that it should not have been found in the analogous climates of the old continent. In this, its range is widely extended, as we can trace it from Labrador westward to Fort de la Fourche, in latitude 56°, the borders of Peace river, and Montague island on the north-west coast, where it was found by Dixon. Round Hudson's Bay it is common and well known, probably extending far to the north-west, as Mackenzie appears to allude to it when speaking of the only land bird found in the desolate regions he was exploring, which enlivened, with its agreeable notes, the deep and silent forests of those frozen tracts. It is common on the borders of Lake Ontario, and descends in autumn and winter into Canada and the northern and middle States. Its migrations, however, are very irregular. During four years it had escaped my careful researches, and now, while writing (in the first week of November 1827), they are so abundant, that I am able to shoot every day great numbers out of flocks that are continually alighting in a copse of Jersey scrub-pine (*Pinus inops*) even opposite my window. It is proper to mention, that owing perhaps to the inclemency of the season, which has so far been distinguished by rains, early frost, and violent gales of wind, there have been extraordinary flights of winter birds. Many flocks of the purple finch are seen in all directions. The American siskin (*Fringilla pinus*, Wils.), of which I never saw a living specimen before, covers all the neighbouring pines and its favourite thistles with its innumerable hosts. The snow-bunting (*Emberiza nivalis*) has also made its appearance in New Jersey, though in small parties, after an absence of several years.*

The white-winged crossbills generally go to Hudson's Bay on their return from the south, and breed there, none remaining during summer, even in the most northern parts of the United States, where they are more properly transient irregular visitors than even winter residents. They are seldom observed elsewhere than in pine-swamps and forests, feeding almost exclusively on the seeds of these trees, together with a few berries. All the specimens I obtained had their crops filled to excess entirely with the small seeds of *Pinus inops*. They kept in flocks of from twenty to fifty, when alarmed suddenly taking wing all at once, and after a little manœuvring in the air, generally alighting again nearly on the same pines whence they had set out, or adorning the naked branches of some distant, high, and insulated tree. In the countries where they pass the summer, they build their nest on the limb of a pine,

* This is the case also with the common crossbills and European siskin, and has hardly yet met with any reasonable solution. See notes to these birds.—ED.

towards the centre ; it is composed of grasses and earth, and lined internally with feathers. The female lays five eggs, which are white, spotted with yellowish. The young leave their nest in June, and are soon able to join the parents in their autumnal migration,

In the northern countries, where these birds are very numerous, when a deep snow has covered the ground, they appear to lose all sense of danger, and by spreading some favourite food, may be knocked down with sticks, or even caught by the hand, while busily engaged in feeding. Their manners are, in other respects, very similar to those of the common crossbill, as described by Wilson, and they are said also to partake of the fondness for saline substances so remarkable in that species.

FEMALE INDIGO FINCH. (*Fringilla cyanea.*)

PLATE XV.—FIG. 4.

See *Wilson's American Ornithology*, i. p. 99, pl. 6, fig. 5, for the male.—Tanagra cyanea, *Linn. Syst.* i. p. 315, sp. 6, adult male in full plumage.—Emberiza cyanea, *Gmel. Syst.* i. p. 876, sp. 54.—*Lath. Ind.* p. 415, sp. 60.—Emberiza cyanella, *Sparm. Mus. Carls.* ii. pl. 42, 43.—*Gmel. Syst.* i. p. 887, sp. 74.— Emberiza cœrulea, *Gmel. Syst.* i. p. 876.—*Lath. Ind.* p. 415, sp. 59, male in moült.—Tanagra cœrulea ? *Gmel. Syst.* i. p. 891, sp. 27.—*Lath. Ind.* p. 427, sp. 27, adult male.—Tanagra Carolinensis cœrulea, *Briss. Av.* iii. p. 13, sp. 6, adult male in full dress.—Emberiza Canadensis cœrulea, *Briss. Av.* iii. p. 298, sp. 12, pl. 14, fig. 2, male moulting.—Passerina cyanea, *Vieill. Nouv. Dict. Hist. Nat.*—Fringilla cyanea, *Nob. Obs.* sp. 112.—*Id. Cat. and Synop. Birds U. S.* sp. 164.—Linaria cyanea *Bartram's Trav.* p. 296.—Linaria cœrula, the Blue Linnet, *Catesby, Carolina,* i. p. 45, pl. 45.—Le Ministre, *Buff. Ois.* iv. p. 86.—L'Azuroux, *Buff. Ois.* iv. p. 369, male moulting.—Passe-bleu ? *Buff. Ois.* iii. p. 495, adult male in full plumage.—Moineau Bleu de Cayenne ? *Buff. Pl. enl.* 203, fig. 2, adult male in full dress.—Blue Linnet, *Edwards, Av.* iv. p. 132, pl. 273, lower figure.—Indigo Bunting, *Penn. Arct. Zool.* ii. sp. 235.—*Lath. Syn.* iv. p. 205, sp. 53.—Blue Bunting, *Penn. Arct. Zool.* ii. sp. 234.—*Latham, Syn.* iii. p. 205, sp. 52, male moulting.—Blue Tanager ? *Lath. Syn.* iii. p. 234, sp. 28.—*Philadelphia Museum,* No. 6002, male ; 6003, female.

SPIZA CYANEA.—BONAPARTE.

Male, see vol. i. p. 99, and note to *F. amœna* of present volume.

THE remarkable disparity existing between the plumage of the different sexes of the common indigo bird renders it almost indispensably requisite that the female, unaccountably neglected by Wilson, as he generally granted this distinction in similar, and often less important cases, should be figured in this work.　Hardly any North American bird more absolutely stands in need of being thus illustrated than the beautiful finch which is now the subject of our consideration.　It could scarcely be expected that the student should easily recognise the brilliant indigo bird of Wilson's first volume in the humble garb in which it is represented in the annexed plate.　But, however simple in its appearance, the plumage of the female is far more interesting and important than that of the male, as it belongs equally to the young, and to the adult male after the autumnal moult, and previous to the change which ensues in the spring,—a large proportion of the life of the bird.

The importance of a knowledge of these changes will also be duly estimated on recurring to the copious synonymy at the head of our article, by which it will be seen that several nominal species have been made by naturalists who chanced to describe this bird during its transitions from one state to another.　Errors of this kind too frequently disfigure the fair pages of zoology, owing to the ridiculous ambition of those pseudo-naturalists, who, without taking the trouble to make investigations, for which, indeed, they are perhaps incompetent, glory in proclaiming a new species established on a single individual, and merely on account of a spot, or some such trifling particular !　The leading systematists who have enlarged the boundaries of our science have too readily admitted such species, partly compelled to it perhaps by the deficiency of settled principles.　But the more extensive and accurate knowledge which ornithologists have acquired within a few years relative to the changes that birds undergo will render them more cautious, in proportion as the scientific world will be less disposed to excuse them for errors arising from this source.　Linné may be profitably resorted to as a model of accuracy in this respect, his profound sagacity leading him in many instances to reject species which had received the sanction even of the experienced Brisson. Unfortunately, Gmelin, who pursued a practice directly the opposite, and compiled with a careless and indiscriminating hand, has been the oracle of zoologists for twenty years. The thirteenth edition of the "Systema Naturæ" undoubtedly retarded the advancement of knowledge instead of promoting it; and if Latham had erected his ornithological edifice on the chaste and durable Linnean basis, the superstructure would have been far more elegant. But he first misled Gmelin, and afterwards suffered himself to be misled by him, and was therefore necessarily betrayed into numerous errors, although he at the same time perceived and corrected many others of his predecessor. We shall not enumerate the nominal species authorised by their works in relation to the present bird, since they may be ascertained by consulting our list of synonyms. On comparing this list with that furnished by Wilson, it will be seen that the latter is very incomplete. Indeed, as regards synonymy, Wilson's work is not a little deficient; notwithstanding which, however, it will be perpetuated as a monument of original and faithful observation of nature, when piles of pedantic compilations shall be forgotten.

We refer our readers entirely to Wilson for the history of this very social little bird, only reserving to ourselves the task of assigning its true place in the system. As we have already mentioned in our "Observations," he was the first who placed it in the genus *Fringilla* (to which it properly belongs), after it had been transferred from *Tanagra* to *Emberiza* by former writers, some of whom had even described it under both in one and the same work. But although Wilson referred this bird to its proper genus, yet he unaccountably permitted its closely allied species, the *Fringilla ciris*, to retain its station in *Emberiza*, being under the erroneous impression that a large bill was characteristic of that genus. This mistake, however, is excusable when we consider that almost all the North American birds which he found placed in it, through the negligence or ignorance of his predecessors, are in fact distinguished by large bills.

The transfer of this species to the genus *Fringilla* renders a change necessary in the name of *Loxia cyanea* of Linné, an African bird, now a *Fringilla* of the subgenus *Coccothraustes*. The American bird belongs to *Spiza*, and, together with the *Fringilla ciris* and the beautiful *Fringilla amœna*, it may form a peculiar group, allied to *Fringilla, Emberiza,* and *Tanagra,* but manifestly nearest the former.

The adult male, in full plumage, having been described by Wilson, may be omitted here. The female measures four inches and three-quarters in length, and nearly seven in extent, The bill is small, compressed, and less than half an inch long ; is blackish above, and pale horn-colour beneath; the irides are dark brown ; above she is uniformly of a somewhat glossy drab; between the bill and eyes, and on the cheeks, throat, and all the inferior parts, of a reddish clay colour, much paler on the belly, dingy on the breast, and strongly inclining to drab on the flanks, blending into the colour of the back, the shafts of the feathers being darker, giving somewhat of a streaked appearance ; the whole base of the plumage is lead colour ; the wings and tail are of a darker and less glossy brown, each feather being edged with lighter, more extended on the secondaries, and especially the wing-coverts ; the wings are two inches and a half long, not reaching, when folded, beyond the tail-coverts ; the first primary is subequal to the fourth, the second and third being longest ; the three outer, besides the first, are greatly attenuated on the outer web, half an inch from the point, where it is extremely narrow ; the tail is two inches in length, and but slightly emarginated ; the feet are dusky, the tarsus measuring three-quarters of an inch.

The male, after his autumnal moult, exhibits pretty much the same dress, except being more or less tinged with bluish. We shall here observe, that we do not believe that the individual kept by Wilson in a cage through the winter, in which the gay plumage did not return for more than two months,

3

2

1

Drawn from Nature by A. Rider

Engraved by W.H. Lizars.

1 Pallas Dipper.   2 Bohemian Waxwing.   3 Female Pine Bullfinch.

Cinclus Pallasii.   Bombycilla Garrula   Pyrrhula Enucleator

formed an exception to the general law, as he supposed. We have no doubt that this circumstance is characteristic of the species in its wild state.

The young strongly resemble the female; the drab colour is, however, much less pure and glossy, being somewhat intermixed with dusky olive, owing to the centre of the feathers being of the latter hue. Consequently, during the progress from youth to adolescence, and even during the two periodical changes, the plumage of this bird is more or less intermixed with drab, blue, and white, according to the stage of the moulting process, some being beautifully and regularly spotted with large masses of those colours symmetrically disposed. In one of these males, but little advanced in its changes, we readily recognise the *Emberiza cœrulea* of authors, *Azuroux* of Buffon, &c.; and in another, which has made further progress towards the perfect state, the shoulders only retaining the ferruginous tinge, we can trace the *Emberiza cyanella* of Sparmann.

PALLAS' DIPPER. (*Cinclus Pallasii.*)

PLATE XVI.—Fig. 1.

Cinclus Pallasii, *Temm. Man. Orn.* i. p. 177.—*Nob. Suppl. Gen. Am. Birds,* sp. 94, *bis, in Zool. Journ. London,* iv. p. 4.—*Id. in Ann. Lyc. New York,* ii. p. 438.—Cinclus Mexicanus, *Swainson, Syn. Birds of Mexico,* sp. 27, *in Phil. Mag. New Series,* i. p. 368.—*Collection of Mr Leadbeater, in London.*

CINCLUS AMERICANUS.—Richardson and Swainson.*

Cinclus Americanus, *North. Zool.* ii. p. 173.

The recent discovery of the genus *Cinclus* in America, furnishes an interesting fact in the history of the geographical distribu-

---

* The Prince of Musignano has considered this identical with the *C. Pallasii* of Temminck; and Mr Swainson, from specimens procured by Mr Drummond near the sources of the Athabasca river, and by Mr Bullock in Mexico, has judged them to be distinct. As far as figures and descriptions can be taken as criterions of species, I should consider that of *Pallas* different, and perhaps the analogue in its own country to that of America. There is great similarity in the few birds that compose this genus, and their locality renders the possession and comparison

tion of birds, this genus being one of the twenty-five European
enumerated in our Observations as not known to inhabit this

of them difficult, and it is probable that in three or four will be comprised the whole of the nearly nominal species at present recorded.
The various generic names that have been given to these birds, with the
changes of place in the different systems, at once pronounce it a form of
no ordinary interest ; and there is perhaps none that shows so much
combination between the truly aquatic birds and those of the incessores.
The peculiar form is familiar to most persons at all conversant with
birds, and has been detailed in the description ; the habits, however,
are not so easily observed.   The present remarks will be confined to the
species of Great Britain, which, I believe, may stand as typical of the
genus.

The common *water crow*, or *pyet*, is abundant on most of the more
alpine and rugged streams of Scotland and North of England, enlivening the picturesque and sometimes solitary scenery by its clean and
cheerful appearance, or associating more sedate recollections with the
low and pensive melody of its song.   They live generally in pairs,
keeping entirely to the line of the brook ; and in their flight fly directly
up or down, with a rapid motion of the wings, uttering a single monotonous alarm note, and when about to alight, fall, or splash as it were,
in the stream, and swim to shore.   Previous to the season of incubation, they become more noisy.   The nest is formed exactly like that of
the common wren, with a single entrance, and is composed of the
ordinary mosses found near the stream, without much lining.   It is
generally placed under a projecting rock, a few yards above the water,
and often where a fall rushes over, in which situation the parent birds
must dash through it to gain the nest, which they do with apparent
facility, and even seem to enjoy it.   At night they roost in similar
situations, perched, with the head under the wing, on some little projection, often so much leaning as to appear hanging with the back
downwards.   I recollect a bridge over a rapid stream, which used to
be a favourite nightly retreat, under the arch ; I have there seen four
at a time sitting asleep in this manner, and used to take them with a
light.   Before settling for their nightly rest, they would sport in the
pool beneath, chasing each other with their shrill and rapid cry, and at
last suddenly mount to their perch ; when disturbed, they would return
again in five minutes.   During winter, they migrate to the lower
stream ;—flowing into the Annan, in Dumfriesshire, there are many
alpine tributary rivulets where these birds are abundant in summer ;
during winter they remove almost entirely to the latter, where they
find a more abundant supply of food, and their aquatic powers are more
easily observed.   On every reach one or two may be seen perched upon

continent. A specimen from the northern countries, communicated by Mr Leadbeater, first enabled us to introduce it into
the American fauna; and, almost simultaneously, Mr Swainson,
in his Synopsis of the birds discovered in Mexico by Mr Bullock,
announced it as occurring in that country, but in no other part,
as he thought, of America. Judging from his short description (and the species does not admit of a long one), we have
no hesitation in affirming that both Mr Swainson's and that
described by Temminck, and supposed to have been found by
Pallas in the Crimea, are identical with ours, notwithstanding the localities are so widely distant from each other, as
well as from that whence ours comes, which, however, it will
be perceived, is intermediate between them.

It has been frequently remarked by us (and the fact is now
well established), that many birds of Mexico, entirely unknown
in the Atlantic territories of the United States, are met with
in the interior, and especially along the range of the Rocky

some projecting stone or stick, or watching by the very edge of the ice,
whence they drop at once on their prey, consisting now almost entirely
of the smaller fishes ; when successful, they return to the edge and
devour the spoil. They are most active in their motions during this
occupation, and dive and return with such rapidity, as to seem constantly dipping and rising, or, as perhaps better expressed by a quaint
poet, it

> Comes and goes so quickly and so oft,
> As seems at once both under and aloft.

In milder weather, when the river was less choked with ice, I have
seen them swimming and diving in the centre of the pools, and so
expertly, that I have mistaken and followed them for the little grebe.
But in all their activity, I have never been able to see them walk upon
the bottom, where the depth exceeded a few inches, and I believe it is
contrary to the habit of any aquatic bird ; the motion has been in all
cases, like all others, performed by the wings.

The species of the genus at present stand nominally as follows :—The
common European *Cinclus aquaticus, C. Pallasii,* Temm., *C. Americanus,*
Swain., *C. Asiaticus ?,* from India, and the *C. septentrionalis* and *melanogaster* of Brehm, mentioned by our author.

Mr Gould has figured a bird, in his beautiful illustrations of Himalaya,
under *C. Pallasii,* which is decidedly different from the American ; but
I do not see any proof why it should be called *C. Pallasii.*—ED.

Mountains, at considerably higher latitudes. But it was not to be expected that a Mexican species should extend so far north as the Athabasca Lake, where our specimen was procured. The circumstance is, however, the less surprising in birds of this genus, as their peculiar habits will only allow them to live in certain districts. The case is similar with the dipper of the old continent, which, though widely dispersed, is only seen in mountainous and rocky countries. Though we do not see any improbability in the American species inhabiting the eastern Asiatic shore, we prefer believing that the specimens on which Temminck established the species, and whose supposed native place was the Crimea, were in fact American. The two species are so much alike in size, shape, and even colour, as to defy the attempts of the most determined system-maker to separate them into different groups.

The single species of which the genus *Cinclus* had hitherto consisted was placed in *Sturnus* by Linné, and by Scopoli, with much more propriety, in *Motacilla*. Latham referred it to *Turdus*. Brisson, mistaking for affinity the strong and curious analogy which it bears to the waders, considered it as belonging to the genus *Tringa* (sandpipers). Bechstein, Illiger, Cuvier, and all the best modern authorities, have regarded it as the type of a natural genus, for which they have unanimously retained the name of *Cinclus*, given by Bechstein, Vieillot alone dissenting, and calling it *Hydrobata*. This highly characteristic name, notwithstanding its close resemblance in sound and derivation to one already employed by Illiger as the name of a family, appears to be a great favourite with recent ornithologists, as they have applied it successively to several different genera, and Temminck has lately attempted to impose it on the genus of ducks which I had named *Fuligula*. In my system, the genus *Cinclus* must take its place in the family *Canori*, between the genera *Turdus* and *Myiothera*.

The dippers, or water-ouzels, are well distinguished by their peculiar-shaped bill, which is compressed-subulate, slightly

bent upwards, notched, and with its edges bent in, and finely
denticulated from the middle; but more especially by their
long, stout, perfectly smooth tarsi, with the articulation exposed, a character which is proper to the order of waders, of
which they have also the habits, nay, are still more aquatic
than any of them. Their plumage also being thick, compact,
and oily, is impermeable to water, as much so as that of the
most decidedly aquatic webfooted birds; for when dipped into
it, that fluid runs and drops from the surface. Their head
is flat, with the forehead low and narrow; the neck is stout;
the body short and compact; the nostrils basal, concave, longitudinal, half covered by a membrane; tongue cartilaginous
and bifid at tip. Their wings are short and rounded, furnished
with a very short spurious feather, and having the third and
fourth primaries longest; the tail short, even, and composed
of wide feathers; the nails large and robust; the lateral toes
are subequal, the outer united at base to the middle one, the
hind toe being short and robust. The female is similar to the
male in colour, and the young only more tinged with reddish.
They moult but once in the year.

These wild and solitary birds are only met with singly, or
in pairs, in the neighbourhood of clear and swift-running
mountain streams, whose bed is covered with pebbles, and
strewed with stones and fragments of rock. They are remarkably shy and cautious, never alight on branches, but keep
always on the border of the stream, perched, in an attitude
peculiar to themselves, on some stone or rock projecting over
the water, attentively watching for their prey. Thence they
repeatedly plunge to the bottom, and remain long submerged,
searching for fry, crustacea, and the other small aquatic
animals that constitute their food. They are also very destructive to mosquitoes, and other dipterous insects, and their aquatic
larvæ, devouring them beneath the surface. They never
avoid water, nor hesitate in the least to enter it, and even precipitate themselves without danger amidst the falls and eddies
of cataracts. Their habits are, in fact, so decidedly aquatic,

that water may be called their proper element, although
systematically they belong to the true land birds.   The web-
footed tribes swim and dive ; the long-legged birds wade as
long as the water does not touch their feathers ; the dippers
alone possess the faculty of walking at ease on the bottom, as
others do on dry land, crossing in this manner from one shore
to the other, under water.   They may be often seen gradually
advancing from the shallows, penetrating deeper and deeper,
and, careless of losing their depth, walking with great facility
on the gravel against the current.   As soon as the water is
deep enough for them to plunge, their wings are opened,
dropped, and agitated somewhat convulsively, and, with the
head stretched horizontally, as if flying, they descend to the
bottom, where they course up and down in search of food.
As long as the eye can follow them, they appear, while in the
water, covered with bubbles of air, rapidly emanating from
their bodies, as is observed in some coleopterous insects.

The dippers run very fast : their flight is direct, and swift
as an arrow, just skimming the surface, precisely in the manner
of the kingfisher.   They often plunge under at once, without
alighting, reappearing at a distance.   When on their favourite
rocks, these birds are constantly dipping in the water, at the
same time flirting their erected tail.   While on the wing, they
utter a feeble cry, their voice being weak and shrill, but somewhat varied ; and they sing from their perch, not loud, but
sweetly, even in the depth of winter.   Early in the spring,
they begin to utter clear and distinct notes, and are among
the first to cheer the lonely and romantic haunts which they
frequent with their simple melody.

These birds, like others that live about the water, pair early,
and have two broods in the season.   The young can leave their
nest before being full fledged ; and, at the approach of danger,
drop from the height where it is generally placed into the
water.   In order that this may be done, they build in some
place overhanging the water, the ledge of a rock, or the steep
bank of a rivulet ; or sometimes, in inhabited countries,

take advantage of mills, bridges, or other works of man. The nest is large, composed of moss, and vaulted above ; the eggs are from four to six, and of a milky white. Though very carefully hid, it may be easily discovered by the incessant chirping of the young.

Having seen nothing but the dried skin of the American dipper, and being utterly unacquainted with its habits, we have been describing as common to the genus those of the European species, which are well known, and which we have stopped to watch and admire among the precipices of the Alps and Apennines, where it struggles with the steepest and most noisy cascades and the wildest torrents. The exceedingly great similarity of form in the two species strongly warrants the belief of equal similarity in habits. The more uniform and cinereous hue of the American, the want of reddish, but especially the striking absence of the white on the throat and breast, are the sole, but sufficient marks of difference between the two species.

Pallas' dipper is longer than the common species, measuring eight and a half inches. The bill is perfectly similar, and three-quarters of an inch long, blackish, paler beneath and on the edges. The whole bird, without any exception, is of a dark greyish slate colour, with the base of the plumage some-what lighter; at the superior orbit is a slight indication of whitish. The uniform general colour is somewhat darker on the head, and a shade lighter beneath. The wings are three and a half inches long, as in the genus; the coverts and ter-tials slightly tipt with dingy whitish; the primaries incline somewhat to brown ; the tail measures one inch and a half, and is perfectly even. The feet are of a flesh-colour, and the nails dusky white ; the tarsus is precisely one inch long.

If we could rely on Brehm, four species of this genus exist, which are all found in the old continent. Two are new ones, proposed by himself, under the names of *Cinclus septentrion-alis,* and *Cinclus melanogaster.* The latter, according to him, is a Siberian species, appearing occasionally on the northern coast of European Russia in winter, and is, perhaps, a genuine species, easily distinguished from the *Cinclus aquaticus* by having but ten feathers in the tail, whilst all others have twelve, in addition to its smaller size, darker colour, and dingy throat ; but the former can hardly be regarded even as a northern variety produced by climate. Mr Brehm is probably quite correct in observing that both his new species are perfectly similar to the old one.

BOHEMIAN WAXWING. (*Bombycilla garrula.*)

PLATE XVI.—Fig. 2.

Ampelis garrulus, *Linn. Syst.* i. p. 297, sp. 1.—*Gmel. Syst.* i. p. 838, sp. 1.—*Lath. Ind.* p. 363, sp. 1.—*Muller,* p. 30.—*Kram. El.* p. 363, sp. 1.—*Borowsk. Nat.* iii. p. 171, sp. 68.—*Meyer and Wolf, Tasch. Deutsch.* i. p. 204.—Lanius garrulus, *Faun. Suec.* ii. sp. 82.—*Scop. Ann.* i. sp. 20.—*Brunn.* sp. 25, 26.— Bombyciphora poliocœlia, *Meyer, Vog. Liv. and Esth.* p. 104.—Bombycivora garrula, *Temm. Man. Orn.* i. p. 124.—*Selby, Ill. Br. Orn.* i. p. 87, pl. 34.— Bombyciphora garrula, *Brehm, Lehr. Eur. Vog.* ii. p. 980.—Bombycilla garrula, *Vieill. Nouv. Dict.*—*Nob. Suppl. Syn. Am. Birds in Zool. Journ. London,* iv. p. 3, sp. 65 ; *bis, Ranz. Elem. Orn.* iv. p. 136, sp. 1.—Bombycilla Bohemica, *Steph. Cont. Shaw's Zool.* x. p. 421.—Garrulus, *Gesn. Av.* p. 703.—*Aldr. Orn.* i. p. 796, pl. 798.—*Mus.* p. 674, pl. 675.—*Raii, Syn.* p. 85, A.—*Will. Orn.* p. 96, pl. 20.—*Alb. Av.* ii. p. 25, pl. 26.—Turdus cristatus, *Wirsing, Vog.* pl. 4.—*Frisch,* pl. 32, fig. 1, male.—*Klein, Stemm.* p. 11, p. 13, fig. 5, *a–c.*—Turdus Bombycilla Bohemica, *Briss. Orn.* ii. p. 333, sp. 63 ; *Id.* 8vo, i. p. 250.—Garrulo di Boemia, *St degli Ucc.* ii. pl. 16.—Le Jaseur, *Buff. Ois.* iii. p. 426, pl. 26.—*Le Vaill. Ois. Para.* i. p. 137, pl. 49.—Le Jaseur de Bohéme, *Buff. Pl. enl.* 261.—*Cuv. Regn. Anim.* i. p. 349.—Euro- paischer Seidenschwanz, *Bechst. Nat. Deutschl.* iii. p. 410, pl. 34, fig. 1.— Rothlichgraver, Seidenschwanz, *Naum. Vog.* pl. 32, fig. 66.—*Meyer und Wolf, Ois. d'Allem. Livr.* 22, pl. 6, fig. 1, male ; fig. 2, female.—Silk-tail, Ray, *Syn.* p. 85, A.—*Phil. Trans.* xv. p. 1165, pl. 1, fig. 9.—Bohemian Chatterer, *Penn. Brit. Zool.* sp. 112, pl. 48 ; *Id.* fol. 7, pl. 1, c.—*Lath. Syn.* iii. p. 91, sp. 1.—*Uebersetz,* iii. p. 86, sp. 1.—*Bell, Travels,* i. p. 98.— *Flor. Scot.* i. sp. 92.—*Mont. Orn. Dict.*—*Lewin, Brit. Birds,* i. pl. 2.— *Bewick, Br. Birds.*—Donovan, *Br. Birds,* i. pl. 11.—*Pult. Cat. Dorsetsh.* p. 11.—*My Collection.*

BOMBYCILLA GARRULA.—Bonaparte.

Bombycilla garrula, *North. Zool.* ii. p. 237.

If the absurd theory advanced by Buffon, that European ani- mals degenerate, or become more or less changed, in other

climates, needed in our time any additional refutation, the
discovery of this bird in the north-western territory, near the
Rocky Mountains, would afford it. By appearing in its full
size and perfection, exactly similar to the European individuals
of its species, it would vindicate its smaller relation, the common and familiar cedar-bird, from the reproach of degeneracy.
But with the more enlightened opinions that now prevail, its
occurrence in that unexplored portion of the globe is important, chiefly as tending to solve the problem of the place of
abode of this mysterious wanderer ; especially as, by a singular
coincidence, whilst we were proclaiming this species as American, it was received by Temminck from Japan, together with
a new species, the third known of the genus, which he has
caused to be figured and distinguished by the appropriate name
of *Bombycilla phœnicoptera,* Boiè. Besides the red band
across the wing, whence its name is derived, the length of its
crest, adorned with black feathers, and the uniform absence,
in all states, of the corneous appendages of the wings, this
new species, resembling more in size and shape the Carolina
waxwing (cedar-bird) than the present, is eminently distinguished from both by wanting the small, closely-set feathers
covering the nostrils, hitherto assigned as one of the characters
of the genus. This example evinces the insufficiency of that
character, though Illiger considered it of such importance as
to induce him to unite in his great genus *Corvus* (comprehending this as well as several other distinct groups) all the species possessing it. It shows especially how erroneous it is to
form two separate families for the allied genera with covered
or naked nostrils. In fact, the genus as it now stands is, not
the less for this aberration, an exceedingly natural one, though
the two species that are now known to inhabit America are
still more allied to each other than either of them to the
Japanese, the present (Bohemian) differing chiefly by its larger
size, mahogany-brown tail-coverts, and cinereous belly, the
first being white and the second yellowish in the cedar-bird,
which also wants the yellow and white markings on the wing.

Of the three species now comprehended in the genus, one is peculiar to America, a second to eastern Asia, and the present common to all the arctic world.

This small but natural group, at one time placed by Linné in the carnivorous genus *Lanius*, notwithstanding its exclusively frugivorous habits, was finally restored by him to *Ampelis*, in which he was followed by Latham. Brisson placed it in *Turdus*, and Illiger in *Corvus*. Ornithologists now concur in regarding it as a genus, disagreeing only as to the name, some calling it *Bombyciphora*, others *Bombycivora*, though they all appear to have lately united in favour of the more elegant and prior termination of *Bombycilla*.

The waxwings, which we place in our family *Sericata*, having no other representative in Europe or North America, are easily recognised by their short, turgid bill, trigonal at base, somewhat compressed and curved at tip, where both mandibles are strongly notched; their short feet, and rather long subacute wings. But their most curious trait consists in the small, flat, oblong appendages, resembling in colour and substance red sealing-wax, found at the tips of the secondaries in the adult. These appendages are merely the coloured corneous prolongation of the shafts beyond the webs of the feathers. The new species from Japan is, as we have mentioned, at all times without them, as well as the young of the two others. The plumage of all is of a remarkably fine and silky texture, lying extremely close; and they are all largely and pointedly crested, the sexes hardly differing in this respect.

The waxwings live in numerous flocks, keeping by pairs only in the breeding season; and so social is their disposition, that, as soon as the young are able to fly, they collect in large bands from the whole neighbourhood. They perform extensive journeys, and are great and irregular wanderers. Far from being shy, they are simple and easily tamed, but generally soon die in confinement. Their food consists chiefly of juicy fruits, on which they fatten, but to the great detriment of the orchard, where they commit extensive ravages. When

fruits are scarce, they seize upon insects, catching them dexterously in the same manner as their distant relatives the flycatchers. No name could be more inappropriate for these
birds than that of chatterers, as there are few less noisy, and
they might even be called mute with much better reason.
They build in trees, and lay, twice in a year, about five eggs.

Whence does the Bohemian waxwing come at the long and
irregular periods of its migrations? Whither does it retire
to pass its existence and give birth to its progeny? These
are circumstances involved in darkness, and which it has not
been given to any naturalist to ascertain. It has been stated,
and with much appearance of probability, that these birds
retire during summer within the arctic circle; but the fact is
otherwise, naturalists who have explored these regions asserting that they are rarer and more accidental there than in
temperate climates. It seems probable that their chief place
of abode is in the oriental parts of the old continent, and, if
we may hazard an opinion, we should not be surprised if the
extensive and elevated tableland of Central Asia were found
to be their principal rendezvous, whence, like the Tartars in
former times, they make their irregular excursions.

As we can only arrive at the truth in this matter by observing facts and collecting localities, we shall endeavour to
do this with the greatest accuracy. In northern Russia and
the extreme north of Norway they are seen in great numbers
every winter, being observed there earlier than in temperate
countries. In northern Asia and eastern Europe their migrations are tolerably regular; very numerous flocks generally
pass through Scania in November, and are again seen on
their return in the spring. But they appear only at very
remote and irregular periods, and merely as occasional and
rare visitants in western, southern, or even central and
northern Europe, and then only in the coldest months of the
most severe winters. Notwithstanding that they at times
invade peculiar districts in vast numbers, so remarkable is
the appearance of these winged strangers then considered,

that we find it placed upon record. However extraordinary it may seem to those who live in this enlightened age and country that the unusual appearance of "cedar-birds of a large kind" should strike terror into the souls of men, such, notwithstanding, was the effect in more ignorant times. They have been looked upon as the precursors of war, pestilence, and other public calamities. One of their irruptions was experienced in Italy in 1571, when flocks of hundreds were seen flying about in the north of that country in the month of December, and were easily caught. A similar visit had taken place in 1530, in February, marking the epoch when Charles V. caused himself to be crowned at Bologna. Aldrovandi, from whom we learn the above particulars, also informs us that large flocks of them appeared in 1551, when it was remarked, that, though they spread in numbers through the Modenese, the Plaisantine, and other parts of Italy, they carefully avoided entering the Ferrarese, as if to escape the dreadful earthquake that was felt soon after, causing the very birds to turn their flight. In 1552, Gesner informs us, they appeared along the Rhine, near Mentz in Germany, in such numbers as to obscure the sun. They have, however, of late years, in Italy and Germany, and in France especially, at all times been extremely rare, being seen only in small companies or singly, appearing as if they had strayed from their way. In England, the Bohemian waxwing has always been a rare visitant, coming only at long and uncertain intervals. In the winter of 1810 large flocks were dispersed through various parts of that kingdom, from which period we do not find it recorded by English writers till the month of February 1822, when a few came under Mr Selby's inspection, and several were again observed during the severe storm in the winter of 1823. Upon the Continent, its returns are subject to similar uncertainty. In M. Necker's very interesting memoir, lately published, on the birds of Geneva, we read, that from the beginning of this century only two considerable flights have been observed in that canton, one in January 1807, and the other in January 1814, when they were very numerous, and spent the winter there, all departing in March. In 1807 they were dispersed over a great portion of western Europe, and were seen near Edinburgh in the first day of that year.

What extent of country they inhabit or frequent in this continent, and whether numerous or not, we are unable to state. The specimen here figured was obtained, together with others, from the north-western range of the Rocky Mountains, and the species appears to spread widely, as we have been credibly informed by hunters that "cedar-birds of a large kind" have been shot a little beyond the Mississippi, at a very great distance from the spot where ours were obtained. Thus does this species extend its range round the whole earth, from the coasts of Europe eastwardly to the Rocky Mountains in America, and we are at a loss to conceive why it should never have been observed on this side of the Mississippi.

Very little is known of the peculiar habits of this elegant bird. It assembles in large flocks, and feeds on different kinds of juicy berries or on insects, which during summer constitute its principal food. In common with many other birds, they are fond of the berries of the mountain-ash and phytolacca, are extremely greedy of grapes, and also, though in a less degree, of juniper and laurel berries, apples, currants, figs, and other fruits. They drink often, dipping in their bill repeatedly. Besides their social disposition and general love of their species, these birds appear susceptible of individual attachment, as if they felt a particular sentiment of benevolence, even independent of reciprocal sexual attraction. Not only do the male and female caress and feed each other, but the same proofs of mutual kindness have been observed between individuals of the same sex. This amiable disposition, so agreeable for others, often becomes a serious disadvantage to its possessor. It always supposes more sensibility than energy, more confidence than penetration, more simplicity than prudence, and precipitates these, as well as nobler victims, into the snares prepared for them by more artful and selfish beings. Hence they are stigmatised as stupid, and, as they keep generally close together, many are easily killed at once by a single discharge of a gun. They always alight on trees, hopping awkwardly on the ground. Their flight is very rapid : when taking wing, they utter a note resembling the syllables, *zi, zi, ri,* but are generally silent, notwithstanding the name that has been given them. They are, however, said to have a sweet and agreeable song in the time of breeding, though at others it is a mere whistle. The place of breeding, as we have intimated, is not known with any certainty, though they are said to build in high northern latitudes, preferring mountainous districts, and laying in the clefts of rocks, which, however, judging from analogy, we cannot believe.

What can be the cause of their leaving their unknown abodes, of their wide migrations and extraordinary irruptions, it is very difficult to determine. That they are not compelled to them by cold is well proved. Are they to be ascribed to necessity from excessive multiplication, as is the case with the small quadrupeds called lemmings, and even with man himself in a savage state or in over-populous countries? or shall we suppose that they are forced by local penury to seek elsewhere the food they cannot be supplied with at home? Much light may be thrown on the subject by carefully observing their habits and migrations in America.

The Bohemian chatterer being so well known, we shall here only give a description of our best American specimen, which is a female shot on the 20th March 1825, on the Athabasca river, near the Rocky Mountains. The sexes hardly differ in plumage.

Length, eight and a half inches ; extent, fifteen ; bill, three-quarters of an inch long, black, paler at the base of the under mandible ; irides, reddish, often quite red ; nostrils, entirely uncovered. From the base of the ridge of the bill arises, on each side, a velvety black line, bordering the forehead, and spreading on the ophthalmic region, and surrounding almost the whole crown ; throat also deep black. The anterior part of the head is bright bay, behind passing gradually into vinaceous drab ; the feathers of the crown are elongated into a crest measuring nearly an inch and a half ; base of these feathers, blackish ; middle, white ; whole neck and hind head and breast, cinereous drab, slightly tinged with vinaceous, and passing by degrees on the posterior parts above and beneath into pure cinereous, slightly tinged with bluish, which predominates on the rump and upper tail-coverts. The black of the throat is somewhat margined with bright bay, and is separated from the black of the eye by a slight obliterated white line. The cinereous of the belly and femorals is paler ; the vent and lower tail-coverts are chestnut rufous, and the feathers very long. The wings measure four and a half inches in length; the second primary is somewhat longer than the first, the others decreasing in succession rapidly. The upper tail-coverts are cinereous drab, like the back, the lower whitish grey ; quills, dusky black, much paler on their inner vane towards the base. The first is unspotted, the second has a slight mark of white on the outer web at tip. This mark increases in size successively on the following, becoming a longitudinal spot, much larger on the secondaries, four of which are furnished with bright red appendages. Each feather of the winglet is broadly white at tip, constituting a remarkable white spot on the wing, which appears to be on the primaries. No yellow whatever is observable on the wing. The tail is three inches long, black, broadly tipt with pale yellow for half an inch, dark bluish grey at base. Tarse, which is three-quarters of an inch long, and feet, black.*

* See vol. i. p. 106, for *B. Americana* and notes.—ED.

FEMALE PINE BULLFINCH.   (*Pyrrhula enucleator.*)

PLATE XVI.—Fig. 3.

See *Wilson's American Ornithology*, Pine Grosbeak, Loxia enucleator, vol. i. p. 79, pl. 5, fig. 2, for the male at the age of one year.—Loxia enucleator, *Linn. Syst.* i. p. 299, sp. 3.—*Faun. Suec.* sp. 223.—*Schœn. Act. Holm.* 1757, p. 139.—*Gmel. Syst.* i. p. 845, sp. 3.—*Brunn.* sp. 239.—*Muller,* sp. 246.—*Borowsk. Nat.* iii. p. 133, sp. 3.—*Lath. Ind.* i. p. 372, sp. 5.—*Retz. Faun. Suec.* p. 234, sp. 211.—*Meyer and Wolf, Taschenb. Vog. Deutschl.* i. p. 142.—Loxia flamingo, *Mus. Carls.* i. pl. 17.—*Gmel. Syst.* i. p. 864, accid. var.—Loxia pyrrhula, var. δ.—*Lath. Ind.* i. p. 388, sp. 56, accid. var. —Coccothraustes Canadensis, *Briss. Orn.* iii. p. 250, sp. 15, pl. 12, fig. 3 ; *Id.* 8vo, i. p. 378.—Pyrrhula enucleator, *Temm. Man. Orn.* i. p. 383.—*Sabine, Zool. App. to Frank. Exp.* p. 675.—*Brehm, Lehr. Eur. Vog.* i. p. 169.—*Ranz. Elem. Orn.* vi. p. 70, sp. 2.—*Selby, Ill. Brit. Orn.* i. p. 256, pl. 53, fig. 1, male; fig. 2, female.—*Nob. Obs. Wils. Nom.—Cat. and Syn. Birds U. S.* sp. 193.—Corythus enucleator, *Cuv. Regn. Anim.* i. p. 392.—Strobilophaga enucleator, *Vieill. Gal. Ois.* i. pl. 53, young male.—Fringilla enucleator, *Meyer, Syst. Taschenb.* iii. p. 250, sp. 2.—Ciufolotto snocciolatore, *Ranz. loc. cit.*— Dur-bec, *Buff. Ois.* iii. p. 457.—Gros-bec du Canada, *Id. Pl. enl.* 135, fig. 1, male a year old.—Haken Kernbeisser, *Bechst. Nat. Deutsch.* iii. p. 28.— *Naum. Vog. Nachtr.* pl. 19, fig. 36, male; fig. 37, female.—Der Fichten Kernbeisser, *Meyer and Wolf, Vog. Deutschl.* xii. pl. 5, fig. 1, young male; fig. 2, old female.—Greatest Bullfinch, *Edwards,* pl. 123, young male ; pl. 24, adult female.—Pine Grosbeak, *Penn. Brit. Zool.* sp. 114, pl. 49, fig. 2.—*Arct. Zool.* ii. sp. 209.—*Ellis, Nar.* ii. p. 15.—*Lewin, Brit. Birds,* ii. pl. 68.—*Lath. Syn.* ii. p. 111, sp. 5 ; *Id. Supp.* p. 148.—*Mont. Orn. Dict.* i.—*Walck. Syn.* pl. 207.—*Donov. Brit. Birds,* i. pl. 17.—*Bewick, Brit. Birds,* i. p. 135.—*Shaw's Zool.* ix. p. 238, pl. 43.—*Ubers.* ii. p. 106, sp. 5.—Flamingo Grosbeak, *Lath. Syn. Suppl.* p. 155, accid. var.—*My Collection,* male, female, and young.

CORYTHUS ENUCLEATOR.—Cuvier.*

Pyrrhula (Corythus) enucleator, *North. Zool.* ii. p. 262.

The female pine bullfinch is eight and a half inches long, and thirteen and a half in extent ; the bill measures more than half an inch, is blackish, with the lower mandible paler at base ; the feathers of the whole head, neck, breast, and rump, orange, tipt with brownish, the orange richer on the crown, where are a few blackish dots ; the plumage at base plumbeous ; the back is cinereous, somewhat mixed with orange ; the shafts darker ; belly and femorals, pure cinereous ; lower tail-

---

* See description of the male, note, &c., vol. i. p. 79.

coverts, whitish, shafted with dusky ; the wings are four and a half inches long, reaching beyond the middle of the tail ; the smaller coverts are similar to the back, cinereous, slightly tinged with orange ; middle and larger, blackish, margined with whitish exteriorly and widely at tip ; the lower coverts are whitish grey ; quills, blackish, primaries margined with pale greenish orange, secondaries and tertials with broad white exterior margins ; the tail is three and three-quarter inches long, blackish, the feathers with narrow pale edges ; feet, dusky ; nails, blackish.

In the young female, the head and rump are tinged with reddish. The male represented and most accurately described by Wilson is not adult, but full one year old ; at which period, contrary to the general law of nature, it is the brightest, as was first stated by Linné, though his observation has since been overlooked or unjustly contradicted. In the adult male, the parts that were crimson in the immature bird exhibit a fine reddish orange, the breast and belly being also of that colour, but paler ; the bars of the wings, tinged with rose in the young, become pure white.

We have nothing to add to Wilson's history of this bird. Although, after the example of Temminck and others, we place this species at the head of the bullfinches, we cannot avoid remarking, that its natural affinities connect it most intimately with the crossbills, being allied to them closely in its habits and in its form, plumage, general garb, and even in its anomalous change of colours. The bill, however, precisely that of a bullfinch, induces us to leave it in that genus, between which and the crossbills it forms a beautiful link ; the obtuse point of the lower mandible, but especially the small porrect, setaceous feathers covering the nostrils, as in these latter, eminently distinguish it from all others of its own genus. These characters induced Cuvier to propose it as a subgenus under the name of *Corythus,* and Vieillot as an entirely distinct genus, which he first named *Pinicola,* but has since changed it to *Strobilophaga.* These authors have

of course been followed by the German and English ornithologists of the new school, who appear to consider themselves
bound to acknowledge every genus proposed, from whatever
quarter, or however minute and variable the characters on
which it is based.

WHITE-CROWNED PIGEON. (*Columba leucocephala.*)

PLATE XVII.—Fig. 1.

Columba leucocephala, *Linn. Syst.* i. p. 281, sp. 14.—*Gmel. Syst.* i. p. 772, sp.
14.—*Lath. Ind.* p. 594, sp. 5.—*Temm. Ind. Col. in Hist. Pig. et Gall.* i. p.
459.—*Vieill. Gal. Ois.* ii. p. 331, pl. 194.—Columba minor leucocoryphas,
Raii, Syn. p. 63, sp. 16, and p. 184, sp. 24.—*Klein, Av.* p. 120, sp. 18.—
Columba saxatilis, Jamaicensis, *Briss. Orn.* i. p. 137, sp. 33 ; *Id.* 8vo, i. p.
34.—Columba capite albo, the White-crowned Pigeon, *Catesby, Car.* i. p.
25, pl. 25.—*Seligman, Saml. Selt. Vog.* 2, col. plate.—Le Pigeon de Roche
de la Jamaique, *Buff. Ois.* ii. p. 529.—*Sonn. Buff.* vii. p. 216.—Colombe à
calotte blanche, *Tem. Hist. Pig. et Gall.* i. p. 204 ; *Id.* folio, pl. 13 of the
second family.—*My Collection.—Edinburgh College Museum.*

COLUMBA LEUCOCEPHALA.—Linnæus.

Columba leucocephala, *Bonap. Synop.* p. 119.

This bird has been already alluded to in our first volume,
when pointing out the difference between it and the new
Columba fasciata of Say. We were then far from supposing
that we should so soon have to become its historian ; but
having ascertained that it inhabits Florida, as well as the
West Indies, we are enabled to give it a place in these pages.
A glance at the plate will now render the difference strikingly
obvious to the American student, who will thus perceive, better
than can be explained by words, how entirely distinct the above-
named species is from the present.

The white-crowned pigeon, well known as an inhabitant of
Mexico and the West Indies, is likewise found in great numbers
on some of the Florida keys, such as Key Vacas, and others,
early in spring, where it feeds almost exclusively on a kind of
wild fruit, usually called beach-plum, and some few berries
of a species of palmetto, that appears to be peculiar to those
keys. It is also extensively spread in Jamaica and St Domingo,

Drawn from Nature by A.Rider

Engraved by W.H.Lizars

1. White-crowned Pigeon. 2. Zenaida Dove.
Columba Leucocephala. Columba Zenaida.

and is very abundant in the island of Porto Rico, frequenting deep woods, and breeding on rocks, whence they are called by some *rock pigeons*. They are very numerous on all the Bahama Islands, and form an important article of food with the inhabitants, particularly when young, being then taken in great quantities from the rocks where they breed. On the Florida keys also they breed in large societies, and the young are much sought after by the wreckers. They there feed principally on berries, and especially on those of a tree called *sweetwood*. When the fruit of this is ripe, they become fat and well-flavoured, but other fruits again make their flesh very bitter.

Buffon, in accordance with his whimsical idea of referring foreign species to those of Europe, considers the present as a variety of the biset (*Columba livia*, Briss.) To that bird it is in fact allied, both in form and plumage, and has, moreover, the same habit of breeding in holes and crevices of rocks ; but it is, at the same time, entirely distinct.

The size of the white-crowned pigeon has been underrated by authors. Its length is fourteen inches, and its extent twenty-three ; the bill is one inch long, carmine red at the base, the end from the nostrils being bluish white ; the irides are orange yellow, the bare circle round the eye, dusky white, becoming red in the breeding season ; the entire crown, including all the feathers advancing far on the bill, is white, with a tinge of cream colour, and is narrowly margined with black, which passes insensibly into the general deep slate colour : on the nape of the neck is a small deep purplish space changing to violet ; the remainder of the neck above, and on the sides, is covered by scale-like feathers, bright green, with bluish and golden reflections, according as the light falls ; the sides of the head, the body above, and whole inferior surface, the wings and tail above and beneath—in short, the whole bird, without any exception but the parts described, is of a uniform deep bluish slate, much lighter on the belly, more tinged with blue on the stout-shafted rump-feathers,

somewhat glossy, and approaching to brownish black on the
scapulars : the quills are more of a dusky black ; the wings
are nearly eight inches long, reaching, when closed, to two-
thirds of the tail ; the first primary is somewhat shorter than
the fourth, and the second and third are longest ; the third is
curiously scalloped on the outer web, which is much narrowed
for two inches from the tip; all are finely edged with whitish;
the tail is five inches long, perfectly even, of twelve uniform
broad feathers, with rounded tips; the feet are carmine red,
the nails dusky ; the tarsus measures less than an inch, being
subequal to the lateral toes, and much shorter than the middle
one. The female is perfectly similar. It is one of this sex,
shot in the beginning of March, that is represented in the
plate, and is perhaps a young, or not a very old bird; for it
would seem that as they advance in age these pigeons become
somewhat lighter coloured, the crown acquiring a much purer
white. This, however, we only infer from authors, our plate
and description being faithfully copied from nature.

The young are distinguished by duller tints, and the crown
is at first nearly uniform with the rest of their dark plumage :
this part, after a time, changes to grey, then greyish white,
and becomes whiter and whiter as the bird grows older. It is
proper to remark, after what has been said under the article
of the band-tailed pigeon in vol. i., that the white colour
extends equally over the whole crown, not more on one part
than another ; thus never admitting of a restricted band or
line, as in that much lighter coloured bird.

Another species closely allied to, and perhaps identical with,
our band-tailed pigeon (though we have equally good reasons
for believing it the *Columba rufina* of Temminck), and of
which we have not yet been able to procure specimens, is also
well known to breed on the Florida keys, whither probably
almost all the West Indian species occasionally resort.

ZENAIDA DOVE. (*Columba Zenaida.*)

PLATE XVII.—FIG. 2.

Columba Zenaida, *Nob. Add. Orn. U. S. in Journ. Acad. Phil.—Id. Cat. Birds U. S.* sp. 198, *in Contr. Macl. Lyc. Ph.* i. p. 22.—*Id. Syn. Birds U. S.* sp. 198, *in Ann. Lyc. Nat. Hist. N. Y.* ii. p. 119.—*Id. Suppl. in Zool. Journ. Lond.* v. p. 6.—*My Collection.*

COLUMBA? ZENAIDA.—BONAPARTE.

THE name of dove is not commonly used to designate a systematic group, but is employed for all the small pigeons indiscriminately, whilst the larger doves are known as pigeons. Even this distinction of size, however, does not seem to be agreed upon, as we find authors calling the larger species doves and the smaller ones pigeons, and sometimes even applying both appellations to different sexes or ages of the same species, as in the case of the common American pigeon, *Columba migratoria.* This extensive family of birds—so remarkable for richness and splendour of colours—so important as contributing largely to supply the wants of mankind— so interesting as forming so perfect a link between the two great divisions of the feathered tribes—has been divided on more philosophical principles into three groups, which some naturalists consider as genera, and others as subgenera or sections. Of these, two only are found represented in America; the third, a very natural group, being confined to Africa and the large eastern islands of the old world. That to which the present bird, and all the North American species but one, belong, is the most typical of all, being characterised by a straight and slender bill, both mandibles of which are soft and flexible, and the upper turgid towards the end; by their short tarsi, divided toes, and long, acute wings, with the first primary somewhat shorter than the second, which is the longest. This group (the true pigeons and doves) is, however, so numerous in species, that we cannot but wonder that it should still remain comparatively untouched by the reforming hand of our contemporaries; especially seeing that as good reasons may be found for subdividing them as the parrots and other large natural groups. We may indicate the differences exhibited in the form of the scales covering the tarsus, and the shape of the tail, &c., as offering characters on which sections or genera could be founded. But as the species of the United States, which are those we are to treat of, are but few, we shall leave the promising task to any one whose researches may lead him to engage in it; and shall only observe, that the two species described by Wilson belong to a different group from the three we have since introduced into the fauna of this country. Of these, the present beautiful dove is the only one hitherto undescribed.

This new and charming little species inhabits the Florida keys with the preceding, but is much more rare. We have also received it from Cuba, and noticed a specimen in a collection of skins sent from that island by Mr MacLeay to the Zoological Society of London. They are fond of being on the ground, where they are most commonly observed, dusting themselves, and seeking for the gravel which, like the gallinaceous birds, they swallow to assist digestion. When flushed, they produce the same whistling noise with their wings as the common turtle-dove, *Columba Carolinensis.*

The Zenaida dove measures ten inches in length. The bill is somewhat more robust than that of the common dove, but otherwise perfectly similar, less than an inch long, black, the corners of the mouth being lake; the irides are dark brown, the pupil of the eye large, and the eye itself full, giving the whole bird a mild and pleasing expression; the naked orbits are of a bluish grey. The whole plumage above is yellowish ashy brown, tinged with vinaceous on the crown, and paler on the sides of the head and neck; under the ears is a small bright rich and deep violaceous spot, rivalling the amethyst in splendour, and above this a similar smaller one, not very distinguishable; the sides of the neck before the bend of the wing exhibit splendid golden violaceous reflections, slightly passing into greenish in different lights; the scapulars are spotted with black, the spots being large and roundish ; the exterior wing-coverts, spurious wing and quill feathers, are blackish ; the primaries are edged with white externally, and, with the exception of the outer ones, at tip also ; the secondaries are broadly terminated with white ; the chin is yellowish white ; the whole inferior surface is bright vinaceous, paler on the throat, and gradually passing into richer on the belly ; the flanks and under wing-coverts are delicate lilac, and the under tail-coverts are mixed with the same colour, some of the longest being entirely lilac, which is also found at the base of the plumage on the belly and rump ; the wings are six inches and a quarter long ; reaching within one inch of the tip of the tail ; the primaries are entire on both vanes ; the first is longer than the fourth, the second longest, though scarcely longer than the third ; the tail is four and a half inches long, composed of twelve broad, full, rounded feathers, extending but one inch beyond their coverts ; it is nearly even, and of the colour of the body, with a broad black band at two-thirds of its length, obsolete on the two middle feathers (which are of the colour of the body), purer on the three exterior ; the lateral feathers are pearl grey for half an inch towards the tip, the outer plume being, moreover, of that colour on the outer vane ; all the tail-feathers are blackish on the inferior surface to within three-quarters of an inch of their tips. The feet are red ; the nails blackish ; the tarsus measures three-quarters of an inch in length.

The female is very similar to the male in size and colour ; the head, however, is but slightly tinged with vinaceous, the golden violet reflections of the neck are not quite so vivid, and the inferior surface of a paler vinaceous, but graduated as in the male. The lateral tail-feathers are also much more uniform with the middle one, and of course with the back, the three outer only on each side being pearl grey at tip. This latter character, however, we should rather attribute to age than sex, if we had not good reason to believe that our female is a perfectly adult bird.

At first sight, the Zenaida dove might perhaps be mistaken for the common turtle-dove (*Columba Carolinensis* and *marginata* of authors), having the same general colour and several common markings ; but, to mention no other differential character, the short even tail, composed of but twelve feathers, all rounded, the outer bluish grey at tip, will at once distinguish it from the latter, which belongs to a different group, having the tail long, cuneiform, and (what is found in no other American species, not even its close relation the passenger pigeon) composed of fourteen tapering and acute feathers, the two middle remarkably so, and the lateral pure white at tip.   If any other distinction should be required, the white tips of the secondaries of our new species will afford a good one, as well as the outer tail-feather, the exterior web of which is blue grey, crossed, as well as the others, by the black band ; whilst in the *C. Carolinensis* it is entirely pure white, the black band being confined to the inner web.

DUSKY GROUSE.   (*Tetrao obscurus.*)

PLATE XVIII.--Female.

Tetrao obscurus, *Say, in Long's Exped. to Rocky Mount.* ii. p. 14.—*Nob. Cat. Birds U. S.* sp. 209, *in Contr. Macl. Lyc. Phila.* i. p. 23.—*Id. Syn. Birds U. S.* sp. 207, *in Ann. Lyc. Nat. Hist. N. Y.* pp. 127, 442.—*Philadelphia Museum*, female.—*Collection of Mr Sabine, in London*, male and female.

TETRAO OBSCURUS.—Say.*

Tetrao obscurus, *North. Zool.* ii. p. 334, plate 59, male ; 60, female.

Linné,. in his genus *Tetrao*, brought together so great a number of species bearing no more than a distant resemblance

---

* The authors of the " Northern Zoology " have given a beautiful figure of this bird, and have quoted the *T. Richardsonii* of Douglas as identical with it.   This I am unable to decide, but should certainly give some weight to the comparisons of Dr Richardson, who thinks that those deposited in the Edinburgh Museum are only younger specimens.

The characters given by Mr Douglas are :—

Drawn from Nature by A.Rider.

Engraved by W.H.Lizars

Dusky Grous Female.
Tetrao Obscurus.

18.

to each other, and differing not only in their external characters, but even in their peculiar habits, that he might, with
almost the same propriety, have included in it all typical gallinaceous birds.   Latham very judiciously separated the genus
Tinamus, as well as that of *Perdix*, which latter he restored
from Brisson.   Illiger likewise contributed to our better knowledge of these birds, by characterising two more natural genera,
Syrrhaptes and *Ortygis*.   Temminck, in his "Histoire des Galli-
nacés," carried the number to seven, but has since reduced it
by reuniting *Coturnix* to *Perdix*.

The true *Tetraones* are divided by Vieillot into two genera,
the *Lagopodes* forming a distinct one by themselves.   These,
however, we regard as no more than a subgenus, of which we
distinguish three in our genus *Tetrao*.   I. *Lagopus*, which represents it in the arctic polar regions, for whose climate they
are admirably adapted by being clothed to the very nails in
plumage suited to the temperature, furnished abundantly with
thick down, upon which the feathers are closely applied.   The
colour of their winter plumage is an additional protection
against rapacious animals, by rendering it difficult to distinguish them from the snows by which they are surrounded.

T. RICHARDSONII, *Mas.*—Pallidé plumbeo-griseus fusco sparsim undulatus :
    gulæ plumis in medio albis : abdomine saturatiore albo parcé maculato :
    maculâ laterali sub nuchâ albâ : rectricibus nigris, apice albicante.
Fæm.—Minor, brunnescenti-grisea, dorso brunneo fasciato ; subtùs albo fre-
    quenter notato rectricibus duobus mediis ferrugineo fasciatis.

That gentleman mentions a trait in their manners, which he thinks
is peculiar to this species.   " On being started from the dark shadowy
pine-trees, their usual roosting-place, they descend, or, more properly,
allow themselves to fall within a few feet of the ground, before they
commence flying, a circumstance which often leads the sportsman to
think he has secured his bird, until the object of his attention leaves
him, darting and floating through the forest."

They were very abundant on the subalpine regions of the Rocky
Mountains, in lat. 52 deg., and still more numerous on the mountainous
districts of the river Columbia, in lat. 48 deg.   They were rare, however, on the north-west coast.

The specimens in the Edinburgh Museum have been accurately figured
by Mr J. Wilson in his "Illustrations of Zoology," under the name of
T. Richardsonii.—ED.

II. *Tetrao*, which is distributed over the more temperate climates; the legs being still feathered down to the toes. III. *Bonasia*, a new division, of which we propose *Tetrao bonasia*, L., as the type, in which only the upper portion of the tarsus is feathered. These occasionally descend still farther south than the others, inhabiting wooded plains as well as mountainous regions, to which those of the second section are more particularly attached. But the entire genus is exclusively boreal, being only found in Europe, and the northern countries of America and Asia. The long and sharp-winged grouse, or *Pterocles* of Temminck, which represent, or rather replace, these birds in the arid and sandy countries of Africa and Asia, a single species inhabiting also the southern extremity of Europe, we consider, in common with all modern authors, as a totally distinct genus. That group, composed of but few species, resort to the most desert regions, preferring dry and burning wastes to the cool shelter of the woods. These oceans, as they might be termed, of sand, so terrific to the eye and the imagination of the human traveller, they boldly venture to cross in large companies in search of the fluid so indispensable to life, but there so scarce, and only found in certain spots. Over the intervening spaces they pass with extraordinary rapidity, and at a great elevation, being the only gallinaceous birds furnished with wings of the form required for such flights. This, however, is not the only peculiarity in which they aberrate from the rest of their order, and approach the pigeons, being said to lay but few eggs, the young remaining in the nest until they are full fledged, and fed in the meantime by the parents.

The grouse dwell in forests, especially such as are deep, and situated in mountainous districts; the *Bonasiæ*, however, and the *Tetrao Cupido*, frequenting plains where grow trees of various kinds. The *Lagopodes* of the arctic regions, or ptarmigans, are also found on the very elevated mountains of Central Europe, where the temperature corresponds to that of more northern latitudes. Here they keep among the tufts of dwarf-willows, which, with pines, form the principal vegetation of these climates. The grouse feed almost exclusively on leaves, buds, berries, and especially the young shoots of trees, pines, spruce, or birch, resorting to seeds only when compelled by scarcity of other food, or when their usual means of subsistence are buried beneath the snow. They sometimes, especially when young, pick up a few insects and worms, and are fond of ants' eggs. Like other gallinaceous birds, they are constantly employed in scratching the earth, are fond of covering themselves with dust, and swallow small pebbles and gravel to assist digestion. No birds are more decidedly and tyrannically polygamous. As soon as the females are fecundated, the male deserts them, caring no further about them nor their progeny, to lead a solitary life. Like perfidious seducers, they are full of attention, however, and display the greatest anxiety to secure the possession of those they are afterwards so ready to abandon. The nuptial season commences when the leaves first appear in spring. The males then appear quite intoxicated with passion ; they are seen, either on the ground, or on the fallen trunks of trees, with a proud deportment, an inflamed and fiery eye, the feathers of the head erected, the wings dropped, the tail widely spread, parading and strutting about in all sorts of extravagant attitudes, and expressing their feelings by sounds so loud as to be heard at a great distance. This season of ardour and abandonment is protracted till June. The deserted female lays, unnoticed by the male, far apart on the ground, among low and thick bushes, from eight to sixteen eggs, breeding but once in a season. They sit and rear their young precisely in the manner of the common fowl, the chicks being carefully protected by the mother only, with whom they remain all the autumn and winter, not separating until the return of the breeding season. It is only at this period that the males seek the society of the females.

The grouse are remarkably wild, shy, and untamable birds, dwelling in forests or in barren uncultivated grounds, avoiding cultivated and thickly inhabited countries, and keeping together in families. The *Lagopodes* only live in very numerous flocks, composed of several broods, parting company when the return of spring invites them to separate in pairs of different sexes, which is always done by the birds of this division. Except in the breeding season, the grouse keep always on the ground, alighting on trees only when disturbed or when going to roost at night; by day retiring to the deepest part of the forest. The flesh of all grouse is delicious food, dark-coloured in some, and white in others, the dark being more compact, juicy, and richly flavoured, as in *Tetrao Cupido;* while the white, though somewhat dry, is distinguished for delicacy and lightness. Such are the *Bonasiæ, T. umbellus* of America, and *T. bonasia* of Europe.

The grouse are distinguished by a short stout bill, feathered at base, and they are, of all gallinaceous birds, those in which the upper mandible is the most vaulted; the feathers of the bill are very thick and close, and cover the nostrils entirely; the tongue is short, fleshy, acuminate, and acute; the eye is surmounted by a conspicuous red and papillous naked space; the tarsi are generally spurless in both sexes, and partly or wholly covered with slender feathers, which in the *Lagopodes* are thicker and longer than in the rest, extending not only beyond the toes, but growing even on the sole of the foot—a peculiarity which, agreeably to the observation of Buffon, of all animals is again met with only in the hare. These feathers in winter become still longer and closer. All the others have the toes scabrous beneath, and furnished with a pectinated row of processes on each side.* This roughness of the sole of the feet enables them to tread firmly on the slippery surface of the ground or frozen snow, or to grasp the branches of trees covered with ice. Their nails are manifestly so formed as to

---

* These processes are liable to fall off, at least in preserved skins. It is owing to this circumstance that we committed several errors in characterising these birds in our "Synopsis of the Birds of the United States."

suit them for scratching away the snow covering the vegetables
which compose their food. The wings of the grouse are short
and rounded; the first primary is shorter than the third and
fourth, which are longest. The tail is usually composed of
eighteen feathers, generally broad and rounded. The red
grouse, *T. Scoticus*, however, and the European *Bonasiæ*, and
T. Canadensis, or spotted grouse, have but sixteen; while our
two new North American species have twenty, one of them
having these feathers very narrow and pointed, the narrowness
being also observed in the sharp-tailed grouse. They have
the head small, the neck short, and the body massive and
very fleshy.

The females of the larger species differ greatly from the
males, which are glossy black or blackish, while the former
are mottled with grey, blackish, and rufous: such are all the
typical *Tetraones* of Europe, and the cock of the plains, the
dusky and the spotted grouse of America. The smaller species,
in which both sexes are mottled, such as *T. phasianellus* and
T. Cupido, exhibit little or no difference in the plumage of the
two sexes; which is also the case in all the *Bonasiæ* and
Lagopodes. The young in their first feathers are in all
respects like the female, and the males do not acquire their
full plumage until after the second moult. All moult twice
a year, and most of the *Lagopodes* change their colours with
the seasons in a remarkable manner.

The genus *Tetrao* is now composed of thirteen species,—
three *Lagopodes*, two *Bonasiæ*, and eight typical *Tetraones*.
This enumeration does not include the *Tetrao rupestris*, which
we do not consider well established, any more than the new
species of Mr Brehm. The species of *Lagopus*, as might be
inferred from their inhabiting high northern latitudes, are
common to both continents, with the exception of the red
grouse, *T. Scoticus*, which is peculiar to the British islands,
and which, from its not changing the colours of its plumage
with the seasons, may be considered as forming the passage
to the true *Tetraones*. Of these, there are five in North

America, each and all distinct from the three European. Of the two *Bonasiæ*, one is peculiar to the old, and the other to the new continent, the former having sixteen, the latter eighteen feathers to the tail. Thus the entire number is seven in Europe, while it is eight in North America. Setting aside the two common to both, and the respective *Bonasiæ*, we may consider the cock of the woods of Europe as the parallel of the cock of the plains of America. The black grouse, *T. tetrix*, will find its equivalent in the dusky grouse, *T. obscurus ;* but the *T. hybridus* has no representative in America, any more than the *T. Scoticus*. These, however, are more than replaced as to number by the *T. phasianellus, T. Cupido*, and *T. Canadensis*, all American species, which have none corresponding to them in the old world.

Perhaps no other naturalist has personally inspected all the known species of this genus of both continents, and having examined numerous specimens even of some of the rarest, and possessing all but one in my own collection, my advantages are peculiar for giving a monography of this interesting genus. Such a work it is my intention hereafter to publish, illustrated with the best figures, and accompanied with further details repecting their habits. In the meantime, I shall merely state, that, being replaced in Africa by *Pterocles,* and in South America by *Tinamus*, all the known species of grouse are found in North America or in Europe, the European also inhabiting Asia, from whose elevated central and northern regions, yet unexplored, may be expected any new species that still remain to be discovered. The extensive wilds of North America may also furnish more, though we do not think so ; for since we have become acquainted with both sexes of the dusky grouse and the cock of the plains, we have been able to refer satisfactorily to known species all those of which any indications occur in the accounts of travellers in this country.

North America is exceeded by no country in the beauty, number, and valuable qualities of her grouse; and she is even perhaps superior to all others in these respects since the discovery of the cock of the plains. Although the careful and accurate researches of Wilson had led him to the belief that there existed but two species of grouse in the territory of the United States, no less than six are now known to inhabit within their boundaries. But we are not aware that any of the subgenus *Lagopus* ever enters the confines of the Union, notwithstanding the pains we have taken to obtain information on this point from the high northern districts of Maine and Michigan, in which, if anywhere, they are most likely to be discovered. It would, however, be very extraordinary if these birds, which are found in the Alps of Switzerland, should not also inhabit the lofty ranges of the Rocky Mountains, which are known to be the resort of the various species of grouse. With the exception, therefore, of the well-known *Tetrao umbellus*, which belongs to *Bonasia*, all the others are true grouse, *Tetraones.*

The spotted and the sharp-tailed grouse were long since known as inhabitants of that part of America north of the United States, but the two others are newly added, not only to our fauna, but to the general system, being found for the first time in the American territory, and not elsewhere. For the history of the discovery, the manners, habitation, and a particular description of each of these, we shall refer the reader to their several articles.

The dusky grouse is eminently distinguished from all other known species by having the tail slightly rounded, and composed of twenty broad and rounded feathers. This peculiarity of the extraordinary number of tail-feathers is only found besides in the cock of the plains, in which, however, they are not rounded, but very slender, tapering, and acute. In size and colour, the dusky grouse may be compared to the black grouse of Europe, so remarkable for the outward curvature of the lateral feathers of the tail.

The figure in our plate is taken from the specimen on which Say established the species; this was killed on a mountain in the great chain dividing the waters of the Mississippi from those which flow towards the Pacific, at a spot where, on the 10th of July 1820, the exploring party of Major Long were overlooking, from an elevation of one or two thousand feet, a wide extent of country. A small river poured down the side of the mountain, through a deep and inaccessible chasm, forming a continued cascade of several hundred feet. The surface of the country appeared broken for several miles, and in many of the valleys could be discerned columnar and pyramidal masses of sandstone, some entirely naked, and others bearing small tufts of bushes about their summits. When the bird flew, and at the unexpected moment of its death, it uttered a cackling note, somewhat resembling that of the domestic fowl.

The female dusky grouse is eighteen inches in length. The bill measures precisely an inch, which is small in proportion ; it is blackish, with the base of the under mandible whitish. The general colour of the plumage is blackish brown, much lighter on the neck and beneath, all the feathers having two or three narrow bars of pale ochreous, much less pure and bright on the neck and breast ; the small short feathers at the base of the bill covering the nostrils are tinged with ferruginous, those immediately nearest the forehead have but a single band, and are slightly tipt, while the larger ones of the neck, back, rump, and even the tail-coverts, as well as the feathers of the breast, have two bands and the tip. These rufous terminal margins on the upper portion of the back and on the tail-coverts are broad, and sprinkled with black, so as to be often blended with the lower band. The sides of the head and the throat are whitish, dotted with blackish, the black occupying both sides of each feather, deepening and taking a band-like appearance on the inferior portion of the upper sides of the neck ; on each feather of the breast is a whitish band that becomes wider on those nearest the belly ; the flanks are varied with rufous, each feather having, besides the small tip, three broad cross lines of that colour, and a white spot at the tip of the shaft, increasing in size as they are placed lower ; the belly feathers are plain dull cinereous, the lower tail-coverts are white, black at their base, with one or two black bands besides, and tinged between the bands with greyish ochreous ; the wings are nine and a half inches long, with the third and fifth primaries subequal ; the coverts, as well as the scapulars, are of the general colour, with about two bands, the second of which is sprinkled as well as the tip, each feather being white on the shaft at tip ; the primaries, secondaries, and outer wing-coverts, including their shafts, are plain dusky ; the secondaries have ochreous zigzag marks on their outer webs, and are slightly tipt with dull whitish ; the primaries themselves are somewhat mottled with dingy white externally, but are notwithstanding entirely without the regular white spots so remarkable in other grouse ; the lower wing-coverts and long axillary feathers are pure white. The tail measures in length seven and a half inches, is very slightly rounded, of twenty broad feathers, of which the lateral are plain blackish, with the exception of a few whitish dots at the base of their outer webs, and the middle ones being varied with rufous dots disposed like the bands across their whole width ; all are thickly dotted with grey for half an inch at tip, which in the specimen figured, but by no means so much so in others, gives the tail an appearance of having a broad terminal band of cinereous sprinkled with blackish. This circumstance evinces the inutility of describing with the extreme minuteness to which we have descended in this instance, as, after all the pains bestowed, the description is only that of an individual. The tail is pure black beneath, considerably paler at tip and on the undulations of the middle feathers. The tarsus is three-quarters of an inch long ; the feathers with which it is covered, together with the femorals, are pale greyish ochreous, undulated with dusky ; the toes are dusky, and the nails blackish.

The male is but little larger, and entirely, but not intensely, black. We can, however, say very little about it, having taken but a hasty and imperfect view of a specimen belonging to

Mr Sabine of London, and writing merely from recollection. The tail-feathers are wholly black, perfectly plain and unspotted; and in the female and young, they are but slightly mottled, as is seen in almost all grouse. Mr Sabine has long had this bird in his possession, and intended dedicating it, as a new species, to that distinguished traveller, Dr Richardson.

SHARP-TAILED GROUSE. (*Tetrao phasianellus.*)

PLATE XIX.

Tetrao phasianellus, *Linn. Syst.* ed. 10, p. 160.—*Gmel. Syst.* i. p. 747.—*Forst. Phil. Trans.* lxii. pp. 394 and 425.—*Lath. Ind. Orn.* p. 635, sp. 2.—*Briss. Suppl.* p. 9.—*Temm. Ind. Gall. in Hist. Pig. and Gall.* iii. p. 702.—*Vieill. Nour. Dict. Hist. Nat.—Sabine, Zool. App. to Frankl. Exped.* p. 681.—*Nob. Cat. Birds U. S.* sp. 208.—*Id. Syn. Birds U. S.* sp. 209.—Tetrao urogallus, var. β, *Linn. Syst.* i. p. 273, sp. 1.—Gelinotte à longue queue, *Buff. Ois.* ii. p. 286.—*Sonn. Buff.* vi. p. 72.—*Bonap. Tabl. Encyc. Orn.* p. 196, pl. 91, fig. 1.—Francolin à longue queue, *Hearne, Voy. à l'Ocean du Nord* (Fr. transl.) p. 386.—Tetras phasianelle, *Temm. Pig. et Gall.* iii. p. 152.—Long-tailed Grouse, *Edwards, Glean.* pl. 117.—*Lath. Syn.* iv. p. 732; *Id. Suppl.* p. 21. —Sharp-tailed Grouse, *Penn. Arct. Zool.* sp. 181.—The Grouse, or Prairie Hen, *Lewis and Clark, Exp.* ii. p. 180, sp. 1.—*Philadelphia Museum*, female. —*My Collection*, male and female.

TETRAO PHASIANELLUS.—Linnæus.

Tetrao (Centrocercus) phasianellus, *Swain. North. Zool.* ii. p. 361.

This species of grouse, though long since said to inhabit Virginia, is, in fact, a recent acquisition to the fauna of the United States; for it was only through an awkward mistake that it was ever attributed to that country. Mitchell, upon an inspection of Edwards's bad drawing of this bird, mistaking it for the ruffed grouse of that and the neighbouring States, declared it to be an inhabitant of Virginia; and upon his authority Edwards gave it as such   This statement, however, led Wilson into the erroneous belief of the identity of the two species, in which he was further confirmed when, after the most careful researches, he became satisfied that the ruffed grouse was the only species to be found in Virginia.

The gallant and lamented Governor Lewis gave the first

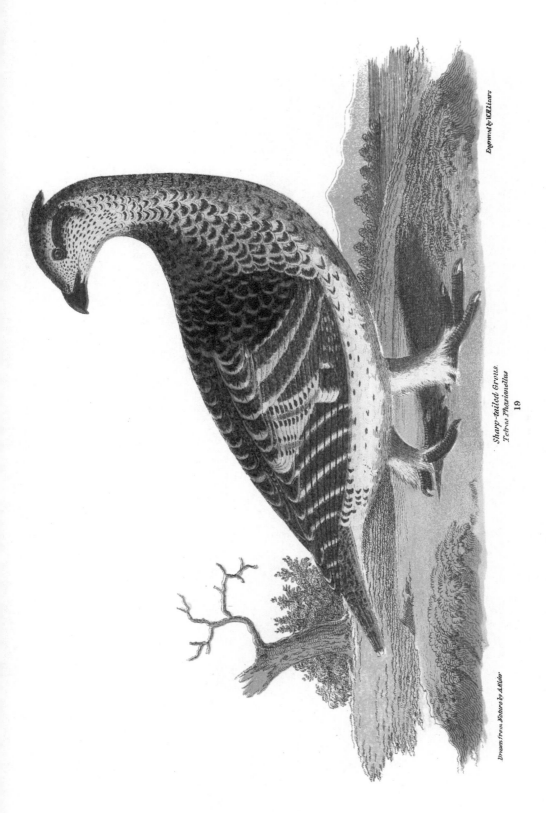

Sharp-tailed Grous.
Tetrao Phasianellus
19

Drawn from Nature by A.Rider.

Engraved by W.H.Lizars

authentic information of the existence of this bird within the limits of these States. He met with it on the upper waters of the Missouri, but observes that it is peculiarly the inhabitant of the great plains of the Columbia. He states also, that the scales, or lateral processes of the toes, with which it is furnished in winter, like the rest of its genus, drop off in summer.

Say introduced the species regularly into the scientific records of his country. The expedition under Major Long brought back a specimen now in the Philadelphia Museum, from which, though a female, and unusually light coloured, we have had our drawing made, on account of its having been procured in the American territory. The bird is never seen in any of the Atlantic States, though numerous in high northern latitudes. It is common near Severn river and Albany Fort, inhabiting the uncultivated lands in the neighbourhood of the settlements, and particularly near the southern parts of Hudson's Bay, being often killed in winter near Fort York; but it does not extend its range to Churchill. Near Fort William, on Lake Superior, the sharp-tailed grouse is also found in spring, and we have seen specimens killed in winter at Cumberland House, and others at York Factory in summer. In collections it is very rare; and Temminck, when he wrote his history of gallinaceous birds, had never seen a specimen, nor did it exist at the time in any European museum.

It is by the shape of the tail that this grouse is eminently distinguished from all others. The English name, which we have, with Mr Sabine, selected from Pennant, is much more applicable than that of long-tailed, given by Edwards; for instead of being long, it is, except the middle feathers, remarkably short, cuneiform, and acute, more resembling that of some ducks than of the pheasant. By the elongated feathers, but in no other particular, this species approaches the African genus *Pterocles*. At Hudson's Bay it is called pheasant, a name which, though inappropriate, seems at least better applied to this than the ruffed grouse.

The original writers that have mentioned this grouse·are
Edwards, who first introduced it, and has figured the female
from a badly stuffed specimen, being, however, the only figure
before ours ; Pennant ; Hearne, who has given the most information concerning its habits, derived from personal observation;
and Forster, who has described it with accuracy. Linné at first
adopted it from Edwards, but afterwards most unaccountably
changed his mind, and considered it as a female of the European
cock of the woods. It was restored by Latham and others to
its proper rank in the scale of beings.

The sharp-tailed grouse is remarkably shy, living solitary
or by pairs during summer, and not associating in packs till
autumn, remaining thus throughout the winter. Whilst the
ruffed grouse is never found but in woods, and the pinnated
grouse only in plains, the present frequents either indifferently.
They, however, of choice, inhabit what are called the juniper
plains, keeping among the small juniper bushes, the buds constituting their principal food. They are usually seen on the
ground, but when disturbed, fly to the highest trees. Their
food in summer is composed of berries, the various sorts of
which they eagerly seek : in winter they are confined to the
buds and tops of evergreens, or of birch and alder, but especially poplar, of which they are very fond. They are more
easily approached in autumn than when they inhabit large
forests, as they then keep alighting on the tops of the tallest
poplars, beyond the reach of an ordinary gun. When disturbed
in that position, they are apt to hide themselves in the snow ;
but Hearne informs us that the hunter's chance is not the
better for that; for so rapidly do they make their way beneath
the surface, that they often suddenly take wing several yards
from the spot where they entered, and almost always in a
different direction from that which is expected.

Like the rest of its kind, the sharp-tailed grouse breeds on
the ground near some bush, making a loose nest with grass,
and lining it with feathers. Here the female lays from nine
to thirteen eggs, which are white spotted with blackish. The

young are hatched about the middle of June ; they utter a piping noise, somewhat like chickens. Attempts have been repeatedly made to domesticate them, but have as constantly failed, all the young, though carefully nursed by their step-mother, the common hen, dying one after another, probably for want of suitable food. This species has several cries : the cock has a shrill crowing note, rather feeble ; and both sexes, when disturbed, or whilst on the wing, repeat frequently the cry of *cack, cack.* This well-known sound conducts the hunter to their hiding-place, and they are also detected by producing with their small, lateral, rigid tail-feathers a curious noise, resembling that made by a winnowing fan. When in good order, one of these grouse will weigh upwards of two pounds, being very plump. Their flesh is of a light brown colour, and very compact, though, at the same time, exceedingly juicy and well tasted, being far superior in this respect to the common ruffed, and approaching in excellence the delicious pinnated grouse.

The adult male sharp-tailed grouse, in full plumage, is sixteen inches long, and twenty-three in breadth. The bill is little more than an inch long, blackish, pale at the base of the lower mandible, and with its ridge entering between the small feathers covering the nostrils: these are blackish, edged with pale rusty, the latter predominating ; the irides are hazel. The general colour of the bird is a mixture of white, and different shades of dark and light rusty, on a rather deep and glossy blackish ground, the feathers of the head and neck have but a single band of rusty, and are tipt with white ; those, however, of the crown, are of a much deeper and more glossy black, with a single marginal spot of rusty on each side, and a very faint tip of the same, forming a tolerably pure black space on the top of the head. The feathers between the eye and bill, those around the eye, above and beneath, on the sides of the head, and on the throat, are somewhat of a dingy yellowish white, with a small black spot on each side, giving these parts a dotted appearance ; but the dots fewer and smaller on the throat. The feathers of the back and rump are black, transversely varied on the margin and at tip with pale bright rusty sprinkled with black, forming a confused mixture of black and rusty on the whole upper parts of the bird; the long loose-webbed upper tail-coverts being similar, but decidedly and almost regularly banded with black, and sprinkled with rusty, this colour being there much lighter, and approaching to white, and even constituting the ground colour. The breast is brown, approaching to chocolate, each feather being terminated by a white fringe, with a large arrow-shaped spot of that colour on the middle of each feather, so that, when the plumage lies close, the feathers appear white with black crescents, and are generally described so. On the lower portion of the breast, the white spots, as they descend, become longer and narrower, the branches forming the angle coming closer and closer to each other, till the spot becomes a mere white streak along the shaft, but, at the same time, the white marginal fringe widens so considerably, that the feathers of the belly may be properly called white, being brown only at their base, but the shaft is white even there, with no more than a brown heart-shaped spot visible on the middle. The heart-shaped brown spots of the belly become so very small at the vent, that this part appears pure white, with a few very small blackish spots; the long flank feathers are broadly banded with black and white, somewhat tinged with ochreous exteriorly; the under tail-coverts are white blackish along the shafts, and more or less varied with black in different specimens; also vary considerably as to the size and shape of all the spots, being in some more acute, in others more rounded, &c. The wings are eight inches long, the third and fourth primaries being the longest; the scapulars are uniform with the back, but, besides the rusty sprinkling of the margins and tip, the largest have narrow band-like spots of a pure bright rufous, a slight whitish streak along the shaft in the centre, and a large white spot at the end. The smaller wing-coverts are plain chocolate brown; the spurious wing

and outer coverts are of the same brown, but each feather
bears at the point a large and very conspicuous pure white
spot; all the other superior coverts are blackish, sprinkled,
and banded with rusty, each furnished with a conspicuous
terminal spot; the under wing-coverts, together with the long
axillary feathers, are pure white, each with a single small
dusky spot, and are marbled with white and brownish on the
outer margin; the quills are plain dusky brown, the primaries
being regularly marked with pure white spots half an inch
apart on their outer webs, except at the point of the first;
the longest feather of the spurious wing, and the larger outer
coverts, have also a pair of these spots; the secondaries, besides the outer spots, which assume the appearance of bands,
are tipt with pure white, forming a narrow terminal margin; those nearest the tertials are also slightly marked with
rusty; the tertials themselves are similar to the scapulars,
that is, they are black, banded and sprinkled with different
shades of rusty; the tail is strongly cuneiform and graduated,
of eighteen feathers, with the middle five inches long, which
is three more than the outer. According to some accounts,
the two middle feathers are by more than two inches longer
than the adjoining, but in all we have examined, the difference
was little more than an inch; the four middle are similar in
shape, texture, and colour, being narrow, flaccid, equal in
breadth throughout, though somewhat dilated and cut square
at the end. In colour, they vary considerably in different
specimens, the ground being generally black, and the tips
white, but more or less varied, in some with white, and in
others with rusty; these colours being at one time pure, at
another sprinkled with blackish, and assuming various tints;
in one specimen they are disposed in spots, in another in
bands, lines, chains, angles, &c.; but generally in a long stripe
on each side of the shaft at base, and in transverse spots at the
point of the two longest, while they are in round spots all
along each side of the two shortest: in one specimen, the
latter are even almost plain, being dingy white, sprinkled with

blackish on the whole of their outer web; all the other lateral feathers, entirely concealed by the coverts, are pure white at the point, but with dusky shafts, and are more or less broadly dark cinereous at base; these feathers are very rigid, and of a curious form, tapering from the base to the point, where they suddenly dilate; they are deeply emarginated at tip, and their inner lobe projects considerably; the tarsus is two inches long; the slender hair-like feathers covering it are, as well as the femorals, of a dingy greyish white, obsoletely waved with dusky; the toes are strongly pectinated, and are, as well as the nails, of a blackish dusky, while the long processes are whitish.

The foregoing minute description is chiefly taken from a handsome male specimen from Arctic America. There is no difference between the sexes, at least we have not been able to detect any in all the specimens of both that we have examined; hence we conclude that the difference generally described by authors, and which we have ourselves copied in our Synopsis, that of the breast being chocolate-brown in the male, and uniform with the rest of the plumage in the female, does not exist. The female is merely less bright and glossy. Both sexes, like other grouse, have a papillous red membrane over the eye, not always seen in stuffed skins, and which is said to be very vivid in the male of this species in the breeding-season. This membrane, an inch in length, becomes distended, and projects above the eye in the shape of a small crest, three-eighths of an inch high. The male at this season, like that of other species, and indeed of most gallinaceous birds, struts about in a very stately manner, carrying himself very upright; the middle feathers of the tail are more or less elongated, in young birds scarcely exceeding the adjoining by half an inch.

The spring plumage is much more bright and glossy than the autumnal, and also exhibits differences in the spots and markings. The specimen we have selected for our plate, on account of its being the only one we had from the United

States territory, is a female in the autumnal dress, and was brought from the Rocky Mountains. We think proper to insert here in detail the description we took from it at the time, thus enabling the reader to contrast it with that made from a northern specimen in spring plumage, rather than point out each and all of the numerous, and at the same time minute and unimportant variations.

The female represented in the figure was fifteen inches long ; its general colour mottled with black and yellowish rufous ; the feathers of the head above are yellowish rufous banded with black, the shaft yellowish ; a line above the eye, the cheeks and the throat, are pure yellowish rusty, with very few blackish dots, and a band of the latter colour from the bill beneath the eye, and spreading behind ; all the lower parts are whitish cream, with a yellowish rusty tinge ; each feather of the neck and breast, with a broad blackish subterminal margin, in the shape of a crescent, becoming more and more narrow and acute as they are lower down on the belly, until the lowest are reduced to a mere black mark in the middle ; the lower tail-coverts and the femorals are entirely destitute of black. All the upper parts, viz., the back, rump, upper tail-coverts, and scapulars, have a uniform mottled appearance of black and rusty, each feather being black, with rusty shafts, spots, bands, or margins, the rusty again minutely dotted with black ; on the rump, but especially on the tail-coverts, the rusty predominates in such a manner that each feather becomes first banded with black and rusty, then decidedly rusty varied with black, which, however, does not change in the least the general effect. The wing-coverts are dusky, each with a large round white spot at tip, the inner gradually taking the markings of the back and scapulars ; the lining of the shoulder is plain dusky, as well as the spurious wing and the primaries, each feather of the spurious wing having about five large round spots of white on its outer web ; the primaries are regularly marked on the same side with eight or ten squarish, equi-distant, white spots, with a few inconspicuous whitish dots on their inner web besides; the secondaries are also dusky, but in them the spots take the appearance of bands continued across the whole feather, of which bands there are three or four, including the terminal; the inner secondaries become darker and darker as they approach the body, the white becomes rufous, the dots are more frequent, and they become confounded with the scapulars, and are banded and mottled with various tints of black and rusty; the lower wing-coverts, and long axillary feathers, are pure white, the outer coverts being marbled with dusky. The tail is composed of eighteen feathers; it is cuneiform, very short, and entirely hidden by the coverts, except the four middle feathers; the two middle feathers are flaccid, narrow, equal in breadth throughout, longer than the others by more than an inch, rusty, crossed by chained bands of black, and dotted with black and whitish at tip; the two next are also longer than the others, nearly whitish, but almost similar in shape, markings, and texture, to the longest; the lateral decrease in size very fast from the centre, but by regular degrees, and are remarkably stiff, somewhat like those of woodpeckers, wider at base and tip than in the middle, pure white at the end and on the inner web, the shaft black, and the outer web dotted with blackish; they are deeply emarginate at tip, the inner lobe being longer, acute, and singularly shaped.

SPOTTED GROUSE.   (*Tetrao Canadensis.*)

PLATE XX.—Male.   PLATE XXI.—Fig. 1, Female.

Tetrao Canadensis, *Linn. Syst.* i. p. 274, sp. 3.—*Gmel. Syst.* i. p. 749, sp. 3.—
Lath. Ind. p. 637, sp. 6.—*Forster, in Phil. Tr.* lxii. p. 389.—*Temm. Ind.
Gall. in Hist. Pig. et Gall.* iii. p. 702.—*Vieill. Nouv. Dict. Hist. Nat.*—
Sabine, Zool. App. Frankl. Exped. p. 683.—*Nob. Cat. Birds U. S.* sp. 207.—*Id.
Syn. Birds U. S.* sp. 208.—Tetrao canace, *Linn. Syst.* i. p. 25, sp. 7, female.
—Lagopus Bonasa, Freti Hudsonis, *Briss. Orn.* i. p. 201, sp. 6; *Id. Suppl.* p.
10; *Id.* 8vo, iv. p. 56, male.—Lagopus Bonasa Canadensis, *Briss. Orn.* i. p.
203, sp. 7, pl. 20, fig. 2; *Id.* 8vo, iv. p. 57, female.—Lagopus, Freti Hudsonis,
Klein, Av. p. 117, sp. 6.—La Gelinotte du Canada, *Buff. Ois.* ii. p. 279; *Id.
Pl. enl.* 131, male, 132, female.—*Sonn. Buff.* vi. p. 58.—*Bonap. Tabl. Enc.*

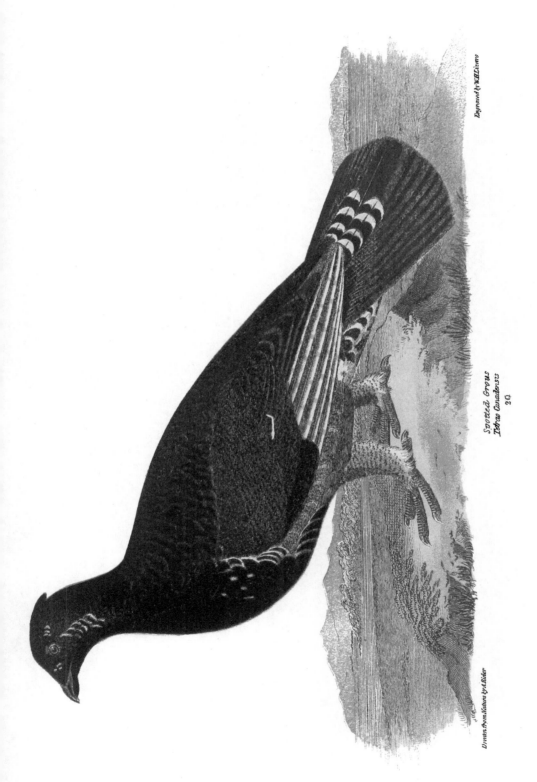

Drawn from Nature by A. Rider

Spotted Grous
Tetrao Canadensis
20

Engraved by W.H.Lizars

Orn. p. 197, pl. 91, fig. 2.—Tetras Tacheté, ou Acaho, *Temm. Pig. et Gall.* iii. p. 160, *bis.*—Black and Spotted Heathcock, *Edw. Glean.* p. 118, pl. 118, male.—Brown and Spotted Heathcock, *Edw. Glean.* p. 71, pl. 71, female. —*Ellis, Hudson's Bay,* i. t. p. 50.—Spotted Grouse, *Penn. Arct. Zool.* sp. 182. —*Lath. Syn.* iv. p. 735, sp. 6 ; *Id. Suppl.* p. 214, accid. var.—The Small Speckled Pheasant, *Lewis and Clark, Exp.* ii. p. 182, male.—The Small Brown Pheasant, *Id.* ii. p. 182, female.—*Philadelphia Museum,* male.—*My Collection,* male and female.

TETRAO CANADENSIS.—Linnæus.[*]

Tetrao Canadensis, Spotted Grouse, *North. Zool.* ii. p. 346, pl. 62. *Temm.*

As may be seen by the synonymy, two separate species have been made of the present, the male and female being taken

---

[*] In this place must be introduced the *Tetrao Franklinii* of Douglas first noticed by that gentleman in the "Transactions of the Linnæan Society." It is a species which has been involved in confusion with the *T. Canadensis,* from the different opinions which those persons who have met with it have formed. By the Prince of Musignano and Mr Drummond, an acute observer, it is thought to be a variety only, while Mr Douglas and the authors of the "Northern Zoology" consider it distinct.

I have added the description of the latter naturalists, and some observations on its habits by Mr Douglas. I cannot decide, not having specimens of both to compare ; but, from the known variation of the markings of those birds, which will stand in the division *Tetrao,* a very accurate comparison indeed of numerous specimens, with the investigation of their habits, would be necessary to distinguish those which are seemingly so nearly allied.

Tetrao Franklinii, Douglas.

Tetrao Franklinii, *Dougl. Trans. Linn.* xvi. p. 139. *North. Zool.* ii. p. 384.

"There is such a close resemblance between this and the common Canadian grouse, that the Prince of Musignano considers it only as a variety ; and this opinion is entertained also by Mr Drummond, a very acute practical observer. The latter had ample opportunities of study-ing the manners of both, and he assures us that he is not aware of any difference between them ; Mr Douglas, on the other hand, who has also seen these birds in their native regions, thinks differently, and although he observes that in habits it assimilates more with *T. Canadensis* than any other, he considers the 'unusually long, square tail,' and its colour-ing, as sufficiently distinctive characters. In our species, the tail is not longer than that in *Canadensis;* and, did we look to the colouring

for different birds.   This error, which originated with Edwards and Brisson, from whom it was copied by Linné, was recti-

alone, however strikingly different it is in the two birds, we should be disposed to class them as varieties.   But a more accurate examination will detect some essential difference in the structure of the feathers themselves.   Those of *Canadensis* are more graduated, giving the tail a rounder appearance ; and they are all slightly, but distinctly emarginate in the middle, from whence arises a little mucro, or point, formed by the end of the shafts.   Now, this emargination is not seen in *Franklinii*, nor is there any appearance of the mucro.   Again, in the latter bird, the tail-feathers are much broader, fully measuring one inch and a fifth across ; whereas those of *Canadensis* are barely one inch broad.   Until, therefore, we become persuaded that distinctions drawn from structure are not to be relied upon, we must coincide with Mr Douglas in considering the present as a distinct species."—W. S.

" I have never heard," says Douglas, " the voice of this bird, except its alarm note, which is two or three hollow sounds, ending in a yearning, disagreeable, grating noise, like the latter part of the call of the well-known *Numida meleagris*.   It is one of the most common birds in the valleys of the Rocky Mountains, from latitude 50 to 54 degs., near the sources of the Columbia River.   It may perhaps be found to inhabit higher latitudes.   Sparingly seen in some small troops on the high mountains which form the base or platform of the snowy peaks ' Mount Hood,' ' Mount St Helen's,' and ' Mount Baken,' situated on the western parts of the continent."

Those grouse known under the name of *ptarmigan*, or the form more familiar as represented by the common red grouse of Britain, typical of the genus *Lagopus*, should be also enumerated here.

The species mentioned by Mr Douglas, and further described in the " Northern Zoology," are all that have yet been authentically described as natives of North America.   Mr Douglas hints at his knowledge of other two, but this was too imperfect to offer any detailed description. The species are :

Lagopus, Ray.

1. *L. mutus*, Leach.—Common ptarmigan.—According to Sabine, inhabits the island on the south-west side of Baffin's Bay.   Dr Richardson never met with it himself in the Fur Countries, and thinks that the only authentic specimens from the New World are in the possession of Lord Stanley, to whom they were presented by Mr Sabine.
2. *L. saliceti*, Swain.—Willow grouse.—Inhabits the Fur Countries from the 50th to the 70th degs. of latitude, within which limits it is partially migratory ; breeds in the valleys of the Rocky Mountains, the barren grounds, and arctic coasts.   It seems identical with the willow grouse

fied by Buffon, Forster, and others ; and in their decision
Gmelin, Latham, and all subsequent writers, have acquiesced.
Both sexes were tolerably well figured by Buffon, as they had
also been previously by Edwards ; but we feel justified in say-
ing that none of their plates will bear a comparison with the
present.

The spotted grouse is well characterised by its much rounded
tail, of but sixteen broad and rounded feathers, and may be
at once distinguished from all others by the large and
conspicuous white spots ornamenting the breast, flanks, and
under tail-coverts. It has been inaccurately compared with
the European *Tetrao bonasia,* from which it differs very
materially, not even being of the same subgenus, and
approaching nearer, if indeed it can be compared with any, to
the *Tetrao urogallus.*

This bird is common at Hudson's Bay throughout the year,
there frequenting plains and low grounds, though in other
parts of America it is found on mountains, even of great
elevation. It inhabits Canada in winter, and was seen by
Vieillot in great numbers during the month of October in
Nova Scotia. Lewis and Clark met with it on the elevated
range of the Rocky Mountains, and brought back from their
western expedition a male specimen, now deposited in the
Philadelphia Museum, where it has been long exhibited under
the name of Louisiana grouse. This, as truly observed by
Say, first entitled it to rank among the birds of the United

of the old continent, which inhabits the greater part of Scandinavia,
Kamtschatka, Greenland, Iceland, and the valleys of the Alps.

3. *L. rupestris.*—Rock grouse.—Closely allied, and long confused with *L.
mutus ;* inhabits Melville Peninsula, and the barren grounds, and descends
along the coast of Hudson's Bay, as far as lat. 58 degs. ; found on the
Rocky Mountains as far south as lat. 55 degs.

4. *L. leucurus,* Swain.—White-tailed grouse.—*Hyeme albus : æstate variegatus, rectricibus semper albis.*—This species is first described and figured
in the "Northern Zoology." The specimens were killed on the Rocky
Mountains, and it is said to inhabit the snowy peaks near the mouth of
the Columbia. Its summer dress is intermediate between *L. mutus* and
rupestris, and it is at once distinguished from all the others by its smaller
size and the pure white colour of the tail at all seasons.—ED.

States.   But the Rocky Mountains are not the only region of
the United States territory where the spotted grouse is found.
We have traced it with certainty as a winter visitant of the
northern extremity of Maine, Michigan, and even of the State
of New York, where, though very rare, it is found in the
counties of Lewis and Jefferson.   On the frontiers of Maine
.it is abundant, and has been seen by Professor Holmes, of the
Gardiner Lyceum, near Lake Umbagog, and others.   In these
countries, the spotted grouse is known by the various names of
wood partridge, swamp partridge, cedar partridge, and spruce
partridge.   The American settlers of Canada distinguish it
by the first.   In Michigan and New York, it goes generally
by the second ; in Maine it bears the third, and in other parts
of New England, New Brunswick, &c., more properly the last.
We have been informed by General Henry A. S. Dearborn
that they are sent from Nova Scotia and New Brunswick to
Boston in a frozen state, as in the north they are known to
be so kept hanging throughout the winter, and, when wanted
for use, they need only be taken down, and placed in cold
water to thaw.   General Dearborn, to whom we are much
indebted for the information which his interest for science has
induced him voluntarily to furnish, further mentions, that he
has heard from his father, that during the progress of the
expedition under Arnold, through the Wilderness to Quebec,
in 1775, these grouse were occasionally shot between the tide
waters of Kennebeck river and the sources of the Chaudiere,
now forming part of the State of Maine.   Fine specimens of
the spotted grouse have been sent to the Lyceum of Natural
History of New York, from the Sault de Ste. Marie, by Mr
Schoolcraft, whose exertions in availing himself of the opportunities which his residence affords him for the advancement
of every branch of zoology merit the highest praise.   He
informs us that this bird is common from Lake Huron to
the sources of the Mississippi, being called in the Chippeway
language, *mushcodasee—i.e.,* partridge of the plains.

The favourite haunts of the spotted grouse are pine woods

and dark cedar swamps, in winter resorting to the deep forests of spruce, to feed on the tops and leaves of these evergreens, as well as on the seeds contained in their cones, and on juniper berries. Hence their flesh, though at all times good, is much better in summer, as in winter it has a strong flavour of spruce. At Hudson's Bay, where they are called indifferently wood or spruce partridge, they are seen throughout the year. Like other grouse, they build on the ground, laying perhaps fewer eggs; these are varied with white, yellow, and black. They are easily approached, being unsuspicious, and by no means so shy as the common ruffed grouse, and are killed or trapped in numbers, without much artifice being necessary for this purpose. When much disturbed, like their kindred species, they are apt to resort to trees, where, by using the precaution of always shooting the lowest, the whole of the terrified flock may be brought down to the last bird.

The spotted grouse is smaller than the common partridge or pheasant, being but fifteen inches in length. The bill is black, seven-eighths of an inch long. The general colour of the plumage is made up of black and grey, mingled in transverse wavy crescents, with a few of greyish rufous on the neck. The small feathers covering the nostrils are deep velvety black. The feathers may all be called black as to the ground colour, and blackish plumbeous at the base; on the crown, upper sides of the head above the eye, and the anterior portion of the neck, they have each two grey bands or small crescents, and tipped with a third; these parts, owing to the grey margin of the feathers being very broad, appear nearly all grey; these longer feathers of the lower part of the neck above, and between the shoulders, are more broadly and deeply black, each with a reddish band, and grey only at tip; the lowest have even two reddish bands, which pass gradually into greyish; a few of the lateral feathers of the neck are almost pure white; all the remaining feathers of the upper parts of the body have two greyish bands, besides a slight tip of the same colour; some of the lowest and longest having even three of these

bands, besides the tip.   The very long upper tail-coverts are
well distinguished, not only by their shape, but also by their
colours, being black brown, thickly sprinkled on the margins
with greyish rusty, and a pretty well defined band of that
colour towards the point, then a narrow one of deep black,
and are broadly tipt with whitish grey, more or less pure in
different specimens ; their shafts, also, are brownish rusty.
The sides of the head beneath the eyes, together with the
throat, are deep black, with pure white spots, the white lying
curiously upon the feathers, so as to form a band about the
middle, continued along the shaft, and spreading at the point ;
but the feathers being small on these parts, the white spots are
not very conspicuous.   The breast, also, is deep black, but
each feather broadly tipt with pure white, constituting the
large spots by which this species is so peculiarly distinguished.
On the flanks, the feathers are at first, from their base, waved
with black and greyish rusty crescents ; but these become
gradually less pure and defined, and by getting confused, make
the lowest appear mottled with the two colours ; all are marked
along the shaft with white, dilating at tip, forming on the
largest a conspicuous terminal spot.   The vent is for a space
pure white, the tips of its downy feathers being of that colour :
the under tail-coverts are deep black, pure white for half an
inch at their tip, and with a white mark along the shaft besides.   The wings are seven inches long, the fourth primary
alone being somewhat longer than the rest.   The upper coverts and scapularies are blackish, waved and mottled with
greyish rusty ; the longest scapularies have a small terminal
spot of pure white along their shaft.   The smaller coverts are
merely edged with greyish rusty, and in very perfect specimens they are even plain ; the under wing-coverts are brownish dusky, edged with greyish, some of the largest, as well as
the long axillary feathers, having white shafts, dilating into
a terminal spot ; the remaining inferior surface of the wing is
bright silvery grey ; the spurious wing and the quills are plain
dusky brown, the secondaries being slightly tipt and edged

externally with paler, and those nearest the body somewhat
mottled with greyish rusty at the point and on the inner
vane ; the primaries, with the exception of the first, are slightly
marked with whitish grey on their outer edge, but are entirely
destitute of white spots. The tail is six inches long, well
rounded, and composed of only sixteen feathers. These are
black, with a slight sprinkling of bright reddish on the outer
web at base, under the coverts, which disappears almost entirely with age ; all are bright dark rusty for half an inch
at their tip, this colour itself being finely edged and shafted
with black. The tarsus measures an inch and a half ; its
feathers, together with the femorals, are dingy grey, slightly
waved with dusky; the toes are dusky; the lateral scales
dingy whitish, and the nails blackish.

The female is smaller than the male, being more than an
inch shorter. The general plumage is much more varied, with
less of black, but much more of rusty. There is a tinge of
rufous on the feathers of the nostrils ; those of the head, neck,
and upper part of the back are black, with two or three bright
bands of orange rusty, and tipt with grey; there is more of
the grey tint on the neck, on the lower part of which above,
the orange bands are broader ; all the remaining parts of the
body above, including the tail-coverts, are more confusedly
banded and mottled with duller rusty, orange, and grey, on a
blackish ground, these colours themselves being also sprinkled
with a little black ; the sides of the head, the throat, and all
the neck below, are dull rusty orange, each feather varied
with black; on the lower portion of the breast, the black
bands are broad and very deep, alternating equally with the
orange rusty, and even gradually encroaching upon the ground
colour ; the breast is deep black, each feather, as well as those
of the under parts, including the lower tail-coverts, are broadly
tipt with pure white, forming over all the inferior surface
very large and close spots, each feather having besides one or
two rusty orange spots, much paler and duller on the belly,
and scarcely appearing when the plumage lies close: the

feathers of the flanks are blackish, deeper at first, and barred
with very bright orange, then much mottled with dull greyish
rusty, each having a triangular white spot near the tip. The
wings and tail are similar to those of the male, the variegation
of the scapulars and upper coverts being only of a much more
rusty tinge, dull orange in the middle of the shaft, all the
larger feathers having, moreover, a white streak along the
shaft, ending in a pure white spot, wanting in the male. The
outer edge of the primaries is more broadly whitish, and the
tertials are dingy white at the point, being also crossed with
dull orange ; the tail-feathers, especially the middle ones, are
more thickly sprinkled with rusty orange, taking the appearance of bands on the middle feathers, their orange-coloured
tip being, moreover, not so pure, and also sprinkled.

The bird represented in the plate comes from the Rocky
Mountains : it is a male, and remarkably distinguished from
the common ones of his species by having the tail-feathers
entirely black to the end. This difference I have observed to
be constant in other specimens from the same wild locality ;
whilst all the northern specimens, of which I have examined
a great number, are alike distinguished by the broad rufous
tip, as in those described, and as also described by Linné and
all other writers, who have even considered that as an essential mark of the species. The Rocky Mountain specimens are,
moreover, somewhat larger, and their toes, though likewise
strongly pectinated, are, perhaps, somewhat less so, and the
tail-coverts are pure white at tip, as represented in the plate.
But Heaven forbid that our statements should excite the
remotest suspicion that these slight aberrations are characteristic of different species! If we might venture an opinion not
corroborated by observation, we would say, that we should not
be astonished if the most obvious discrepancy, that of the tail,
were entirely owing to season, the red tip being the full spring
plumage ; though it is asserted that this species does not vary
in its plumage with the seasons. However this may be, we
have thought proper to give a representation of the anomalous

Drawn from Nature by A. Rider

1. Cock of the Plains. Female. 2. Female Spotted Grous.
Tetrao Urophasianus.   Tetrao Canadensis.

21

Engraved by W.H.Lizars

male bird from the Rocky Mountains in our plate, whilst the
female, placed with the cock of the plains, that its reduced
size may be properly estimated, has been chosen among the
ordinary specimens having the tails tipt with red; the red
tip being still more conspicuous in the common males, from
which, in order to comprehend all, our description has been
drawn up.

COCK OF THE PLAINS.   (*Tetrao urophasianus.*)

PLATE XXI.—Fig. 2.

Tetrao urophasianus, *Nob. in Zool. Journ. Lond.—Id. App. to Syn. Birds U. S.*
p. 442, *in Ann. Lyceum Nat. Hist. N. York.*—The Cock of the Plains,
Lewis and Clark, Exp. ii. p. 180, sp. 2.

TETRAO UROPHASIANUS.—Bonaparte.[*]

Tetrao (Centrocercus) urophasianus, *Swain. North. Zool.* ii. p. 358, pl. 53.

It is with the liveliest satisfaction that we are enabled finally
to enrich the North American fauna with the name, portrait,
and description of this noble bird, which must have formed
from the earliest periods a principal ornament of the distant
wilds of the west. Hardly inferior to the turkey in size,
beauty, and usefulness, the cock of the plains is entitled to
the first place in the beautiful series of North American
grouse, in the same rank that the cock of the woods so justly
claims among those of Europe and Asia.

This fine bird, like its European analogue, seems to be
restricted within certain bounds, and is probably nowhere
numerous, owing to its bulk, limited powers of flight, and the
eagerness with which it is pursued; but chiefly to its polyga-

---

[*] This fine species, with the *Tetrao phasianellus,* have been made into
a subgenus by Mr Swainson. I have provisionally retained these birds
in *Tetrao,* but have little doubt that this form will show its own and
separate station among the *Tetraonidæ.* Mr Swainson thinks that it
will represent the scansorial form among these birds, from the structure
of the tail; while the blackcock of Europe is placed by him to show
the fistirostral form, by its lyre-shaped tail and glossy colour bearing a
faint resemblance to the Drongo Shrikes.—Ed.

mous habits, which are the cause of desperate combats between
the males for the possession of the females. However long
the period since it was first heard of in the accounts of hunters
and travellers, no more was known than that there existed
in the interior of America a very large species of grouse,
called by the hunters of the west the prairie turkey. We
have little to add, it is true, to what is known of its habits,
but we have it in our power to say that we have seen it; we
can determine its place in the system ; and now give a faithful
representation of at least one sex.

We have again to acknowledge ourselves indebted, no less
to the industry and sagacity, than to the liberal views of Mr
Leadbeater, for the present opportunity of representing this
bird. His invaluable collection contains the only specimen
known to be anywhere preserved.

The name of cock of the plains was given by Lewis and
Clark, and we have retained it, as being not only appropriate,
but at the same time analogous to that of the large European
species called cock of the woods. Similar reasons have influenced us in selecting the scientific name, which, though
perhaps too long, and ill compounded, has nevertheless the
advantage of combining analogy in meaning with the indication of a most remarkable characteristic of the bird. This
species is in fact distinguished from all others of its genus,
and especially from its European analogue, by its long tail,
composed of twenty narrow, tapering, acute feathers ; thus
evincing the fallacy of the character erroneously attributed to
all the grouse, of having broad and rounded tail-feathers. It
is a singular fact, that both of the newly-discovered species
from the north-western part of America, and they only, should
be distinguished by the extraordinary number of the feathers
of the tail. In the dusky grouse, however, they are broad
and rounded. The cock of the woods, like the greater part
of the species, has but eighteen, which are also broad and
rounded. The only grouse in which they are found narrow
is the sharp-tailed, though without being either acute or taper-

ing, but, on the contrary, square at tip, and of equal breadth throughout, or, if anything, the lateral rather broader at the tip.

Lewis and Clark first met with this bird on their journey westward, near the fountain of the Missouri, in the heart of the Rocky Mountains. They inform us that it is found on the plains of the Columbia in great abundance, from the entrance of the south-east fork of the Columbia to that of Clark's river. It appears also to extend to California, for there can be but little doubt that it is the bird erroneously called bustard by the travellers who have visited that country. Lewis and Clark state, that in its habits it resembles the grouse (meaning probably *T. phasianellus*), except that its favourite food is the leaf and buds of the pulpy-leafed thorn. The gizzard is large, and much less compressed and muscular than in most gallinaceous birds, and perfectly resembles a maw. When the bird flies, he utters a cackling note, not unlike that of the domestic fowl. The flesh of the cock of the plains is dark, and only tolerable in point of flavour, and is not so palatable as either that of the pheasant or grouse. It is invariably found in the plains.

The cock of the plains is precisely equal in size to the cock of the woods ; at least such is the result of a comparison of the female with the corresponding sex of the European bird, both lying before us. Each part exactly coincides in form and dimension, excepting that the tail rather gives the superiority to the American, so that if the male bears the same relative proportion to his female, the cock of the plains must be proclaimed the largest of grouse. The two females are strikingly similar. The cock of the plains is, however, a much more greyish bird, wanting entirely the reddish that mottles and occupies so much of the plumage of its analogue. This, the total want of beard-like appendages, and the singular shape of the tail, are the prominent discriminative features ; to which may be added, that the under wing-coverts, marbled with black in the European, are pure white in our new species, though this, as well as the want of reddish, might be ascribed to the youth of our specimen. However this may be, the remaining differences will be better estimated by attending to the following minute and accurate description.

The female of the cock of the plains, represented in the plate of one-half the natural size, is from twenty-eight to thirty inches in length. The bill is one inch and a quarter long, perfectly similar to that of *T. urogallus*, perhaps a trifle less stout, and with the base (if this remarkable character be not accidental in our specimen) farther produced among the feathers of the front; the whole plumage above is blackish, most minutely dotted, mottled, and sprinkled with whitish, tinged here and there with very pale yellowish rusty, hardly worth mentioning; on the head, and all the neck, the feathers being small, minutely crossed transversely with blackish and whitish lines, gives the plumage quite a minutely dotted appearance; the superciliar line is slightly indicated by more whitish; on a spot above the eye, in the space between the bill and eye, and along the mouth beneath, the black predominates, being nearly pure: on the throat, on the contrary, it is the white that prevails, so as to be whitish dotted with black; on the lower portion of the neck, the black again is the prevailing colour, the black feathers there being nearly tipt with greyish; the sides of the neck are pure white for a space; from the lower portion of the neck to the upper tail-coverts inclusively, the back, scapulars, wing-coverts, and secondaries, the blackish feathers have each two or three yellowish white bands, which are broader, especially on the upper part of the back and are moreover sprinkled with white somewhat tinged with rusty; the scapulars and wing-coverts are besides shafted with white, somewhat dilating towards the point, the scapulars being of a deeper black; the spurious wing and primaries are plain dusky, with paler edges, the outer with some indications of whitish dots (generally found in grouse) on the outer vane, but no regular white spots; the secondaries are tipt with white, and those which are next to the primaries nearly plain on their inner web; the primaries are rather slender, the inferior surface of the wings is of a very pale silvery grey; the under wing-coverts and long axillary feathers being pure silvery white, excepting on the lining of the wing, which is dusky blackish; the wings are twelve inches long; the breast is greyish, somewhat mottled with black; on each side below is a pure white space, some of the feathers of which are tipt or banded with black; the large feathers of the flanks are blackish, shafted with white, crossed by several whitish bands, and sprinkled with yellowish; a broad oblong patch of deep brownish black occupies the whole of the belly and vent, the outer feathers being shafted with white, and broadly white at the point of their outer webs; the femorals and small feathers of the tarsus extending between the toes are yellowish grey, minutely waved with blackish; the tarsus measures two inches; the toes are dusky black, and the pectinated row of processes long, strong, and dingy whitish; the nails, blackish; the whole base of the plumage, with the exception of that of the neck beneath, which is white, is of a dusky grey. The tail is ten inches long, and in colour is, as well as its coverts, in harmony with the rest of the plumage; the ground colour is blackish, and crossed, or rather mottled, with bands of whitish spots disposed irregularly, between which are small additional darker spots; the two middle ones are mottled all over, but the others are almost immaculate on their inner vane, and at the point; hence the lower surface of the unexpanded tail is of a silvery grey, much darker than that of the wings; at the very tip of the tail-feathers, the middle excepted, appears a very small whitish spot, the two outer pairs being rather broadly yellowish white, dotted with blackish on that part; the tail is composed of twenty feathers, the highest number ever met with in any tribe of birds. Although it appears strongly cuneiform, owing to the remarkable shape and curve of the feathers, it is, when expanded and properly examined, nothing more than much rounded; the two in the middle, which are the longest, reaching but a trifle beyond the adjoining, and so on in succession, the difference in length increasing progressively, but very gradually at first, and more and more as they are distant from the centre, there being nearly an inch difference between the third and second, and full that between the second and the outer, which is only six inches long, while the middle is ten. All the twenty are narrow, tapering, acute, and falciform, turning inward. Those towards the middle are less curved, but more conspicuously acuminate and narrow for nearly two inches, all but the middle ones being slightly square at their narrow tips.

Though we have reason to believe that the specimen described and figured is a female, yet, from the broad patch upon the belly, and other marks unnecessary to be specified, we should not be surprised at its being a young male just beginning to change. In that case, and supposing him to have attained his full growth, this species would prove to be inferior in size to the cock of the woods, as its male would only be equal to the female of the latter.

CONDOR.   (*Cathartes gryphus.*)

PLATE XXII.—YOUNG MALE.

Vultur gryphus, *Linn. Syst.* i. p. 121, sp. 1.—*Gmel. Syst.* i. p. 245, sp. 1.—*Lath. Ind. Orn.* i. p. 1, sp. 1.—*Encycl. Brit.* xviii. p. 695, pl. 510.—*Humboldt, Hist. Nat. in Obs. Zool.* i. p. 26, pl. 8, 9.—Vultur Magellanicus, *Lever. Mus.* p. 1, pl. 1, female.—Vultur condor, *Daud. Orn.* ii. p. 8.—*Shaw, Zool.* vii. p. 2, pl. 2, 3, 4.—Cathartes gryphus, *Temm. Ranz. Nob. Cat. and Syn. Bds. U. S.* sp. 2.—Gypagus griffus, *Vieill. Enc.* iii. p. 1174.—*Id. Nouv. Dict.*— Sarcoramphus cuntur, *Dumeril.*—Sarcoramphus gryphus, *Goldfuss, Nat. Atlas,* pl. 107, adult male.—Sarcoramphus condor, *Less. Orn.* i. pl. 7, adult male.—Vultur gryps gryphus, *Klein, Av.* p. 45.—*Briss. Av.* i. p. 473 ; *Id.* 8vo, p. 137.—*Borowski, Nat.* ii. p. 62.—Cuntur, *Laet. Am.* p. 401. —*Ray, Av.* p. 11.—Catarte condoro, *Ranz. Elem.* vii. p. 24, sp. 2, tab. xxii. fig. 2, adult male.—Il Condoro, *Molina, St. Nat. Chili,* p. 223.—Manque, *Molina, Chili,* p. 236 (French edition).—Condor, *Frezier, Voy.* p. iii.—*La Condamine, Voy. Amaz.* p. 175.—*Briss. Orn.* i. p. 463, sp. 12.—*Buff. Ois.* i. p. 184 ; *Id.* (ed. 1770) i. p. 143, v.—*Martinet, Hist. Ois.*—Le Condor, ou Grand Vautour des Andes, *Cuv. Regn. An.* i. p. 306 ; *Id.* 2d ed. p. 316.— Catharte Condor, *Temm. and Laug. Pl. Col.* 133, adult male, 494, head of the adult living male, 408, young female.—Condur Vulture, *Lath. Syn.* p. 4 ; *Id. Suppl.* p. 1 ; *Id. Suppl.* ii. p. 1, pl. cxx.—*Id. Gen. Hist.* i. p. 4, pl. 1, adult male.—*Hawkesw. Voy.* i. p. 75.— *Wood's Zoography,* i. p. 371.— *Stevenson, Voy. Am.* ii. p. 59.—Der Condor Geier, of German authors.— *Cabinet of the Academy of Natural Sciences.*

To such a degree has its history been exaggerated by fable, that the mention of the condor immediately recalls to mind

Drawn from Nature by A. Rider.

Young Male Condor.

Cathartes Gryphus.

22

the roc of Marco Polo and the Arabian Tales. Some authors
have indeed referred this name to it, and even go so far as to
make it the subject of one of the labours of Hercules, the
destruction of the Stymphalian birds. Such, in fact, were the
stories related by the early travellers, that, even when reduced
to what in the judgment of Buffon was their real value, it
cannot but now appear unaccountable that they should ever
have found credence, and still more so that compilers should
have gone on accumulating under the condor's history, not
merely the tales told of it, but others collected from every
quarter of the globe, however remote or different in climate,
not hesitating to give currency to the most revolting absurdities. The accounts of Father Feuillée, who was the first
describer, Frezier, and especially Hawkesworth's, appear, however, to be tolerably correct; while the ardent imagination
of Garcilasso led him to indulge in the wildest extravagances
in relation to this bird. Abbeville and De Laet, no less than
Acosta in his " History of the Indies," ascribed to this cowardly
vulture the strength, courage, and raptorial habits of an eagle,
and even in a higher degree, thus doing him the honour to
represent him as formidable to every living creature, and the
dreaded enemy of man himself. Desmarchais improves if
possible upon these stories, giving the condor still greater
size and strength, and stating that it is well known to carry
off in its prodigious talons a hind, or even a heifer, with as
much ease as an eagle would a rabbit ! Such a creature could
not of course dwell in forests, for how could it among trees
display its enormous wings ? They were therefore limited to
savannahs and open grounds. Antonio de Solis, Sloane in the
" Philosophical Transactions," and even the learned La Condamine, who saw the bird himself, and certainly witnessed no
such exploits as had been related of it, indulged in wild theories
depending on popular tales and superstitions. The obscurity
created by so much misrepresentation could not, however, conceal its true vulture-like nature from the acuteness of Ray,
who pointed out its appropriate place in the system. His

opinion was adopted by Brisson and Linné, and it became among naturalists generally a settled point, notwithstanding the eloquently expressed doubts of Buffon, who wanted rather, on account of its supposed great strength and agility, to elevate the condor to the rank of an eagle, these qualities not permitting him to degrade it so low as the vultures. But a still greater error of the French Pliny, as he may be on every account so appositely styled, was to consider the condor as not peculiar to America, but as a genuine cosmopolite, of which happily there were but few, however, for otherwise the human race would not have been able to stand against them. But it was only in its imaginary character that the condor of Buffon was truly cosmopolite, having no other existence than what was based upon absurd and ridiculous fictions gathered in all parts of the globe ; for no living bird could be placed in competition with one for whose powers of flight distance was no impediment, and whose strength and swiftness united would have rendered him lord of creation.

We should, however, make some allowance for the credulity of our forefathers in believing, upon the reports of weak or lying travellers, all the romantic and extravagant tales related of this wondrous condor. They had not, as we have, the means of personally ascertaining the sober truth. But it is almost incredible, and remarkably illustrates the force of preconceived opinions, that in the year 1830 a traveller could be found with assurance enough to impose upon us, and journals, even of respectable standing, to copy as positive and authentic, a description of a condor of *moderate size,* just killed, and actually lying before the narrator, so large that a single quill-feather was twenty good paces long ! This indeed might have lifted an elephant, and it is quite unfortunate that Peru and Chili should no longer produce them for prey for such a bird, and that the mastodon is now extinct. So much for human credulity, which is often exercised upon more serious occasions, with equal impudence and much worse results.

As in so many other instances of power based upon prejudice

or great reputation unjustly usurped, a near and close examination has shown the falsity of these pretensions. The wonderful condor now proves to be nothing more than a rather large vulture. The same has happened, as Humboldt observes, with its countrymen, the gigantic Patagonians, who are found at last not to exceed the stature of ordinary men.

Notwithstanding the faithful accounts of a few of the older authors, the true history of the condor had remained involved in the obscurity created by mingling it with so many childish tales, when the celebrated Humboldt, studying it living with the sober eye of truth and philosophy, furnished a correct description, a good drawing, and an excellent memoir upon it. Since that time several stuffed as well as living specimens have reached the menageries and museums of the United States and Europe, which, with the three plates published by Temminck, have rendered it familiar to all. It is but just, however, to mention that Latham had, long before Humboldt, given in his second Supplement a tolerably correct description of both sexes, with a figure of the adult male, and taken also from the identical specimens, now at Vienna, and originally brought to England by Captain Middleton from the Straits of Magellan, that furnished the subjects of Temminck's plates.

The adults of both sexes, and a young female, having been tolerably well represented, it is the young male that we have preferred to figure in this work, in order thus to complete the iconography of so interesting a species. And we trust that, through the exertions of our artists, our figure, which is reduced three and a half times from nature, will be found for minuteness of accuracy much superior to all, owing to the extraordinary pains taken by Mr Lawson, who, besides being furnished with a correct drawing, made repeated visits to the living bird, carefully verifying its form and dimensions in all their details.

The genus *Vultur* of Linné, now the family *Vulturini* (or *Vulturidæ*), a family first established by Duméril under the appellation of *Ptilodéres,* or *Nudicolles,* though much less

numerous as well as less intricate in the characters of the
species than the *Falconidæ*, of which we have treated under
the head of *Falco Cooperii*, has nevertheless much exercised
the ingenuity of ornithologists, who nearly all disagree both
as to its limits and its subdivisions.   With respect to the
former, those recognised by us will be clear and well defined,
this family being constituted of the two modern genera *Vultur*
and *Cathartes* of Illiger, which we adopt with some modifications, as will be seen hereafter.   Contrary to the general
practice, we discard from it the aberrant genera forming the
passage to other groups, in which we prefer arranging them.
The groups towards which a direct passage is the most
obvious are the family of *Rapaces* or *Falconidæ*, and some
typical *Gallinæ* and aberrant waders.   With neither the
Passeres nor the webfooted orders (unless it may be with
the frigate-bird) do we perceive any immediate relations.
The passage to these takes place through the intervention of
the three other orders, in the first of which the genera
Gypaëtus and *Gypogeranus* approach so near them as even
to have strong claims to be included in the same family,
being almost exactly intermediate between *Falconidæ* and
Vulturidæ.

Although the *Vulturidæ* are far from exhibiting the same
diversity of conformation, habits, and appetites as the numerous
tribe of the falcons, and form indeed as a whole a much more
compact mass, and much less numerous in species, yet even
those naturalists, with Illiger at their head, who have left
untouched the great genus *Falco*, have joined unanimously
with the reformers in dividing that of *Vultur* into two great
equivalent genera.   This course, though we imitate it ourselves,
we must confess to be more expedient than consistent, and it
is probable that, for the very reason that differential traits are
less numerous and complicated in the different species, the
divisions have been more easily made and admitted.   Let
us analyse them.   Illiger was the first to separate the species
into his two genera *Cathartes* and *Vultur :* we say the first,

excluding Storr and Lacépède, who long since, with so much
reason, withdrew *Gypaëtus* from the genus, and not adverting
to the artificial section made by Duméril in the year 1806,
under the name of *Sarcoramphus*, for the stout-billed carun-
culated species indiscriminately. The characters assigned by
Illiger were precise and natural, and the species he cited as
examples correctly typical. But Temminck, in adopting
Illiger's two genera, misapplied the characters and rendered
them unnatural by declaring the *Vultur percnopterus* a
Cathartes, whilst it is in fact a slender-billed *Vultur*, as
the condor is a stout-billed *Cathartes*. Deceived by Temminck, we at first adopted this erroneous view, which we have
finally rectified in our observations on the second edition of
the "Règne Animal" of Cuvier. In returning to what we consider the principles of Illiger, as they certainly are the dictates
of reason, it so happens that this genus *Cathartes*, as is often
the case, is found to correspond to a geographical division,
being exclusively American, whilst that of *Vultur* is in like
manner confined to the old continent. The other genera which
have been proposed among the *Vulturidœ* may be considered
as groups of secondary importance.

Thus the three European species * belong, according to
Savigny, to as many separate genera, namely, *Gyps*, *Ægypius*,
and *Neophron*. The last, restricted to its proper limits, is a
very well-marked subgenus, which we adopt under the name
of *Percnopterus*, Cuvier. It contains to my knowledge but
two well-ascertained species, which are the *slender-billed vultures of the old continent.*

The other European vultures, with stout bills, are comprised
in my subgenus *Vultur*, composed often well-known species.
But we must confess that the *Vultur cinereus* and *Vultur
fulvus* differ materially, and that even their skeletons present
differences that in other cases might be considered as even
more than generic, while one uniform osseous structure is

---

* Ruppel reckons four. He makes two of *V. fulvus*, considering the
Chassefiente of Le Vaillant a distinct species.

found to prevail throughout the numerous species of falcons. This observation I believe has never before been made. Savigny founded his groups, which are excellent as subdivisions, on the different conformation of the nostrils, on the tongue, aculeated on its margin in *Gyps*, and not in *Ægypius*, and on the number of tail-feathers, which is twelve in the latter, as in the American genus, and fourteen in his genus *Gyps*, as well as in *Neophron*.

Thus are the twelve species constituting my genus *Vultur* divided into two very natural subgenera, corresponding to the two genera of Vieillot, *Vultur* (comprising ten species) and *Neophron* (comprising but two), the first being subdivisible into the two minor groups of Savigny. The three might indeed be considered as co-ordinate subgenera.

As for the genus *Cathartes*, it is by no means so easy to divide, and the two groups or subgenera which we admit are perhaps artificial and blended too much together. The first, comprising the condor, the Californian condor, and the king vulture, that is, the *stout-billed American vultures*, may be called *Sarcoramphus*, a name confined by Duméril and Cuvier to those that have caruncles or fleshy appendages on the head, but to which Vieillot very justly added *C. Californianus*, calling the group *Gypagus*.

The second subgenus of *Cathartes* may be called *Catharista*, Vieillot, or the *slender-billed American vultures*, analogous in a parallel series, where the strength of the bill is considered, to the *Percnopteri*, but having no immediate affinity with them. The only known species are the two of Wilson's work, *Cathartes aura*, and *Cathartes iota* of my Synopsis, the former of which is a link between its own group and the preceding.

The best discriminating mark between the two principal genera of this family, one which is obvious and easily understood, is the striking character of the perviousness of the nostrils in *Cathartes*, through which light appears broadly from one side to the other, while in the *vultures* they are separated by an internal cartilaginous partition. This will make it at once evident that it was for want of proper examination that the *Percnopterus*, merely on account of its slender bill, was ever considered a *Cathartes.* The remaining characters being more of a relative than a positive kind, we shall not here notice them, except remarking that the hind toe being much shorter and set on higher up in the American genus, shows a greater affinity with the gallinaceous birds, an affinity which may be traced in other features of their organisation. The number of tail-feathers is fourteen in several species of *vultures,* whilst no *Cathartes* has ever been found to have more than twelve. The principal traits, both moral and physical, are the same in all the birds composing this highly natural family.

All, in fact, are distinguished by having their head, which is small, and their neck, more or less naked, these parts being deprived of feathers, and merely furnished with a light down, or a few scattered hairs. Their eyes are prominent, being set even with the head, and not deep sunk in the socket, as in eagles and other rapacious birds. They have the power of drawing down their head into a sort of collar formed by longer feathers at the base of the neck: sometimes they withdraw the whole neck and part of the head into this collar, so that the bird looks as if it had drawn its whole neck down into the body. They have a crop covered with setaceous feathers, or sometimes woolly or entirely naked, and prominent, especially after indulging their voracious appetite. Their feet are never feathered, like those of an eagle, although they have been unnaturally so represented in the plates of some authors. The tarsus is shorter than the middle toe, which connected at its base by a membrane with the outer one. The claws are hardly retractile, comparatively short, and from these birds' habit of keeping much on the ground, instead of always perching, as the *Falconidæ*, they are neither sharp-pointed nor much curved. Their wings are long and subacuminate, the third and fourth primaries being longest: they are lined beneath with a thick down of a peculiar and very soft nature.

The young birds have their head entirely covered with down, which gradually falls off as they advance in age. The female is larger than the male : their plumage varies greatly with age, and they moult but once a year. The young are easily distinguished by their downy head and neck, these parts in the adult being naked, and by the absence of the caruncles which in some species are found on the adult. These fleshy appendages are of the same nature as the wattles, &c., of gallinaceous birds.

No part of ornithology has been more confused in its details than that relative to the vultures, and their synonymy, especially the European species, is almost inextricable : the old authors have heedlessly multiplied and even composed species, whilst the modern have brought together the most confused citations under those which at last they founded on the actual observation of nature. We congratulate ourselves that the task of pointing out all these errors, from which no writer without exception appears free, does not belong to us.

Vile, gluttonous, and pre-eminently unclean, the *Vulturidæ* are the only birds of prey that keep together in flocks all the year round : as cowardly as they are indiscriminately voracious, they are too pusillanimous, notwithstanding their numbers, to attack living prey, and content themselves with the abundant supply of food which is offered by the putrid carcasses of dead animals. In fact, they appear to give the preference to these, with all their disgusting concomitants, and only resort to freshly slaughtered animals when impelled by extreme hunger. Their want of courage is denoted by their crouching attitude and the humility of their demeanour. Creatures with such dispositions did not require from Nature strength or powerful weapons; nothing was needed but perfection of smelling, that they might from a distance discover their appropriate food, and this faculty they possess in an eminent degree. Their nostrils have two large external apertures, and an extensive olfactory membrane within.

Though regarded with disgust for their filthy habits, these

well-known birds are extremely serviceable in hot climates,
by devouring all sorts of filth and impurities, and thus pre-
serving the atmosphere from the contamination of noxious
effluvia. On this account their cowardice is protected by
man, who in civilised as well as savage life always looks to
his own advantage, and does not disdain to make use of those
for whom he cannot help feeling contempt. Besides their
usefulness during life, the vultures have an additional security
in their utter uselessness when dead. In consequence of their
food, their body exhales a disgusting effluvium, and their flesh
is so rank, stringy, and unsavoury, that nothing short of absolute famine can bring any one to taste of it. No skill nor
precautions in cooking can overcome its natural bad odour,
which prevails over the most powerful spices. But though
not eatable themselves, they excel in picking clean to the very
bones the carrion they feed upon, leaving them as bare as if
they had been carefully scraped. With this food they gorge
themselves to such a degree as to be incapable of flight, and
hardly able to move for some time, and then allow of a very
close approach. In fact, their indolence, filthiness, and voracity are almost incredible.

They are birds of slow flight at all times, and raise themselves from the ground with difficulty, though, when surprised
and closely pursued after overfeeding, when they are almost
helpless, they can lighten themselves by vomiting up their
superfluous meal, sometimes to the great annoyance of the
pursuer, and then at once take flight. Their sight is exceedingly keen, and is only inferior in power to their sense of
smelling, which enables them to discover their peculiar food
at great and incredible distances. They are dull and heavy,
fond of assembling in flocks upon trees, where they may be
seen perched for hours together, sitting with their wings
open as if ventilating their plumage. They walk with the
body inclined forward, the wings drooping, the tail brushing
the ground. When they wish to take flight, they are obliged
to run a few paces, and then contract the body violently.

Their flight, though slow, is protracted for a greater length of time than even perhaps that of the eagle, though more laborious and heavy. They elevate themselves to such wonderful heights, that as they describe circle after circle they gradually appear no larger than a swallow, next a mere speck is visible, then disappearing altogether from the limited power of human vision. Not, however, beyond their own, for as they hover over the country beneath, they can discover a carcass or carrion anywhere over a very wide district. In the East they are well known to follow the caravans; in Africa and South America they accompany and wait upon the hunter's steps. If a beast is flayed and abandoned, calling to each other with shrill but resounding voice, they pour down upon the carcass, and in a short time, so dexterously do they manage the operation, nothing remains but the naked skeleton. If the skin should be left on the prey they discover, an entrance is soon made through the belly, by which they extract all but the bones, which are left so well covered by the skin as hardly to show that they have been at work there. Should a sickly ox or smaller animal be accidentally exposed defenceless, or from any cause unable to resist, the vultures fall upon and devour him without mercy in the same manner. Thus in the mountainous districts of hot countries, in which they are very numerous, the hunter who wishes to secure his game dares not quit an animal he may have killed, for fear of its immediately becoming their prey. Le Vaillant, while in Africa, met with frequent losses through the rapacity of these parasites, which, immediately notified by the calling of the crows, flocked around in multitudes, and speedily devoured large animals that he had killed, depriving him not only of his own meal, but of many a valuable specimen intended as a contribution to science. They may be frequently seen tearing a carcass in company with dogs or other ravenous quadrupeds, such associations producing no quarrel, however lean and hungry both may be. Harmony always subsists, so long as they have plenty, among creatures of dispositions so congenial. But the eagle drives them to a distance till he is satisfied, and only permits them to enjoy the fragments of the prey he has conquered. With the same expectation of feeding upon the leavings, they attend upon the ferocious quadrupeds of the cat-kind, and may thus indicate the vicinity of these dangerous beasts. That it is cowardice which prevents them from attacking animals capable of making any defence is evident. The innate cruelty of their disposition is often manifested towards the helpless. To a deserted lamb they show no mercy, and living serpents and whatever other minor animals they can overpower are their usual food. They are also, it is said, extremely fond of crocodiles' and alligators' eggs, to obtain which they keep watch unseen in the adjacent forest while the female is laying, and as soon as she is gone descend, and removing the sand where they are buried, greedily devour them.

The vultures are mostly found in warm climates, although by no means afraid of cold, as they prefer the vicinity of lofty mountains; those which inhabit in the north retiring southward in winter in the northern hemisphere. Their favourite abodes are rocks and caverns among broken precipices, where they retire to sleep and to digest their meals when overfed, which happens as often as an opportunity offers : in such retreats they may be often observed in great numbers together, enjoying the exhilarating air of the morning. Their nest is made, with hardly any preparation, on inaccessible cliffs or other places where they can seldom be found by man. They reside generally where they breed, seldom coming down into the plains, except when frost and snow have driven all living things from the heights : they are then compelled to brave danger in pursuit of food. The vultures generally lay but two eggs at a time, sometimes three or four, especially the North American species; and are faithfully monogamous. In their mode of supplying their young with food, there is a striking difference between them and other rapacious birds. The latter place before their progeny the quivering limbs of their prey, that they may learn to employ their beak and talons. The vultures, whose claws are not fitted for seizing and bearing off their spoils, disgorge into the mouth of their young the contents of their crop, from the nature of which, this operation, so interesting when performed by a dove or a canary, becomes in this case one of the most disgusting imaginable.

According to Belon, the Latin name *Vultur* is but a contraction of *volatu tardo:* the name *Cathartes*, imagined by Illiger, means in Greek, *purger.* *Condor* is a corruption of *cuntur*, the true appellation of our species in the *Qquichua* language, derived, according to Humboldt, from the verb *cunturi*, to smell.

Although the largest of American vultures, the condor is inferior in size to several of those which inhabit the old continent, and even to the large golden vulture of Eastern Europe. Both sexes are very nearly of equal size; but the superiority, if any, is found, as usual, upon the side of the female; so that the common statement of writers, that this sex is of less size, has no foundation in fact.

The adult male is always more than three feet long, and measures nine feet from tip to tip of the extended wings. Some gigantic individuals are met with four feet long and twelve in extent.   The bill is dark brown colour at the base, somewhat of a lemon white at tip.   The tongue is entire, cartilaginous, membranous, ovate-cuneate, concave beneath, serrated with spines on the margin.   A longitudinal compressed caruncle, or firm fleshy crest, extends from the top of the head to the front, and to the brown portion of the bill. It is rounded before and behind, a sinus on the upper border, the lower free for a short space at each extremity, papillous, or strongly wrinkled, and, as well as the cere, of a bluish colour.   The nostrils are oval-linear, and with no hairs surrounding them.   The skin of the neck and crop is bare, with the exception of some scattered short and rigid hairs; it is reddish, and has two short pear-shaped processes depending from it.   Two intertwisted fleshy strings arise from

the bill, pass over the auditory region, and descend along the
sides of the neck: these fleshy cords acquire by desiccation,
in stuffed specimens, the appearance of a series of tubercles
or wrinkled protuberances : a double string of a similar substance passes above the eye, which is small, much lengthened,
and lateral, being set far back from the bill : the irides are of
an olive grey. Their cavernous structure enables the bird to
swell out all these appendages at pleasure, like the turkey :
the crest, however, must be excepted, which is very dissimilar
to the flaccid, pendulous cone of the turkey, and incapable of
dilatation. The orifice of the ear is very large, subrounded,
but hidden under the folds of the temporal membrane. The
occiput exhibits a few short brown bristles. Around the
lower part of the neck above is a beautiful half collar of silky
and very soft down, as white as snow, which separates the
naked parts from the feathered body. In front this collar is
interrupted, and the neck is bare down to the black plumage :
this gap in the collar can however only be discovered on close
inspection. The whole plumage is of a very deep blue black ;
the tips of the secondaries and the greater wing-coverts on
the outer web only being of a whitish pearl-grey : the first
seven outer quills are wholly black, twenty-seven being white
on their outer web : the third quill is the longest. The wings
are three feet nine inches long, reaching nearly to the tip of
the tail, but not passing beyond, as in the closely related
species the Californian condor. The tail is very slightly
rounded at the end, rather short in proportion to the bird,
measuring thirteen inches. The feet are bluish : toes connected at their base by a membrane.

The female is entirely destitute of crest or other appendages.
The skin which covers the head is uniformly blackish, like
the plumage, in which there is only a little cinereous on the
wings : in this sex the wing-coverts, which in the male are
white at tip from the middle, are of a blackish grey. This
circumstance is very conclusive, inasmuch as the white forms
a very conspicuous mark on the wings of the male, which has

occasioned it to be said that some condors had a white back.

For several months during the early part of their life, the young are covered with very soft whitish down, curled, and resembling that of young owls: this down is so loose as to make the bird appear a large shapeless mass. Even at two years old the condor is by no means black, but of an obscure fulvous brown, and both sexes are then destitute of the white collar.

The following description and admeasurements are from a pair of young living birds, said to be nine months old, caught on the Peruvian Andes. One of these (which are precisely alike) was captured by an Indian, who discovering two in the nest, ran up at great speed, fearing to be overtaken by the old ones, and succeeded in securing it by putting it in his pocket, not larger than a full-grown chicken. I have carefully compared this with, and found perfectly similar to it, a bill and a quill-feather brought from the Columbia river by Lewis and Clark, and preserved in the Philadelphia Museum. These remains prove the existence of the condor within the United States, and sufficiently authorise its introduction into this work.

Length, three feet nine inches; breadth, nine feet. Bill to the corner of the mouth, two inches six-eighths; to the cere, one inch and a half; to the down, three and a quarter inches. Bill curved and hooked, with several flexures; upper mandible passing over the lower, which is rounded and scalloped: nostrils pervious, rounded-elliptical, cut in the cere. Bill outside, cere, and all the surrounding naked parts, black; ears without any covering; the skin rugose: inside of the bill yellowish white, margined with black, palate furnished with a fleshy skin, having the appearance of a row of teeth in the middle, then of a hard ridge looking like a file, and two marginal rows: tongue broadly concave, and serrated on the turned-up edges with sharp-pointed cutting serratures: an elevation of the skin indicating the frontal caruncle; the place where the bristles begin to appear is also indicated by an elevation. Eye full and rounded; iris blackish; membrane of the throat very dilatable; head and neck covered by a thick silky down of a brownish black colour, on the front more dark and bristly; general colour dark brown, each feather having a banded appearance, tipt with more or less of umber; quill and tail feathers black, with a gloss of blue. The number of tail-feathers is twelve, the closed wings not reaching beyond, though very nearly to the tip. Feet black; acrotarsus beautifully colligate; acrodactylus scutellated: the whole leg measures one foot in length, of which the tarsus is five and a quarter inches, and the middle toe and nail six, the nail being one and a half: lateral toes connected with the middle as far as the first joint by a membrane; the inner, two and a half inches long without the nail, which is one and a half; the outer, with the nail, a quarter of an inch shorter; hind toe articulated inside, bearing on the ground only with the point of the nail, an inch and a half long, the nail one inch more, and much incurved; sole of the foot granulated; fat part of the heel large and rough. The feet have been generally described as white or whitish, owing to their being commonly stained with the excrements, which the bird throws much forward, but they are in fact of a fine blue horn-colour when washed clean; and these birds seemed to be fond of washing themselves.

The condor is diffused over the continent of South America from the Straits of Magellan, extending its range also to Mexico and California, and the western territory of the United States beyond the Rocky Mountains. It was not seen by Lewis and Clark until they had passed the great Falls of the Columbia, and it is by no means common or numerous anywhere in the northern parts of America; those individuals that have been observed here appear to have been stragglers from their native country, which is no doubt South America. It might even be limited to the great chain of the Andes, especially their most elevated ranges, being plentiful in Quito, Peru, New Granada, and Antioquia, and much more rare where they are less lofty ; the condor inhabiting pretty nearly the same altitude with the *Cinchonæ* and other subalpine plants.   It is, moreover, according to the observations of Humboldt, the invariable companion of the guanaco for an extent of nearly three thousand miles of coast, after which this animal is no longer seen ; but the condor continues to be met with much beyond this, as if quite indifferent to climate, or because it can regulate it by varying its elevation with the change of latitude.   In the eastern or even southern United States a condor has never been seen, though the king vulture of South America has been occasionally observed.   The chief abode of the condor is indeed on the highest summits of the Andes, some of which are covered with perpetual snow, and is fixed by Humboldt at between three thousand one hundred and four thousand nine hundred metres.   Every time, says he, that I have been herborising near the limits of perpetual snow, we were sure to be surrounded by condors.   These mountains and the forests that clothe their sides are the condor's home, and from these their excursions are extended over the whole neighbourhood to the very sea, from which they may be often seen hovering at prodigious heights and describing vast circles, but always ready to lower themselves by degrees whenever they espy a chance of satisfying their voracious appetite. They are only known, however, to descend towards the seashore during the rainy season, corresponding to our winter, when they come in search of food and warmer weather : they then obtain the bodies of large fishes or marine animals, such as whales or seals, and the prospect of finding these is their principal attraction to the shore : they arrive here at evening, and as a journey of several hundred miles requires for them but little time or exertion, as soon as their meal is digested, and they begin to feel lighter, they return to their favourite rocks, often during the following day.   They have sometimes been killed at sea, floating on the dead body of a whale which they were tearing for food.   They exhibit the common propensity of their tribe for carrion, and nothing but the urgent stimulus of hunger can bring them to attack living creatures, and then their cowardice will not allow them to meddle with any but the feeble or diseased which are incapable of defending themselves. They will also combine together to overpower their prey, if they see the least danger of resistance. A single cougar, or even a courageous bird, will drive from their prey a whole troop of condors, which, however, seldom amounts to more than five or six, as they do not collect in such numerous bodies as their fellow-vultures. When feeding on a cow, a guanaco, or a paco, they first pick out the eyes, then tear away and devour the tongue, and next the entrails, at last picking the flesh from the bones. Smaller animals they generally swallow whole. Guided by their amazingly acute faculty of smell, the condor will arrive, performing circular evolutions, from the highest regions of the atmosphere upon a carrion, and often, trusting to their powers of digestion, they swallow bones and flesh together. The Indians, too indolent to keep clean their butchering or similar places, and often neglecting to bury their dead with sufficient carefulness, have a great veneration for this bird and others of its kind, to which they trust to rid them of such nuisances. The regard with which they are treated makes them so familiar, that Humboldt relates his being able to approach within two yards of a troop of condors before they retreated, though he had other persons in his company. When full-fed the condor will remain motionless on a projecting rock, and has then a sinister appearance ; if on the ground, however, he allows of a close chase before having recourse to his ample wings, hopping along before his pursuer. When, on the contrary, he is pressed by hunger and light from emptiness, he will soar to extreme heights in the atmosphere, especially in clear weather, whence he can discover prey at any possible distance. They lay in the most inaccessible parts of the Andes, near the limit of perpetual snow, on the most broken and terrific precipices, where no other living creature can dwell. Nests have been found at the extraordinary elevation of fifteen thousand feet.

Their eggs are usually laid on the naked rock, or with very little preparation, and never on trees, which they even avoid alighting on, unlike their congeners in this respect, and always on rocks or the ground, the straightness of their nails making this easier for them. The eggs are white, and three or four inches long. The young are entirely covered with very soft whitish down, and the mother is said to provide for them during a long time. The facts relative to their propagation are not, however, sufficiently ascertained, for how are we to verify assertions relating to operations performed so much beyond the reach of ordinary observation?

Authors describe various modes that have been resorted to for destroying the condors in their native countries, where they sometimes become a nuisance; such as poisoning carrion, seizing them by the legs by hiding under the skin of a calf, and by building narrow enclosures in which is placed putrid flesh, when the birds flying down and feeding greedily, are unable to take wing again for want of space to get a start by running. But we scarcely see any advantage in such stratagems, since they may be caught with running knots while disabled by repletion, or even, as it is reported, knocked down with clubs; and in any case, we are at a loss to reconcile such persecutions with the protection so wisely granted them both by civilised and savage man.

In captivity, the condor is easily tamed if taken young, and does not refuse any kind of animal food whatever, nor do they appear to dread or suffer in the least from the extreme changes of the climate in Europe and the north-eastern parts of America. But it is almost impossible to keep the adults, which are always exceedingly wild and mischievous. They are incredibly tenacious of life: the bones are so hard as to resist a musket-ball, to which also the thick down of their plumage is impenetrable. They can resist strangulation for hours, even when hanged and hard pulled by the feet. A remarkable fact is that in domesticity they will not refuse water, drinking it in a very peculiar manner, by holding their

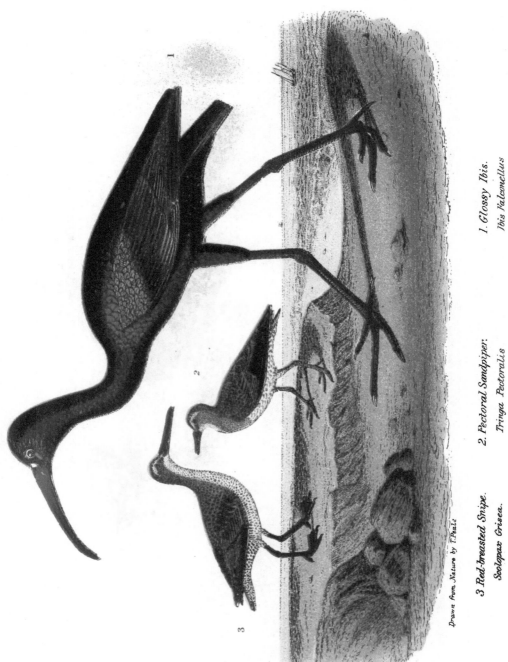

Drawn from Nature by T.Peale

3 Red-breasted Snipe.
Scolopax Grisea.

2. Pectoral Sandpiper:
Tringa Pectoralis

1. Glossy Ibis.
Ibis Falcinellus

lower mandible in it for some time, and using it as a spoon to throw the liquid into their throat. The individual represented in our plate was remarkable for playfulness and a kind of stupid good nature. During Mr Lawson's almost daily visits for the purpose of measuring and examining accurately every part for his engraving, he became so familiar and well acquainted that he would pull the paper out of the artist's hands, or take the spectacles from his nose, so that Mr Lawson, seduced by these blandishments, and forgetting its character in other respects, does not hesitate to declare the condors the gentlest birds he ever had to deal with.

GLOSSY IBIS. (*Ibis falcinellus.*)

PLATE XXIII.—Fig. 1.

Tantalus falcinellus, *Linn. Syst.* i. p. 241, sp. 2.—*Gmel. Syst.* i. p. 648, sp. 2.— *Lath. Ind.* ii. p. 707, sp. 14.—*Brunn. Orn.* sp. 167.—*Scop. Ann.* i. sp. 131.— *Kram. Austr.* p. 350.—*Borowsky,* iii. p. 72.—*Faun. Helv.*—*Retz, Faun. Suec.* p. 171, sp. 135.—*Nilss. Orn. Suec.* ii. p. 43, sp. 160.—Ibis falcinellus, *Vieill. Nouv. Dict.* ; *Id. Enc. Met. Orn.*—*Temm. Man. Orn.* ii. p. 598.— *Savi. Orn. Tosc.* ii. p. 327.—*Nob. Obs. Nom. Wils. Orn.* note to No. 199 ; *Id. Syn. Birds U.S.* sp. 241 ; *Id. Cat.* ; *Id. Spec. Comp.*—*Wagler, Syst. Av. Ibis,* sp. 1.—*Roux, Orn. Prov.* pl. 309.—*Goldfuss, Nat. Atlas,* pl. 95. —Ibis sacra, *Temm. Man. Orn.* 1st ed. p. 385.—Tantalus igneus, *Gmel. Syst.* i. p. 649, sp. 9.—*Lath. Ind.* ii. p. 708, sp. 16, very old individual.— Tantalus viridis, *Gmel. Syst.* i. p. 648, sp. 8.—*Lath. Ind.* ii. p. 707, sp. 15.— *Montagu, in Linn. Trans.* ix. p. 198.—Tantalus Mexicanus? *Ord, in Journ. Ac. Philad.* i. p. 53 (and perhaps of other authors ; in that case the Acalot of Ray and Buffon, and its derivatives).*—Numenius castaneus, *Briss. Orn.* v. p. 329, sp. 5. ; *Id.* 8vo, ii. p. 294 (old indiv.)—Numenius viridis, *Briss. Orn.* v. p. 326, sp. 4, t. 27, fig. 2. ; *Id.* 8vo, ii. p. 293, two years old.—

* The following are the indications of that obscure species, the Mexican Ibis :—Tantalus Mexicanus, *Gmel. Syst.* i. p. 652, sp. 18.— *Lath. Ind.* ii. p. 704, sp. 4.—Ibis Mexicana of later compilers ; Ibis acalot of French compilers and dictionaries.—Numenius Mexicanus varius, *Briss.* v. p. 335, sp. 7 ; *Id.* 8vo, ii. p. 295.—Acacolotl, *Ray, Syn.* p. 104, sp. 5.—*Will.* p. 218 ; *Id. Engl.* p. 296.—Acalot, *Buff. Ois.* viii. p. 45.—Mexican Ibis, *Lath. Syn.* iii. pt. 1, p. 108, sp. 5 ; *Id. Gen. Hist.* ix. p. 146, sp. 5. This bird is said to inhabit Mexico : it will not be superfluous to remark that the proportions assigned to it are much larger than those of our *Ibis falcinellus.*

Numenius viridis, *S. G. Gmel. Reis.* i. p. 167. ; *Id. in Nov. Comm. Petrop.* xv. p. 462, t. 19, young.—Numenius igneus, *S. G. Gmel. Reis.* i. p. 166; *Id. in Nov. Comm. Petrop.* xv. p. 460, t. 13, old specimen.—Tringa autumnalis, *Hasselquist, Iter Palest.* ii. p. 306, sp. 26, 27, two years old.—Falcinellus, *Gesner, Av.* p. 220.—Falcata, *Gesner, Ic. Av.* p. 116, with a bad figure.—Falcinellus, *sive Avis falcata, Aldrov. Av.* iii. p. 422 and 423. —*Johnston, Av.* p. 105.—*Charleton, Excit.* p. 110, sp. 7; *Id. Onomatz,* p. 103, sp. 7.—Falcinellus Gesneri et Aldrovandi, *Willoughby, Orn.* p. 218.— Arcuata minor, &c., *Marsigli, Danub.* v. p. 42, tab. 18, adult ; 20, young.— Numenius subaquilus, *Klein, Av.* p. 110, sp. 8.—Falcinellus or Sithebill, *Ray, Av.* p. 103, sp. 3.—*Will. Orn.* p. 295, tab. 54; *Id. Engl.* p. 295, tab. 54, fig. 4.—Le Fauconneau ! Falcinellus, *Salerne, Orn.* p. 322.—Courlis Vert, *Buff. Ois.* viii. p. 29 (ed. 1783), viii. p. 379.—*Vers. Germ. Otto,* xxii. p. 170.—Courlis d'Italie, *Buff. Pl. enl.* 819, adult male.—Courlis Brillant, *Sonnini, Buff. Ois.* xxii. p. 238, old female.—Ibis Vert, *Cuv. Regn. Anim.* i. p. 485 ; *Id.* 2d ed. i. p. 520.—*Roux, loco citato.*—Savigny, *Egypt. Ois.* tab. vii. left-hand fig., two years old.—Ibis Noir, *Savign. Hist. Nat. et Mythol. de l'Ibis,* p. 36, tab. 4.—Ibis Sacré, *Temm. Man. Orn.* 1st ed., but not of Cuvier.—Ibis Falcinelle, of most French authors and of the Dictionaries.—Chiurlo, &c., *Storia degli Uccelli,* ix. p. 439, old male.—Ibi Falcinello, *Ranzani, Elem.* iii. pt. viii. p. 185, sp. 3.—Mignattajo, *Savi. loco citato.*— Bay Ibis, *Penn. Arct. Zool.* ii. p. 460, A.—*Lath. Syn.* iii. pt. i. p. 113, sp. 13; *Id. Suppl.* p. 67, *Germ. trans. by Bechst.* v. p. 67, tab. 81, young.— *Lath. Gen. Hist.* ix. p. 152, sp. 15.—*Brit. Miscell.* tab. 18.—*Montag. Orn. Dict. Suppl.*—Green Ibis, *Lath. Syn.* iii. pt. 1, p. 114, sp. 13, young.—*Linnean Trans.* ix. p. 198.—*Montag. Orn. Dict. Suppl.*—*Lath. Gen. Hist.* ix. p. 154, sp. 18.—Glossy Ibis, *Lath. Syn.* iii. pt. 1, p. 115, sp. 14, old specimen ; *Id. Gen. Hist.* ix. p. 154, sp. 17.—*Brit. Zool.* 1812, ii. p. 30.—*Montag. Orn. Dict. Suppl.*—Brazilian Curlew, *Nat. Miscell.* tab. 705 ?—Sichelsnabliger Nimmerzatt, *Bechst. Nat. Deutschl.* iv. p. 116.—*Meyer and Wolf, Tasch. Deutschl. Vog.* ii. p. 352.—*Naum. Vog. Nacht.* t. 28, adult male.—Braune Ibis, *Brchm, Lehrb. Eur. Vog.* ii. p. 528.—*Philadelphia Museum.*

Though it may appear very extraordinary, it is not less true, that one of the two species of ibis worshipped by the ancient Egyptians, their black ibis, has a claim to be included in our work as being an occasional visitant of the eastern shores of these States. This fact, which we would be among the first to disbelieve were we to read of it in the eloquent pages of Buffon, is authenticated by the specimen here figured, which, moreover, is not a solitary instance of the kind. Thus, instead of being limited to a peculiar district of Egypt, as stated by Pliny, Solinus, and others, and reiterated by the host of compilers, this celebrated bird is only limited in its irregular wanderings by the boundaries of the globe itself.

The credit of having added this beautiful species to the fauna of the United States is due to Mr Ord, the well-known friend and biographer of Wilson, who several years ago gave a good history and minute description of it in the Journal of the Academy of Philadelphia, under the name of *Tantalus Mexicanus?* His excellent memoir would have been sufficient to establish its identity with the species found so extensively in the old world, even if the specimen itself, carefully preserved in the Philadelphia Museum, did not place this beyond the possibility of doubt.

Among the natural productions which their priests had through policy taught the superstitious Egyptians to worship, the ibis is one of the most celebrated for the adoration it received, though for what reason it is not easy to understand. The dread of noxious animals, formidable on account of their strength or numbers, may induce feelings of respect and veneration, or they may be felt still more naturally for others that render us services by destroying those that are injurious, or ridding man of anything that interferes with his enjoyments, or by ministering to his wants. We can conceive how a sense of gratitude should cause these to be held sacred, in order to insure their multiplication, and that this sentiment should even be carried to adoration. But why grant such honours to the wild, harmless, and apparently useless ibis? It is perfectly well proved at this day that the ibis is as useless as it is inoffensive, and if the Egyptian priests who worshipped the Deity in his creatures declared it pre-eminently sacred; if, while the adoration of other similar divinities was confined to peculiar districts, that of the ibis was universal over Egypt; if it was said, that should the gods take mortal forms it would be under that of the ibis that they would prefer to appear on earth, and so many things of the kind, we can assign no other reason than the fact of their appearing with the periodical rains, coming down from the upper country when the freshening Etherian winds began to blow, when they were driven in search of a better climate by the very rains that produce the inundation of the Nile, doing Egypt such signal benefit. The ibis, whose appearance accompanied these blessings, would disappear also at the season when the south desert winds from the internal parts of Africa brought desolation in their train, which could be averted only by the periodical return of the *circumstances represented by the ibis*, which seemed like Providence to control them, and was therefore declared the real providence of Egypt, though merely the concomitant, and by no means the cause of those blessings by which they profited in common with all. It thus became so identified with the country as to be used as its hieroglyphic representative, and was said to be so attached to its native land that it would die of grief if carried out of it, and it was on account of its fidelity to the soil that it was honoured as its emblem. So good a citizen could not of course from selfish motives migrate periodically, and its absence must have been for its country's sake ! Hence the ridiculous tale current throughout antiquity, and strengthened by the testimony of Herodotus, Ælian, Solinus, Marcellinus, copied by Cicero (who went so far as to assign to the ibis proper instruments for the purpose, such a strong bill), by Pliny and others, and credited in our days to a certain extent by Buffon, who thus accounted for the divine honours it received. I allude to the story of their attacking and destroying periodically on the limits of civilisation immense flocks of small but most pernicious winged serpents generated by the fermentation of marshes, which, without the generous protection afforded by the ibis, would cause the utter ruin of Egypt.

Still more unaccountable is it that naturalists and philosophers should have been so long in finding out the true meaning of this oriental figure. How could the ibis with its feeble bill, whose pressure can be hardly felt on the most delicate finger, and which is only calculated for probing in the mud after small mollusca and worms in places just left bare after an inundation, how could such a weapon cut to pieces and destroy so many monsters if they had existed ? How could these learned men (notwithstanding that Herodotus relates his seeing heaps of their bones or spines) believe for an instant in the existence of these winged serpents? and why try to reconcile truth with a barefaced falsehood, or with expressions manifestly figurative? We are aware that some modern translators of Herodotus, by forcing the Greek original to meet their own views, have attempted to write, instead of winged serpents, the word *locustæ*, which insects are known to come in vast swarms, causing periodically great devastation even in some parts of Europe. But nothing is gained by this plausible and apparently learned supposition, since the conformation of the ibis would prevent it from making any havoc among these enemies, whose being winged would not, moreover, save their author from the difficulty, locusts having certainly neither bones nor spines. The figure intended is still plainer, and Savigny, who first pointed it out, could in my opinion have saved himself many a page of his classical dissertation, and without any recourse to the idea of the *Cerastes*, for to me it is evident that by the winged serpents were originally signified the exhalations from the marshes, so noxious in Egypt when brought by the south-easterly or *Typhonian* winds, against which the ibis was observed to direct its flight and to conquer, aided, it is true, by the powerful sweeping *Etherian* winds.

Be this as it may, no animal was more venerated by the Egyptians than the ibis: there was none whose history was more encumbered with fictions. Notwithstanding the ridicule thrown upon it by Aristotle, the ibis was believed to be so essentially pure and chaste, as to be incapable of any immodest act. The priests declared the water to be only fit for ablutions and religious purposes when the ibis had deigned to drink of it. Yet by some unaccountable contradiction Roman authors made of it an unclean animal. It is needless here to repeat all the fanciful and extravagant things said of the ibis among a people whose credulity, superstition, and wildness of imagination knew no bounds. It was represented by the priests as a present from Osiris to Isis, or the fertilised soil, and as such was carefully brought up in the temples, those first menageries of antiquity. It was forbidden under pain of severest punishment to kill or injure in the least these sacred beings, and their dead bodies even were carefully preserved in order to secure eternity for them. It is well known with what art the Egyptians endeavoured to eternise death, notwithstanding the manifest will of Nature that we should be rid of its dreaded images, and that many animals held sacred shared with man himself in these posthumous honours. In the Soccora plains many wells containing mummies are rightly called birds' wells, on account of the embalmed birds, generally of the ibis kind, which they contain. These are found enclosed in long jars of baked earth, whose opening is hermetically closed with cement, so that it is necessary to break them to extract the mummy. Buffon obtained several of these jars, in each of which there was a kind of doll enveloped in wrappers of linen cloth, and when these were removed, the body fell in a blackish dust, but the bones and feathers retained more consistence, and could be readily recognised. Dr Pearson, who received some of these jars from Thebes, gives a more minute description, as does also Savigny. E. Geoffroy, and Grobert, also brought from Egypt some very perfect embalmed ibises, and I have availed myself of every opportunity to examine such as were within my reach, and especially those preserved in the Kircherian Museum at Rome, one of which, containing a most perfect skeleton, is now before me.

By far the greater part of the jars contain nothing but a kind of fat black earth, resulting from the decomposition of the entrails and other soft parts buried exclusively in them. Each bird is enclosed in a small earthen jar with a cover used for the purpose. The body is wrapped up in several layers of cloth, about three inches broad, saturated with some resinous substance, besides a quantity of other layers fixed in their place by a great many turns of thread crossed with much art, so much, indeed, that it is by no means easy to lay the parts bare for inspection without injuring them. Space appears to have been considered of much value in preparing these mummies, and every means was used to secure them within the least possible.compass, by bending and folding the limbs one upon another. The neck is twisted so as to bring the crown of the head on the body, a little to the left of the stomach, the curved bill with its convexity upward is placed between the feet, thus reaching beyond the extremity of the tail : each foot with its four claws turned forward, one bent upward and elevated on each side of the head ; the wings brought close to the sides, much in their natural position. In separating them to discover the interior, nothing of the viscera nor any of the soft parts remain, the bones exhibit no traces of muscle or tendon adhering to them, and the joints separate at the least touch. Most of these mummies, it must be admitted, are not of the species of which we are writing (and which also is but seldom represented hieroglyphically), but of the white kind, which was more venerated, the *Ibis religiosa* of Cuvier ; and some authors even' deny that a well-authenticated black ibis has ever been unwrapped. Complete birds even of the white species are extremely rare. Cuvier obtained the entire skeleton from an embalmed subject, and Dr Pearson was so fortunate as to discover the perfect bird in two brought among other mummies from Thebes. They have been accurately described in the scientific journals of England under the name of true Egyptian or Theban ibis. The Egyptian ibis of Latham is, however, nothing but the *Tantalus ibis.*

Buffon by means of his mummies was enabled to verify the real size of the ibis, and as he found two bills entire among those he examined, he settled the genus to which the sacred bird belonged, and stated very correctly that its place was between the stork and the curlew, where later naturalists have arranged it. But it is to be regretted that a preconceived opinion should have so blinded him that he could not see the furrows of the upper mandible, which do exist in a very

eminent degree, as I have personally ascertained, notwithstanding his statements to the contrary, in making which he
must have had before him the bill of the tantalus, which he
mistook for the ibis. These furrows it is of the more consequence to note, inasmuch as they form the principal discrimination between the genera *Tantalus* and *Ibis,* and serve to
put an end to a controversy to which the sacred ibis has given
rise.

Although every traveller in Egypt has used his exertions
to collect all the facts relative to a bird which plays such a
part in the sacred legends of that country, a bird associated
with so many of the wonders of antiquity, yet it was for a long
period a question among naturalists and scholars to what
species the name of ibis was properly to be applied. As,
however, contrary to the general practice of the ancients, the
description of the bird did exist, and even a representation,
tolerably good, among their sculptured hieroglyphics, it could
only be because it was supposed that divine honours must
have been the reward of signal service that any dispute could
ever arise on the subject. A sacred bird must of course, it
was concluded, be a great destroyer of venomous animals,
which the timid ibis *is not;* hence the misapplication of the
name. To such an extent did this idea prevail, and predominate over all others, that Buffon, who could only feel contempt
for the idle tales related of the ibis, so involved their true
history as to attribute to them the most violent antipathy to
serpents, on which he supposed they fed, and destroyed them
by all possible means, and assigns to them the habits of a
species of vulture. Others maintained, notwithstanding its
long and falcate bill, that it was in fact a vulture, which was
indeed the most natural conclusion after they had begun by
giving it such habits. Cuvier himself, who cleared up and
rectified everything else in relation to the ibis, because he
found in a mummy some skins and scales of serpents, most
probably embalmed as companions, which was frequently done
with different kinds of animals, declared it a true snake-eater.

Two different kinds of ibis were known to the ancients, and looked upon by the Egyptians as sacred—the white, common throughout Egypt, and the black, which was said to be found only in a peculiar district. It is the latter of which we are now to treat, a bird long known to, but not recognised by, naturalists; whilst the white was only rediscovered, in later times, by the courageous Abyssinian traveller, Bruce, who first among the moderns obtained correct notions respecting it. Bruce's ibis has been since proclaimed by Cuvier and Savigny the true ibis, in place of the *Tantalus ibis* of Linné, which he so called for want of knowing the real ibis, believing this to be it, though but very seldom even found in Egypt. This opinion, which though more plausible than that which it superseded, was still erroneous, originated with Perrault, and was adopted and maintained by Buffon, Brisson, Linné, Blumenbach, and all others until lately, when Colonel Grobert returning from Egypt presented Fourcroy with mummies which enabled Cuvier first to perceive that the ibis was not a *Tantalus*, but a true *Ibis*, which genus he did not then distinguish from *Numenius*. Savigny in the year 1806, by an admirable work on the ibis, put the question at rest.

The sacred white ibis, though not in reality peculiar to Egypt, where it is seen only at certain seasons of the year, does not, however, migrate to far distant countries: it is spread throughout Africa, and species extremely similar to it are found in India and Ceylon. But it is not our province to treat of it, and it has already formed the subject of several volumes.

We have already remarked that Buffon justly indicated the natural relations of the ibis by stating that it was intermediate between the stork and the curlew. What he said of the species we shall extend to the three families to which the three birds belong in our system. In the transition from one group to another, Nature seems often to make the passage by insensible intermediate steps, and it sometimes happens that the species placed on the limits of two groups belong decidedly to one or the other, and even when it may be impossible to say to which they ought to be referred, we still cannot admit them as types of an intermediate group. At other times the inter-mediate species form a small group by themselves, and although a portion of such a connecting group shows great affinity to that which follows it, while another portion is equally con-nected with a preceding group, yet the two parts are still more related between themselves. So it is with the family of *Tantalidæ* or *Falcati*, formed from the genus *Tantalus* of Linné, and composed of but two very natural genera, *Tantalus* and *Ibis*, the former of which retains a resemblance to the *Ardeidæ* or *Cultrirostres*, while the latter claims a stronger affinity with the *Scolopacidæ* or *Limicolæ*. Nothing, in our opinion, shows more the propriety, and even necessity, of distinguishing this small intermediate group from those which touch upon it.

Buffon and Brisson, who used as a character the artificial one of the curved bill, did not separate the *Tantalidæ* from the curlews, which are real *Scolopacidæ*, though somewhat allied to *Ibis*. Linné, whose philosophical tact was seldom at fault, and who crowded all the *Scolopacidæ* into his arbitrary genera *Tringa* and *Scolopax*, did not, however, confound the two families, for he employed as a distinguishing mark of his genus *Tantalus* the important character of the naked face. He was followed by Latham and others. The *Ibis* of Lacépède is equivalent to the *Tantalus* of Linné, though by giving the genus this name, (which Latham had done in English), he obtained the credit of being the founder of the genus *Ibis*, but unjustly, as he included in it all the smooth and thick-billed *Tantali*. To Illiger belongs the merit of having first made the distinction between them, and Cuvier, Vieillot, Temminck, and most others have followed his course, though some German authors call the restricted genus *Falcinellus*. The present family was instituted by Illiger under the name of *Falcati*. Vieillot and Ranzani adopted it under the name of *Falcirostres*. Boiè called it " of the *Ibides*," but Cuvier and Latreille placed

the two genera of which it is composed within the respective
limits of the two families which they connect, and which they
called *Cultrirostres* and *Longirostres.* Although Mr Vigors
and the modern English school have not adopted it (probably
because it interfered with their whimsical quinary arrangement), they do not dismember it, but force the whole into
their family *Ardeidæ*, with which even *Ibis* has, it is true,
more real though less apparent affinity than with *Scolopacidæ:*
as for *Tantalus*, there could be no doubt. Goldfuss has done
the same.

The *Tantalidæ* all have a very long bill, stout at the base,
subulate, falcate, and cylindrical at tip, the edges bent in and
sharp. Their corneo-membranous tongue is remarkably short,
flat, cuneate-acuminate, entire, posteriorly furcate-emarginate.
Their face is destitute of feathers, and their throat somewhat
dilatable into a pouch. Their neck is long. Their feet long,
equilibrate, and always four-toed : the naked space of the tibia
considerable : the toes long, bordered with a narrow membrane
connecting the fore toes at base. The hind toe is articulated,
with the tarsus low down, and is half as long as the middle,
bearing with its whole length on the ground. The wings are
moderate, obtuse, tubercular. The tail short, composed of
but twelve feathers. The falcate shape of the bill will at once
distinguish them from any of the *Ardeidæ*; and the nakedness of the face from the *Scolopacidæ.*

The *Ibis* may be known from the true *Tantalus* by having
a comparatively slender bill, depressed and curved from the
base; instead of being very stout at the base, compressed, and
curved only towards the tip. In *Ibis*, the upper mandible is
deeply furrowed its whole length, and entire. In *Tantalus* it
is not furrowed, and is notched. The nostrils are pervious
and wide open in the latter; half-closed by a membrane in
the former. The head is warty and entirely bald in *Tantalus*,
while in *Ibis* the nakedness generally extends over the face
and throat merely.

Tantalus only contains four species, one in each of the five

divisions of the globe, Europe excepted. In *Ibis* there are
about twenty well-ascertained species, three inhabiting the
United States, of which the present is the only one that ever
visits Europe. In South America are found several beautiful
species. The true ibises may be subdivided into two secondary
groups ; those with the tarsi reticulated, and those which, like
the present species, have them scutellated. The former have
shorter feet, and by their stouter bill, and the more extended
nudity of the face, approach nearest to the *Tantali.* Temminck
wishes to divide them into the sections *Sylvains* and *Riverains.*
Dr Wagler distributes them into three sections, which he
calls *Ibides lepopodiœ, Ibides aspido-lepopodiœ,* and *Ibides
aspidopodiœ !* this last section being formed for our species
alone, principally on account of its having the middle toe-
nail pectinated.

In the ibises, as in their kindred *Tantali,* the females are
considerably smaller than the males, but perfectly like them
in colours. The young differ greatly from the adults until
the third year. Their moult is annual and regular.

They are dull and stupid birds, fearless, and allowing of a
very close approach, so that they are easily shot. They frequent inundated places, the shores of lakes and rivers, and
particularly grounds just left bare by floods, where their
favourite food abounds. They live in flocks, but when once
paired the sexes remain united for life. They feed on insects,
worms, mollusca, and the ibises also on vegetable substances :
they search their food in mud, and often throw it up with
their bill, catching it as it descends in their throat. Shells,
even of considerable size, they swallow entire, trusting to the
muscular power of their stomach to crush them, for which
their bill is too weak. The *Tantali* are also well known to
use their powerful bills against fishes and reptiles, but the
true ibis never, notwithstanding the popular belief to the contrary. When satisfied with feeding, they retire for digestion
to the highest trees, where they stand in an erect posture,
resting their heavy bill upon their breast. The *Ibides* more

than the *Tantali* migrate periodically and to vast distances.
The habit of resting upon trees, as indeed the whole animal
economy (a thing never sufficiently considered in the formation of natural families of the ibis), separate them from the
Scolopacidæ. They are monogamous ; build on high trees,
both sexes assisting in the construction of the nest : the female
lays two or three whitish eggs, which she alone incubates, but
is then fed by the male, and both feed the young, which
require for a long period the care of the parents, and do not
leave the nest till able to flutter. They walk slowly, often
sinking deeply in the mud while watching for prey : their
gait is measured, and they never run rapidly. Their flight
is heavy, but high and protracted. Their voice is loud and
monotonous. In domesticity, like many other birds, they
become omnivorous. As to anatomical conformation, the
ibises resemble the genera of *Scolopacidæ :* a very thick
muscular stomach occupies nearly two-thirds of the anterior
capacity of the abdomen : the swelling of the œsophagus at
its origin is considerable and very glandulous : the intestines
form an elliptic mass, composed of a double spiral, besides first
a turn bordering the gizzard ; they measure upwards of three
feet in length in the species we treat of. There are two rather
short and obtuse cæcums.

The bay or glossy ibis is twenty-six inches in length, and
more than three feet in extent. The bill is of a greenish
lead colour, somewhat reddish at tip, and varies much in
length in different specimens,—the longest we have measured
was five and a half inches from the corners of the mouth : in
many it is but four inches : it is slender, thicker at base, and
higher than broad, rather compressed and obtusely rounded
at tip, and less arcuated than in the other North American
species ; the upper mandible is somewhat longer than the
lower, thickened and subangulated at base, and flattened
at its origin : two deep furrows run from the nostrils to
the extremity, dividing it into three portions ; the edges
of both mandibles are quite entire, and being bent in, they

form together when closed another deep channel: the upper mandible is filled inside to a great extent with the bony substance of the bill, so as to be hardly concave. The under mandible follows exactly the curve of the upper, and is but half as high on the sides: it is strongly canaliculated below from the base to the tip; the channel from the tip to the middle is narrow, but then widens considerably, and is extremely wide at base, where it is filled by a naked membrane forming a kind of jugular pouch. The nostrils are placed near the base of the mandible, at the origin of the lateral furrows, and are oblong, narrow, longitudinal, furnished in the upper part with a naked membrane. The tongue is sagittate and less than three-fourths of an inch from the acute point of its lateral lobe to its tip: the jugular pouch is dusky: the small naked part of the face, the lora and region around the eyes are of a greenish grey, which passes into whitish on the limits of the feathers: the irides are dark brown. The crown of the head and cheeks are of a brownish black with purplish reflections; the throat immediately below the pouch is of the same colour, though somewhat less brilliant, and with more green reflections; the feathers of the head are pointed, those of the occiput being, moreover, suberectile: the whole base of the plumage is of a pale sooty grey. The feathers of the back and wing-coverts are compact and rounded; those of the inferior parts are rather loose in texture at their margins; hind head, neck, upper portion of the back, inner wing-coverts to the shoulder of the wing, and all the internal parts of the body, together with the thighs, of a vivid brownish chestnut, very brilliant and purplish on the interscapular region: lower portion of the back, rump, vent, tail and wings entirely, including the upper and lower coverts and the long axillary feathers, glossy golden green, with purple reflections, except the primaries, which are pure golden green. The wings are one foot long, and when closed reach precisely to the tip of the tail, which is four and a half inches in length, and even at the tip: the first primary is hardly shorter than the third, the second longest. The feet are rather slender, and the tarsus much longer than the middle toe : their colour is greenish lead, somewhat reddish at the joints : tarsus scutellated, four inches long ; the naked part of the tibia nearly three inches ; the toes are slender, the middle without the nail is two and a half, and the hind toe one inch long : the nails are long and slender, but truncated and of a dark horn colour : the middle one is the longest, and slightly curved outwards, dilated on the inner side to a thin edge, which is irregularly and broadly pectinated. This character is particularly worthy of remark, inasmuch as none of the genus but this exhibit it, and it may be of great use in deciding at once whether mummies belong to this species or not, though we regret that no one appears ever to have thought of having recourse to it to determine this controverted question.

The adult female is perfectly similar to the male in all except size, being very sensibly smaller.

Under two years of age they resemble the adult, but the head and neck are of a much darker colour, the chestnut having nothing vivid, but rather verging upon blackish brown, and all speckled with small dashes of white disposed longitudinally on the margins of the feathers, and disappearing gradually as the bird advances in age : the under parts and the thighs are of a blackish grey, more or less verging upon chestnut according to age, the back acquiring its brilliant colours in the same manner. It is in this state that most authors, Brisson especially, have described their *Numenius viridis*, which for a long time usurped the privilege of somewhat representing the type of the species.

The young has these white lines longer and more numerous, and the lower parts of a darker blackish grey.

This bird does not appear in its full plumage until the third year, and is so different from the adult as to furnish an excuse for those who in that state have considered it as a distinct species. The bill is brown : the feathers of the head and of the throat are dark brownish with a whitish margin, wider in proportion as the bird is younger : the breast, belly, vent, under tail-coverts, and thigh-feathers are greyish brown or slate colour : the lower portion of the back, wings, and tail of a somewhat golden green, passing into reddish, with but very little gloss in specimens under one year old, and richer as they advance in age.   The feet are wholly blackish.

No bird ranges more widely over the globe than the glossy ibis : it has long been known to inhabit Europe, Asia, Oceania, and Africa, where it gained its celebrity.   It is now proclaimed as American, though we are not able to tell how numerous or extended the species may be on this continent.   We can hardly doubt, however, that it is found along almost all the shores of North and South America, though far from common in any of these States.   From the fact of this bird having been known to stray occasionally from Europe to far distant Iceland, we may infer that the individuals met with in the United States are merely stragglers from that part of the world, just as the *Scolopax grisea* of the same plate is an American bird well known to push its accidental migrations as far as the old continent.

Lest the discovery of the glossy ibis on the continent of America should give weight to an erroneous supposition of Vieillot, we think proper to mention that the Cayenne ibis of Latham, *Tantalus Cayanensis*, Gmel., represented by Buffon, pl. enl. 820 (Vieillot's own *unseen Ibis sylvatica*), is by no means this bird, but a real species examined by us, and which must be called *Ibis Cayanensis*.

Let it come whence it may, the glossy ibis is only an occasional visitant of the United States, appearing in small flocks during the spring season at very irregular periods, on the coasts of the middle states.   The specimen Mr Ord described, and which produced a strong sensation even among experienced gunners and the oldest inhabitants as a novelty, was shot on the 7th of May 1817, at Great Egg Harbour, and we have seen others from the same locality and obtained at the same season, as also from Maryland and Virginia.   A beautiful specimen preserved in the American Museum at New York, was shot a few miles from that city in June 1828. In Central Italy they arrive periodically about the middle of April, or the beginning of May, and pass a month among us, after which they disappear entirely, and a pair of the glossy ibis is of very rare occurrence, though they have been known to remain here so late as August. A few pairs are brought every year in spring to the market of Rome, and in Tuscany and near Genoa they are more plentiful. The Italian and United States specimens that have come under my observation were all adults. During their stay among us they occupy places near marshes and grounds subject to be overflowed, where there are no trees, but abundance of grass, and plenty of their favourite food. They search for this collected in flocks of from thirty to forty, and explore the ground with great regularity, advancing in an extended line, but closely side by side: when they wish to leave one side of the meadow for another, they do not take wing, but walk to the selected spot. When they have alighted on a newly-discovered rich spot of ground, they may be observed on it for hours, continually boring the mud with their bill. They never start and run rapidly like the curlew and sandpiper, but always walk with poised and measured steps, so that Ælian says the ibis's motions can only be compared to those of a delicate virgin. The body is kept almost horizontal, the neck much bent, like the letter S, and lifting their feet high. If alarmed, or when about to depart, they rise to wonderful heights, ascending first in an inclined but straight flight, and then describing a wide spiral, the whole flock are heard to cry out in a loud tone, their voice resembling that of geese: finally having reached what they consider the proper height, taking a horizontal direction, they soon disappear from the sight: their flight is vigorous and elevated, their pectoral muscles being very thick: they fly with the neck and legs extended horizontally, like most waders, and as they float along, send forth from time to time a low and very hoarse sound. Their food consists chiefly of small aquatic testaceous mollusca, and they do not disdain such small worms and insects as they may meet with : they are supposed to live chiefly on leeches (whence their Tuscan name *Mignattajo*), but erroneously, none of these having ever been found in their stomachs either by Professor Savi or myself. From what is observed in Europe, the regular migration of these birds appears to be in the direction of south-west and north-east. Every circumstance leads to the belief that they come to us in Central Italy through Sardinia and Corsica from the coasts of Barbary, and continue their journey hence to the vast marshes of Eastern Europe and the Caspian Sea, where they are well known to breed, though nothing is yet ascertained of their mode of propagation. Be this as it may, the glossy ibis in the north and west of Europe is a very rare bird, and merely a straggler, whilst it is common at its passage in Poland, Hungary, Southern Russia, Turkey and Greece, especially the islands of the Archipelago. It is found also in Austria and Bavaria, and in other parts of Germany, especially on the Danube ; and occasionally near the lakes of Switzerland, but hardly ever in Holland, the north of France, or England. In Sweden it is also met with, though extremely rare : it has been observed in Gothland, along the marshes of the interior parts of the island, and been killed in Scania : it is registered among the rare birds that visit Iceland. It has been found common along the rivers of the islands of Java and the Celebes ; is periodically known during seven months in Egypt, coming in October and disappearing in March : it is later in coming, and disappears after, and in quite a different direction from the white sacred kind : like this, they follow the overflowing of the Nile, retiring gradually as the water becomes too deep. It is very common about the Black, and especially the Caspian Seas, the great rendezvous and breeding-place of waders, where appears to be their chief quarters, and whence they spread into Siberia and Tartary. Great numbers are also met with in the Ural Desert. The Arabs in Egypt kill the glossy ibis by shooting them, and catch many in nets, so that in autumn the markets of the cities of Lower Egypt, that of Damietta especially, are abundantly furnished with ibises of this species, as well as the white, now no longer sacred, which are exposed for sale with the heads cut off. When taken alive, these birds appear really very low spirited, and reject food : they stand upright, the body horizontal, the neck much bent, the head inclined, moving it from right to left, advancing or withdrawing it, and striking the ground with the point of their bill. They often stand on one leg like the stork : are by no means shy, and will open their bill to defend themselves if you stretch out your hand, but their bite is scarcely felt.

It should be mentioned that although this is the black ibis of antiquity, it is by no means that of systematical writers, which they describe as really black, with a red bill and feet. Such a species is very seldom if ever seen in Egypt.

PECTORAL SANDPIPER. (*Tringa pectoralis.*)

PLATE XXIII.—FIG. 2.

Pelinda pectoralis, *Say, in Long's Exp.* i. p. 171.—Tringa pectoralis, *Nob. Cat. Birds, U. S. ; Id. Synops.* sp. 250 ; *Id. Speech. comp.*—Tringa campestris? *Cat.* ii.—*Vogel.* p. 74, sp. 764.—Tringa cinclus dominicensis? *Briss. Licht. Av.* v. p. 219, sp. 12, pl. 24, fig. 1.—Chorlito a cou brun? *Azara,* iv. p. 284, sp. 404.—Alouette de mer de St Domingue, *Briss. loc. cit.*—*Philadelphia Museum.*

THIS humble species, well marked, though closely allied to several other sandpipers, is, as well as I can judge, accurately described and figured by Brisson ; but since then unnoticed even by compilers. His description had become obsolete, when Say found the bird in the western territory, and we replaced it in the records of the science. We have since shot it repeatedly on the shores of New Jersey, where it is common. The species appears to be spread throughout the States, extending farther into the interior than most of its family : beyond the Mississippi it is very common ; many flocks of them were seen by Major Long's party both in the spring and

autumn at Engineer Cantonment, and it is often met with in
small parties on the coasts of the middle states in the latter
part of autumn.   It also inhabits the West Indies, and, if we
are correct in our reference to Azara, is found in Brazil and
Monte Video.

Unlike other sandpipers, this is not addicted to bare sandy
places, but, on the contrary, is fond of damp meadows, where
it shows some of the habits of the snipe.   Solitary individuals
are often seen, starting up from before the sportsman's feet
much in the manner of that bird.

The family to which this bird belongs has been admitted
by all authors, under various names, and comprehending more
or less aberrant genera.   It was first established by Illiger,
but he excluded from it those which by an unimportant deviation are destitute of a hind toe, which he placed in his artificial family of *Littorales,* while he included in it some true
Charadriadæ on account of the presence of a rudiment of this
member.   Vieillot took the same view, calling the two artificial families *Helionomi,* and *Ægialites ;* as did Ranzani
and Savi under the names of *Limicole* and *Tachidrome ;* and
Mr Vigors erred in like manner by distributing the genera
between his too extensive families of *Charadriadæ* and *Scolopacidæ.*   The arrangement of Cuvier and Latreille is in this
instance much more consonant to nature : these authors called
their better composed, though still far from perfect family,
Longirostres.

This family, which we shall call *Limicolæ* or *Scolopacidæ,*
is strictly natural, especially since we have still farther re-
formed it by withdrawing the genus *Himantopus,* with which
we had encumbered it in our synopsis.   The family now
comprises the six genera *Numenius, Tringa, Totanus, Limosa,
Scolopax,* and *Rhynchæa,* all possessing the most marked
affinity in form and habits.

The *Scolopacidæ* have either a moderate or generally a
long bill, slender, feeble, and extremely soft, being partially
or entirely covered with a nervous and sensitive skin : it is

nearly cylindrical, and mostly obtuse at the point. Their face is completely feathered, and their neck of a moderate length and size. The feet, though rather long, are moderate and quite slender; the tarsus is scutellated: but the chief character which, combined with the bill, will always distinguish them from the allied families, consists in the hind toe, which is short, slender, articulated high up on the tarsus, and the tip hardly touching the ground: in some quite typical species this toe is entirely wanting, and this fact corroborates what we have so often repeated in our writings, that the mode of insertion, or use made of this toe is of more importance than its being absent or present. In all the *Limicolæ* the wings are elongated, falciform, acute, and tuberculated; and the tail rather short.

The females are generally larger than the males, but, luckily for naturalists, similar to them in colour. I say luckily, for as the young differ greatly from the adults, and as the moult which takes place twice a year produces additional changes in the confused plumage of most of these birds, sexual diversity, if it existed, would render the species still more difficult to determine.

All the *Scolopacidæ* inhabit marshy, muddy places, and around waters; and never alight on trees. On the ground they run swiftly. Their food consists of insects, worms, mollusca, and other aquatic animals, which they seek in the mud, feeling and knowing where to seize their prey without seeing it, by means of the delicacy of touch of their bill. They are monogamous; breed on the ground in grassy marshes, or on the sand; and lay mostly four pyriform eggs, both parents sitting upon them, and afterwards attending their young with care, though these latter leave the nest, run about, and pick up food as soon as hatched. All these habits contrast strongly with those of the ibis, which can only be forced into this family on account of the softness of the bill, and its great similarity to that of the curlews.

Our genus *Tringa* is much more extensive than that of

most modern, though much less so than that of former writers;
for we arrange in it all the *Scolopacidæ*, whose bill, short, or
moderately so, straight or slightly curved, is soft or flexible
for its whole length, and with the point smooth, depressed,
somewhat dilated and obtuse; not taking into consideration
the feet, especially the hind toe, which we think in this case
hardly proper to represent subgeneric divisions. Ornithologists will perceive at once from this that our genus thus
constituted reunites in a natural group species that were dispersed by Linné in his genera *Tringa Scolopax* and *Charadrius*, and even some that Latham placed in his restricted
genius *Numenius*. It coincides more nearly with the better
formed genus *Tringa* of Brisson, and especially of Vieillot,
Temminck, and Ranzani, but with the addition of their
Arenaria or *Calidris;* and with the same addition, is wholly
included in the *Actitis* of Illiger; although that learned
systematist does not cite under his comprehensive genus a
single typical *Tringa*, and probably never examined one, as
they do not possess the character he assigns to the group
"pedes colligati." Our *Tringa* embraces and is formed of
the groups *Calidris, Pelidna, Falcinellus, Machetes, Eurynorhyncus* and *Arenaria* of Cuvier; and we subdivide it
pretty nearly into these very groups, which we regard as
subgenera, adding, moreover, to them another which we call
Hemipalama.

All our *Tringæ* have a bill compressed at the base, with
both mandibles furrowed each side their whole length, the
lower a little the shorter; the nostrils are in the furrows,
basal, linear, and pervious, but half closed by a membrane:
their tongue is moderately long, slender, subfiliform, canaliculated above, entire, and acute. The tarsus is slender, longer,
or subequal to the middle toe, and always scutellated: the
fore toes rather elongated and slender, the hind toe when
present is extremely short, slender, much elevated, and
hardly reaching the ground: the nails are moderate, compressed, curved, and acute. The wings moderate for this

family, though in reality long, with the first primary longest; the tertials and scapulars shorter than the primaries. The tail is rather short, subequal to the wings when closed, and always of twelve feathers and no more.

With the exception of the subgenera *Falcinellus*, distinguished by an arched bill, and *Calidris*, by a short, straight one, and, both three-toed, all our *Tringæ* are tetradactyle, having the short hind toe. With the exception of my subgenus *Hemipalama*, whose character is to have the fore toes all connected at base by a membrane, and of *Machetes*, which has only the outer ones connected, all the *Tringæ* have the feet cleft to the base. Of the species that remain after the separation of these four well-marked groups, and which are still the most numerous, we form our subgenus *Tringa*. We must not, however, pass by unnoticed the *Eurynorhyncus* of Nilsson, a group so important as perhaps to merit generic distinction: it is the *Platalea pygmœa* of Linné, of which a single specimen of uncertain nativity is known.* In this, by an extreme development of the *Tringa* character, the bill is remarkably flattened and widened at tip, somewhat in the manner of the spoonbill.

In the sandpipers the female is similar to the male, being only somewhat larger. The young differ from the adult, and they moult twice a year, changing greatly the colours of their plumage. These are a mixture of white and cinereous, changing in summer to rufous and black.

The sandpipers are maritime birds that live in flocks, oftentimes composed of different species, on sandy beaches or muddy shores, preferring mostly salt water. They migrate with the changes of the seasons along coasts and rivers, and are seldom seen in the temperate climates of North America and Europe, except during autumn, winter, and more especially in spring, when they are the most numerous. They retire to the north to breed, which they do socially among the grass near the water, but never in our climates. They feed on

* See Thunberg, Av. Sv. Holm. 1816, p. 194, tab. vi.

insects, mollusca, and other small animals, which they seek in soft ground by thrusting in their flexible bill, or among the rejectamenta of the sea. They run rapidly, and generally fly near the surface of the water in a straight line, and during the day, only short distances. Their flesh, though esculent, is by no means palatable, being too fishy: they grow amazingly fat in autumn, though their fat is not firm, but very oily. They are caught, however, in Italy by spreading nets on their feeding-grounds, and in the United States great numbers are destroyed by the gun.

Spread over all the globe, some of the species even, the sandpipers are very difficult to distinguish from one another, marked traits being few, and detailed descriptions applying mostly to individual specimens. The species have been wantonly multiplied by superficial observers, and too much reduced perhaps by scientific men. We must chiefly rely on the relative dimensions of the bill and the length of the tarsus in fixing them. In North America are found at least ten of the subgenus *Tringa,* most of which likewise inhabit Europe, that has eight: the pectoral sandpiper is the only one beside the *T. pusilla* of those American registered in our synopsis that is not found in Europe.

This new species, though it is quite as large, if not larger than the *Tringa alpina,* has a shorter bill; which is besides reddish at base, distinguishing it at once from all the species it could be confounded with, since each of them has the bill entirely black: the *T. maritima* and *T. platyrhynca* have a similarly-coloured bill, but are otherwise too well marked to be mistaken ; the former by the restricted naked space of the tibia, and the latter by the depressed form of its bill.

The pectoral sandpiper is eight and a half inches long, some females being nearly nine: the bill is little more than an inch long, compressed throughout, reddish yellow at base, the rest black, and with a few snipe-like punctures near the tip. The crown of the head is black, each feather margined with rufous: the orbits, a line over the eye, and the forehead narrowly are whitish, minutely dotted with blackish; the irides are dark: a very distinct brown line passes from the eye to the upper mandible: the cheeks, neck above, sides of the neck, and beneath down to the breast are greyish with a rufous tinge, and beautifully streaked with blackish, occupying the middle of each feather, along the shaft: surrounded and well defined (in perfect specimens) by these markings, the throat and chin are of a purer white than in other *Tringæ:* the remaining lower parts from the breast to the lower tail-coverts, including the flanks and long axillary feathers, are white, the base of the plumage dark plumbeous, and a few blackish streaks along the shafts of some of the flank and vent feathers: the feathers of the neck above, owing to the circumstance of the blackish central line widening considerably, become gradually dusky, the feathers there being merely bordered with the greyish buff. The interscapular region, the scapulars, and small wing-coverts are shining black with greenish reflections; they are margined with ferruginous, and near the exterior tips with whitish: the lower part of the back, the rump, and the upper tail-coverts are jet black and without margins. The wings are five inches long, lined with white, which predominates on the under wing-coverts: these are, however, a little varied with blackish and grey: the primaries are dusky as well as the outer wing-coverts, and are slightly edged with whitish: the shaft of the outer quill is white; of the others entirely dusky: the first primary is longest, and after the second they decrease rapidly. The tail is two inches to the tip of the lateral feathers, and a quarter of an inch more to the tip of the middle ones, which are longest by that much, and somewhat tapering, and are black edged with rufous, while the others are pale dusky, margined with white all around the tip. The feet are greenish yellow, the bare space above the knee five-eighths of an inch: the tarsus very nearly one inch, and equal to the middle toe; the outer toe is connected at the very base with the middle by a very small membrane hardly visible in young individuals, which is also the case with *T. platyrhynca:* the nails are of a blackish horn colour. Such is this bird as it appears in the end of summer and early in autumn on the New Jersey coasts, still apparently in its perfect nuptial dress, or nearly so. Mr Say informs us that all the individuals of the many flocks observed at Engineer Cantonment both in the spring and autumn were of equal size; and we have also found the sexes to agree in this respect, perhaps more than is usual in other sandpipers : in the spring dress, according to the same author, the colour of the upper part of the bird is much paler, almost destitute of black, and the feathers margined with pale cinereous. The upper part of the head is always darker than any part of the neck, and margined with ferruginous : the plumage of the neck beneath and the breast does not appear to undergo so much change as that of the upper part of the body. We have not seen the bird in this plumage, but it will be evident to every ornithologist conversant with the sandpipers that the specimens described by Say were still in the winter dress, and we may conclude that the changes in this species are analogous to those of its allies.

Several specimens of both sexes that we shot in New Jersey, evidently young birds, as they were killed at the same season as the adults described, are considerably paler and duller, the tints being blended and ill defined : the white even of the throat is dingy, the quills and tail-feathers almost uniformly dusky and destitute of margins : they have not the least trace of the outer toe membrane.

RED-BREASTED SNIPE. *(Scolapax grisea.)*

PLATE XXIII.—Fig. 3.

Scolopax grisea, *Gmel. Syst.* i. p. 658, sp. 27.—*Lath. Ind.* ii. p. 724, sp. 33. *Suppl.* i. p. 444, sp. 42, winter dress.—*Temm. Man. Orn.* ii. p. 679.—*Nob. Obs. Wils.* sp. 205 ; *Id. Cat. and Syn. Birds U. S.* sp. 267 ; *Id. Specch. Comp.* sp. 206 ; *Id. Monog. Scolop. in Obs. Cuv.* p. 115, sp. 2.—Scolopax noveboracensis, *Gmel. Syst.* i. p. 658, sp. 28.—*Lath. Ind.* ii. p. 723, sp. 32, summer dress.—Scolopax Paykulli, *Nilss. Orn. Suec.* ii. p. 106, sp. 186, tab. xi.—

Scolopax leucophæa, *Vieill. Gal. Ois.* ii. p. 110, tab. 291, changing to the summer dress.—Scolopax grisea, *Vieill. Nouv. Dict.* winter dress.— Totanus griseus, *Vieill. Nouv. Dict.* winter dress.—Totanus ferrugineicollis, *Vieill.* summer dress.—Totanus noveboracensis, *Sabine, Zool. App. Franklin's Exp.* p. 687, summer dress.—Macroramphus griseus, *Leach, Cat. Mus. Brit.* —Limosa scolopacea, *Say, in Long's Exp.* ii. p. 170, winter plumage.—Bec- caccia grigia, *Ranz. Elem.* iii. pt. viii. p. 162, sp. 5.—Becassine grise, *Vieill. Nouv. Dict.* iii. p. 358.—Becassine ponctuée, *Temm. loc. cit.*—Brown Snipe, *Penn. Arct. Zool.* ii. sp. 369.—*Lath. Syn.* v. p. 154, sp. 26 ; *Id. Gen. Hist.* ix. p. 216, sp. 25.—*Mont. Orn. Dict. with a good fig. in the Suppl.* winter dress.—Red-breasted Snipe, *Penn. Arct. Zool.* ii. sp. 368.—*Lath. Syn.* v. p. 153, sp. 26 ; *Id. Gen. Hist.* ix. p. 215, sp. 24, summer dress.—Graubraune Schnepfe, *Meyer and Wolf, Tasch.* iii. p. 46. —*Philadelphia Museum.*

WE can add nothing to the excellent account given by our predecessor of this remarkable species, but as he only figured it in its summer and more familiar dress, our representation of the winter plumage will not be thought superfluous upon referring to our elaborate synonymy, and still less if we bear in mind that even a distinct genus has been instituted for it in this vesture, when it chanced to come under more critical inspection. We shall therefore merely dwell upon the literary and systematical history of the species, referring the reader to Wilson for its natural one.

In its winter plumage the adult red-breasted snipe, then called brown snipe, is so different from the young and from the perfect bird in summer dress, that it is no wonder that it should have been considered a distinct species, especially as it is the only snipe that undergoes such changes, and analogy could therefore no longer serve to guide us. While passing gradually from one plumage to another, the feathers assume so many appearances as to excuse in some degree even the errors of those who have been led to multiply the nominal species by taking a wrong view of the genus to which it belonged.

Pennant, soon followed by Latham, was the first to make known our snipe, which they described in both vestures, and the bird was registered accordingly in the ill-digested com- pilation of Gmelin. Wilson perceived that the two supposed species were one and the same, retaining for it the name of

Scolopax noveboracensis, which appertained originally to the summer dress alone. That given to the winter dress is now, however, with more propriety adopted by all modern ornithologists. As some birds of the old continent are known occasionally to stray to the American shores,* so this common American bird visits accidentally the north of Europe, and especially its islands. There are several instances of its having been killed in the British Isles, where more than one English specimen is preserved, small parties even of these birds having been seen there at different periods and in their different dresses. But these instances are by no means so frequent as reported in authors, the *Limosa rufa* and *Tringa islandica* having been mistaken for it. A specimen in ambiguous plumage, straying into Sweden from the marshes of Lapland (where they may be more common), afforded Nilsson the opportunity of contributing his part to the confusion; but as he gave a figure, besides describing the bird with his characteristic accuracy, it was at once detected. Since Temminck, it is only wilful obstinacy or gross ignorance that can persist in regarding as species the different states of a bird so well marked in its natural genus as to deserve a subgenus for itself, and still more on account of its habits than its conformation (notwithstanding Temminck's statements to the contrary), as will be evident from the following generalities on the genus *Scolopax.*

This genus, as instituted by Linné, and adopted by authors from Latham to Wilson, was, like *Tringa,* a great receptacle, though with the advantage of not containing a single species that is not still admitted as of at least the same family. But however extensive it may have been, had Linné been consistent in arranging under it all the species that possessed the character he assigned to it, he ought to have added to it the

---

* The *Tringa pugnax* of Europe, we are informed by Mr Cooper, who has compared the specimen with one of the species from Austria in analogous plumage, has been shot on Long Island, in the State of New York.

greater part of his *Tringæ,* many of which took rank unperceived in both genera. Cuvier rectified this course, thus
forming a vast genus *Scolopax,* more extensive than our whole
family of *Scolopacidæ.* His subgenus *Scolopax* corresponds,
however, exactly to my genus of that name, which I subdivide
into three natural subgenera, *Rusticola, Scolopax,* and *Macroramphus,* which is the present bird. Illiger first reduced
the genus *Scolopax* within proper limits, but including, it
is true, *Rhynchæa,* since established by Cuvier as a genus.
Modern ornithologists in general agree with us, except that
some, as Vieillot and Savi, consider *Rusticola* a true genus,
leaving the name of *Scolopax* to the rest. *Macroramphus*
and *Scolopax* are, in fact, more closely related than is *Rusticola* to any of them.

All the species of our genus *Scolopax* are very similar as to
the bill, which in all is long, slender, straight, compressed,
especially at base, where it is elevated, soft and flexible its
whole length, with the point depressed, dilated, tumid, and
obtuse : owing to the desiccation of the delicate nervous
apparatus of this part, it becomes wrinkled after death,
exhibiting at the point a dorsal groove and numerous inden-
tations. Both manibles are furrowed to the middle on each
side ; the upper, serrated inside along the palate with spine-
like processes pointing backwards, is terminated by an internal
knob; the lower being shorter, channelled, and somewhat
truncated : the nostrils are in the furrows, basal, marginal,
linear, and pervious, but half closed by a membrane. The
tongue is moderate, filiform, and acute. The head is in all
large, compressed, and angular, low forward and high behind :
the eyes are very large, placed high and far back, but perhaps
less so in the bird which is more immediately the subject of
our remarks : the neck is of moderate length, and stout ; the
body compressed and very fleshy.

But if they have all these traits in common, the feet, tail,
and wings present material differences. The feet are in all,
it is true, moderately long, slender, and four-toed, there being

to this no exception as in the *Tringæ*. But in *Rusticola*
there is no naked space on the tibia, whilst it exists, though
small, in *Scolopax*, and is considerable in the present subgenus. In this the tarsus is much longer than the middle
toe, while in the true snipes it is subequal, and in the woodcocks decidedly shorter. In the present the outer toe is
connected to the first joint with the middle by a membrane,
whilst in the two others all the toes are cleft : in this and
Scolopax the hind nail is falculate and acute, as well as
the others, and projects beyond the toe, which is not the case
in the woodcocks, which have that nail quite blunt and drawn
back. On the other hand, *Macroramphus* agrees with *Rusticola* in the tail, that part having the regular number of
twelve feathers, whilst in the typical snipes the number of
feathers as well as their shape varies amazingly in the different, and otherwise strikingly similar species. Some have it of
twelve, others of fourteen, sixteen, eighteen, and one even of
twenty-four feathers, a number before unheard of in any other
bird whatsoever. In all these groups the tail is nevertheless
short, equal, or more or less rounded. In both the groups
of snipes the first quill is the longest ; but in the species of
woodcocks the quills vary in length and shape, affording the
same anomalies and useful marks as the tail-feathers in the
true snipes. In the European woodcock the primaries are
of equal breadth and the two first longest, while in the
American the three outer quills are very narrow, linear, and
the fourth and fifth longest.

The females in this genus are similar in colour to the
males, but larger, considerably so in the American woodcock. They moult twice in the year, but the present is the
only one that varies much with age or season.

It will not be wronging any to call them all stupid birds,
though the present is less so : this only of its genus is gregarious, associating and flying in numerous flocks. Like the
snipes, and contrary to the woodcocks, they do not dwell in
damp woods or forests, but frequent open marshy grounds

and morasses ; but, unlike the snipes, they prefer the vicinity
of the sea. They might indeed be called salt-water snipes,
in contradistinction to the others, which are fresh-water
snipes. Their flight is high, rapid, and irregular, having
nothing of the heaviness of the woodcocks. The flesh of all
these birds is exquisite food, and much sought after.

The *Rusticola* of Vieillot, which we adopt as a subgenus
for the woodcocks, is distinguished, even from most water
birds, by the want of nudity of the tibia, which is completely
covered with feathers, as in land birds. It contains but the
two species alluded to, that are closely allied, though they
have specific traits that might constitute genera in other cases.
This shows the difficulty in our science of knowing where to
seek for generic and specific traits in the different groups.
The two species of woodcocks vary greatly in their respective
habits, one being a summer, the other a winter visitant in
temperate climates, and one of course retiring south, the other
northward from them. Some authors prefer for this group
the name of *Scolopax,* because it is to its type that the Greeks
gave this name.

Our subgenus *Scolopax,* of which we have published a
monograph in our observations on the second edition of
Cuvier's "Animal Kingdom," is composed of nine or ten species,
all of which, with their characteristic details, will be carefully
figured in our inedited work " Lithographic Monography of
Obscure Genera of Aquatic Birds." In these the tail-feathers
furnish the specific characters. The number, shape, and
disposition of these afford a sure clue, as in *Numenius* it is
the rump, under wing-coverts, and long axillary feathers
which are our best guide to a knowledge of the species.
Without this clue they cannot well be distinguished, and
those who undertake to make phrases with this object in a
group to which they have not the clue, will only make
pedantic nonsense, as is done every day. This very natural
group is called *Telmatias* by Boie, and *Gallinago* by the
English.

As for *Macroramphus*, as we have observed, it forms the transition to *Totanus*, which would be enough to show the impropriety of Boie's course in considering the genus *Scolopax* as a family of itself. Temminck's name of *Becassine chevalier* is peculiarly descriptive, and alone contradicts his unjust censure of Dr Leach's genus, a group whose scientific characters were first laid down by our friend Mr Say, though he referred the species to *Limosa*.

In its winter plumage the red-breasted snipe, instead of the mottled garb in which it is familiar, is of an uniform dusky cinereous : the specimen lying before us is eleven and a half inches long and nineteen in extent. The bill is two and a quarter inches long, of a dull greenish; the tip is black, and obtains the strongly marked dorsal groove that so well distinguishes a *Scolopax* from the allied genera. The prevailing dusky cinereous colour extends over the head, neck, and wing-coverts, the back and scapulars being of a lighter dusky cinereous, and each feather darker on its margin and tip : a broad line from the upper mandible passing over the eye, and the lower orbit, are white : between the eye and bill is a dusky line; the irides are brown : the cheeks, throat, and upper portion of the breast are pale cinereous, each feather being margined with whitish : the lower part of the back, the rump, and upper tail-feathers are white, beautifully and closely fasciated with black : the breast, belly, and thighs are white, the sides being spotted and waved with blackish : the lower tail-coverts are white with short black bands, narrower than those of the upper parts. The wings are six inches long : the lesser wing-coverts of the colour of the body, but they are margined with whitish; the middle and greater wing-coverts are darker, with pure white margins and a little white along the shafts : the primaries are plain blackish dusky, the inner one slightly edged with white : the secondaries are broadly margined and narrowly shafted with white : the first quill is longest, the shaft white : the under wing-coverts and long axillary feathers are white, fasciated with black. The tail is two and a half inches long, composed of twelve feathers, all full and rounded, the two middle a little longer, and marked like the coverts already described, that is, white and densely fasciated with black bands. The feet are of a dull green : naked space on the tibia one inch long : tarsus nearly one inch and a half : middle toe without the nail hardly an inch : hind toe more than a quarter : the toes webbed at base, the outer web reaching to the first joint of the outer toes, the inner being hardly visible.

Wilson's description of the summer plumage being sufficient, we omit it here, though admitting of much more detail. In few words it may be stated, that however great the apparent difference, it may be reduced to this : 1. All those parts that are plain cinereous in winter take on a mottled appearance, being strongly tinged with reddish, and varied with black and yellowish. 2. The anterior parts that are white, such as the superciliar line, and breast, become reddish. The strongly characteristic marks of the other parts remain unchanged.

The young birds of the year have the plumage above generally black, the back of the head dusky, and the feathers broadly margined with bright rufous ; the superciliar line and the inferior parts are of a dingy white, inclining to rufous ; this colour predominates on the breast, where the feathers, as well as on the flanks and the superciliar line, have numerous dusky dots : the middle tail-feathers are terminated by reddish.

Notwithstanding the statements of Wilson, we do not perceive any difference in plumage in the female, which is merely of a larger size. As the species breeds in high northern latitudes, visiting the temperate regions of America in spring and autumn, on its passage to and from its winter quarters, it is the more extraordinary that it should not equally extend these regular migrations to Europe.

WILSON'S PHALAROPE. (*Phalaropus Wilsonii.*)

PLATE XXIV.—FIG. 1, ADULT.　PLATE XXV.—FIG. 1, YOUNG.

See Wilson's American Ornithology, Grey Phalarope, Phalaropus lobatus, vol. iii. p. 94, pl. 73, fig. 2, for a very bad figure and imperfect account; and a much better one illustrating the same figure in the second edition of the same volume, called by Mr Ord, Supplement to the American Ornithology of Wilson, under the name of Brown Phalarope, Phalaropus lobatus, p. 232.—Phalaropus Wilsonii, *Sabine, Zool. App. Franklin's Exp.* p. 691.—*Nobs. Obs. Wils.* sp. 233; *Id. Add. Orn. U. S. in Ann. Lyc. N. Y.* ii. p. 159; *Id. Suppl. Syn. in Zool. Journ. Lond.*; *Id. Cat. and Syn. Birds U. S.* sp. 279; *Id. Specch. Comp.*—Phalaropus frœnatus, *Vieill. Gal. Ois.* ii. p. 178, pl. 271.—Phalaropus fimbriatus, *Temm. Pl. Col.* 370.—Lobipes fimbriatus, *Selby and Jardine, Orn. Ill.* 1, *Syn.* sp. 2, adult.—Lobipes incanus, *Selby and Jard. Orn. Ill.* 1, *Syn.* sp. 3, tab. 16, young.—Phalarope liseré, *Temm. loc. cit.*—Phalarope bridé, *Vieill. loc. cit.*—American Phalarope, *Sabine, loc. cit. Lath. Gen. Hist.* x. p. 4, sp. 2.—*Philadelphia Museum*, adult.—*Leadbetter's Collection*, young.

THIS beautiful, and, as regards system, so remarkable bird, was first discovered by Wilson, who, had he lived to publish the species himself, would doubtless have fixed it on the same firm basis as in other instances of the kind. But death put an end to his labours, and to the advantage which science daily realised from them, when, among other important materials, this phalarope remained in his portfolio. It became the task of friendship to publish a few rough notes and unfinished sketches, the present among the rest, and a figure was thus produced impossible to be recognised except upon actual reference to the specimen itself. The description which accompanied it was as defective as the figure, the author's pencil-notes having been found partly illegible, and it was marked by him as a *Tringa*. In a second and much improved edition, which it has pleased the author to call an original work, though the plates are identical with the former, Mr Ord's description and personal observations are very correct and ingenious, but the name and synonyms are altogether misapplied, through his mistaking it for the *Phalaropus hyperboreus*. In a paper published in the Annals of the Lyceum of New York, I availed myself of the first opportunity

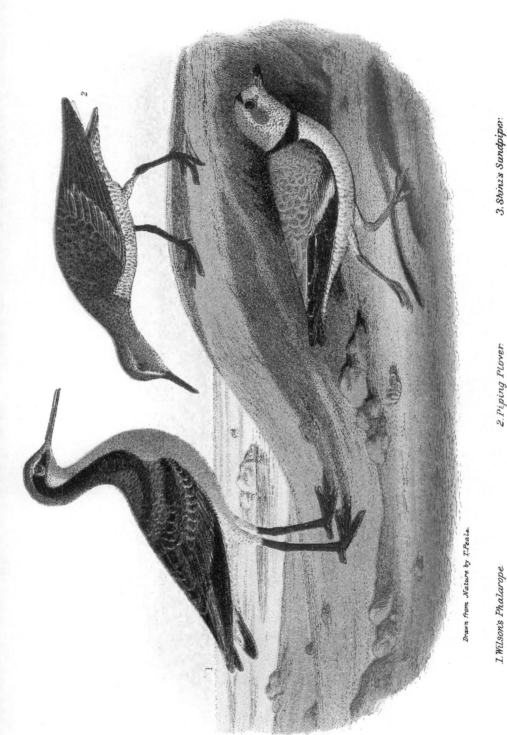

Drawn from Nature by T. Peale.

1. Wilson's Phalarope.

Phalaropus Wilsoni.

2. Piping Plover.

Charadrius Melodius

24.

3. Shinz's Sandpiper.

Tringa Schinzii

that offered to explain the confusion respecting the three species, and finally distinguished among them three groups, which were exemplified in my Synopsis.

Mr Sabine was not aware, when he applied to this bird the name of our predecessor, that he was performing not merely an act of courtesy and respect, but one of justice also towards its first discoverer. It was only by actual inspection of the specimen examined by Wilson, and preserved in the Albany Museum, that we could identify the species, and it does not appear surprising to us that some who have not thus verified the fact for themselves should still express doubts, as Baron Cuvier has done by implication in the new edition of his " Règne Animal." We ourselves, when we first procured the bird, had not the least suspicion that it was contained in Wilson's work. Every one will therefore be sensible of the propriety of publishing a new figure, more needed, in fact, in this case than if the species had been new. The description in Sabine's Appendix to Franklin's Expedition could not, however, be misunderstood, and Temminck and Vieillot by its perusal would have spared this bird two synonyms, as they simultaneously figured and described it in their respective works under the different names quoted in our list, though Vieillot perceived it to be the species intended by Wilson. The authors of the " Illustrations of Ornithology " did not recognise in their *Lobipes incanus* the young of this, which is not much to be wondered at; but it is rather extraordinary that writers so justly scrupulous about the rights of priority should adopt, though greatly posterior, Temminck's name instead of Sabine's, thus slighting over one of the best of the few positive zoological labours of their own countrymen, and after it had been already sanctioned by strangers.

That the *Lobipes incanus* is the young of this species, which any one familiar with the changes of plumage of the phalaropes might have suspected, will, it is hoped, be placed beyond future question by the figure we now give also of it.

If the bill only were considered, this species might with

some propriety be united subgenerically with the *P. hyperboreus,* but as by its feet it differs considerably from both the other phalaropes, which agree in this particular, we have instituted for it a peculiar subgenus under the name of *Holopodius,* which we regard as in all respects more essentially different from the old groups than they are from each other. In what respect Mr Sabine found this species, which he so well established, intermediate between the two, we are at a loss to imagine.

In fact, in *Holopodius,* so opposite to Cuvier's *Lobipes* both in name and character, the toes have a narrow border formed by a subentire membrane; the outer connected to the first joint only; the inner almost cleft, and the hind toe long and resting on the ground: the two other groups having the toes broadly bordered with a deeply scalloped membrane, and semipalmated: the hind toe is very short, the nail only touching the ground. The *Lobipes* of Cuvier differs from the *Crymophilus* of Vieillot only in the shape of the bill, stout, flattened, and carinated in the latter, slender and cylindrical in the former, as well as in ours.

Edwards first brought the phalaropes into notice, and it was from his works that Linnæus and Brisson registered these singular birds in their general works: the former, however, thrust them into that storehouse of species, his *Tringa,* whilst the latter established for them the genus *Phalaropus,* than which no group is more natural, and, in our opinion, equivalent to a family.

Latham and all modern authors have retained very properly this genus in their systems. But if they are so far unanimous, they are greatly at variance when they come to assign it a place, some referring to one order or family, and some to another. That these birds belong to the *Grallæ* or waders, though still more aquatic in their habits even than some of the webfooted birds, does not, in my opinion, admit of doubt.

Before the recent discovery of the species now under consideration, *Phalaropus* contained but two real species, out of which as many had been formed as their changeable plumage exhibits phases, and, what is worse, the nominal species founded on the one had been confounded with those taken from the other, and the different plumage of each taken for varieties of its relative, so that not even the two real species were accurately known apart, though so different as to form each of them the type of a peculiar group, in the same manner as we have observed is the case with the *P. Wilsonii.* They are found in the north of both continents, the present being peculiar to America, which possesses them all. Cuvier, losing sight of the strong common tie that connects the phalaropes, has separated his two groups, *Phalaropus* and *Lobipes,* and has placed the one near *Tringa,* and the other near *Totanus,* on account of the analogy of the bill, regarding the *Phalaropus* as a pinnate-footed *Tringa,* and the *Lobipes* as a pinnate-footed *Totanus.* Vieillot, in adopting these groups as genera, placed them adjoining each other in a separate family, but he changed Cuvier's names into *Crymophilus* and *Phalaropus,* transposing the latter name to the other group, the *Lobipes* of Cuvier. All the three known phalaropes are distinguished by a moderate, slender, straight, and subcylindrical bill : both mandibles are furrowed each side nearly their whole length, and the upper somewhat curved at the point ; the lower is hardly shorter, quite straight, and the point subulate. The nostrils are in the furrows, basal, longitudinal, linear, half closed by a membrane. Their head is small, completely feathered, compressed and rounded above ; the eyes are small, the neck well-proportioned, and the body roundish. The feet are moderately long, four-toed ; the naked space on the tibia rather extensive ; the tarsus as long as the middle toe, moderate, robust, somewhat compressed, and scutellated ; the toes are moderate and rather slender, the three anterior bordered by a festooned membrane, and the outer at least is always connected at base to the middle one ; the hind toe is short, bordered only on the inside with a small entire membrane, articulated rather high and internally, touching the ground at tip: the nails are short, curved, and acute: the wings long, falciform, and acute, the first primary being the longest: the quills twenty-five in number. The tail is short, and consists of twelve feathers, with its under-coverts extending quite to the tip.

The female is but little different from the male, but larger and handsomer in full plumage. The young are very different from the adults, and they vary much with age. They moult twice in the year, their colours changing strangely, which has occasioned the wanton multiplication of species. Their plumage is close, thick, abundantly furnished with down, and impermeable to water. Their colours are principally brownish and reddish, changing in winter to grey and white, which is always to be found on their under parts.

Their habits are essentially aquatic. They inhabit the sea-coasts, the shores of lakes and occasionally of rivers; are gregarious, but never collect in large flocks. Probably from being so seldom met with, they show little dread of mankind, and allow of the nearest approach; and not being alarmed at the report of a gun, it is easy to kill several without moving from one spot. Their food consists of aquatic insects and other small animals that are found in the water. They are strictly monogamous, and are generally seen in pairs, carrying fidelity to an extreme: delighting in their peculiar element, they even copulate on the sea, and reluctantly leave it to build their nest on shore, among grasses: they lay from four to six eggs, which both sexes incubate, the male being even more strongly marked on the belly by the naked places which this causes: they share between them all the parental duties, and the young leave the nest, run about and swim, as soon as they are hatched. The phalaropes are hardly ever seen on dry ground, where, however, they walk and run swiftly, without the embarrassment of some other birds of less aquatic propensities. Though certainly the smallest of swimmers, they perform this operation with great dexterity, resisting the heaviest waves, or rising over their top, but are never known to dive : they notwithstanding swim with perfect ease, when they have all the appearance of a miniature duck, with their head carried close to their back. While swimming, they dip their bill often in the water, frequently turning round, with much elegance in all their motions. Their flight is rapid. Their flesh is oily and unpalatable.

The abode of these diminutive swimmers is the arctic and polar regions, to which their thick coat of feathers is well adapted. Hence they migrate in autumn to the temperate regions of both continents, where they are also seen in spring. They are essentially arctic birds, and breed in the most northern parts of the world, and although they retire more to the south in winter, yet their visits to our temperate climates are rare and casual. From such a combination of traits as are above related, it will be evident that, though much restricted in the number of species, the phalaropes are entitled to a conspicuous rank in classification. They can only be compared with the allied genera *Himantopus* and *Recurvirostra,* and we see how materially they differ from them. They may be said to connect the *Scolopacidæ* with the *Laridæ,* forming a beautiful link between the order of waders and that of the webfooted birds.

Our subgenus *Holopodius,* which resembles *Lobipes* in the bill, while *Crymophilus* resembles it in the feet, is furnished with a long, very slender, smooth, flexible, and cylindrical bill, of equal breadth throughout, subulate to the tip, with the point narrow, sharp, and slightly curved : the nostrils are quite basal, and linear-elongated : the tongue is filiform and acute. The tarsi are elongated, and much compressed, in which it comes nearer to the ANSERES, and compensates for the other traits which remove it further from them than the other phalaropes. Thus do we find ourselves baffled in all attempts at a regularly symmetrical or mathematical arrangement. Nature acknowledges no artificial nor contracted limits. The toes are long, and by no means semipalmated, the outer being connected to the middle only as far as the first joint, and the inner almost divided; the bordering membrane narrow and subentire; the hind toe long, and resting on the ground. The wings are long, even for the genus, and the tertials very long, reaching nearly to the tip of the primaries when the wings are closed. The tail is moderate, being neither so long as in *Crymophilus,* nor so short as that of *Lobipes.* The general form is slender, and, together with the bill and other traits, gives this bird a strong resemblance to the *Totani*—a bare analogy, however, which we should not with Cuvier mistake for affinity.

The American or Wilson's phalarope has been so well described from the recent specimen, by Mr Ord, as not to be susceptible of improvement, and the following description is merely intended to elucidate our figure, which represents of the natural size a beautiful female in the perfect plumage of spring. This individual was nine and a half inches long, and sixteen in extent of wings. The form of the bill we have described above: it is black, and more than an inch and a quarter long, though only a line in thickness: the irides are dark brown. The upper part of the head is of a bluish delicate pale ash colour; the hind head and that part of the neck adjoining it are whitish; a white stripe passes over the eye, and beneath it is a spot of the same colour: a large curving band of black includes the eye and spreads out towards the nucha, descending a good space down the neck, and gradually passes into a reddish brown, which becomes the colour of the sides of the neck; this tint deepens into bright chestnut on the back part of the neck, and descends on each side, thus mingling with the plumage of the back and scapulars, which are dark ash, each feather slightly tipt with whitish: the upper tail-coverts are ash colour. The throat and sides of the head to the black mark, and all beneath, including the lower tail-coverts, are pure white, somewhat tinged with rufous on the lower part of the neck beneath. The wings are five inches long, and in colour dark ash; larger coverts and secondaries very slightly edged with white; under coverts white, most of the smaller wing-coverts being marked with ferruginous; the upper tail-feathers are tinged with reddish at their tips, and the under marked with white on their inner webs. The feet are dark plumbeous; the claws of a dark horn-colour; the naked part of the tibia is nearly an inch long; the tarsus more than one inch and a quarter, and sharpish; the middle toe without the nail is scarcely one inch, and the remarkably long hind toe five-sixteenths without the nail.

There are fewer variations caused in this phalarope than in the others by sex and season: the young, however, is surprisingly different, for which reason we have figured it also of the full size. The bill is like that of the adult, somewhat gaping beyond the middle: the face is whitish mixed with dusky, and with a dusky stripe from the bill to the eye: the crown, neck above, back and wings are dusky brown, darker on the middle of the feathers: the rump, upper tail-coverts, and flanks are broadly white; the throat is pure white: the sides of the neck are tinged with rusty: the neck beneath and breast are white, slightly tinged with reddish-dusky; the belly of a purer white, with a little dusky; the vent, and long lower tail-coverts, which reach to the tip of the tail, are pure white; the wings are four and three-quarter inches long, the lower coverts white; the scapulars blacker, with pale rusty edges: the primaries are blackish, with pale brown shafts, of which the outer is white. The tail is broad and rounded, the middle and outer feathers somewhat longest, all of a pale dusky grey with white shafts, the exterior being also white on the best part of the inner web. All the tail-feathers are also edged with white. The feet are reddish black, the tarsus an inch and a quarter long.

We are acquainted as yet with no peculiarity of this fine phalarope, and even the few facts registered concerning it have been obscured by the heedlessness of compilers. Though it appears to extend its migrations more to the south than its congeneric species, it is decidedly, like them (notwithstanding Temminck's supposition to the contrary), an arctic bird, and the only remarkable circumstance about it is that it should not also be found in Europe. As far as we know, it is exclusively North American, for the specimen of the young inadvertently said by the authors of the "Ornithological Illustrations" to have come from South America, was found in the Vera Cruz market, as appears from their own account. As for Senegal, it was merely a gratuitous supposition on the part of Temminck, too rashly converted by the same English authors into certainty, and it therefore remains strictly North American, for which country we have, besides Wilson's and our experience, the unquestioned authorities of Vieillot and Sabine.

SCHINZ'S SANDPIPER.   (*Tringa Schinzii.*)

PLATE XXIV.—Fig. 2.

Tringa cinclus, var. *Say, in Long's Exp.* i. p. 172.—Tringa Schinzii, *Brehm, Lehrb. Eur. Vog.* ii. p. 571.—*Nob. Obs. on Wils.* before sp. 213 ; *Id. Cat. and Syn. Birds U. S.* sp. 249.—Scolopax pusilla? *Gmel. Syst.* i. p. 663, sp. 40 ?—Tringa cinclus, var. A, minor ? *Briss.*—Tringa alpina ? *Vieill.* (not of authors).—*My Collection.*

In Mr Say's valuable notes to Long's Expedition, he describes as follows the bird which we have had carefully represented in the annexed plate, in order that naturalists may judge whether or not we are right in referring it to the new European species hitherto confounded with *Tringa alpina*, and lately separated by Brehm in his work on the birds of Europe, under the name of *Tringa Schinzii*. It is so difficult to say what is a species and what a variety in this most intricate genus, that we shall not undertake to decide from a single specimen, especially when, as in this case, it involves the identity of the bird in the two continents.

"*Pelidna cinclus, var.* Above blackish brown, plumage edged with cinerous or whitish ; head and neck above cinereous with dilated fuscous lines ; eyebrows white ; a brown line between the eye and the corner of the mouth, above which the front is white ; cheeks, sides of the neck, and throat cinereous lineated with blackish brown; bill short, straight, black; chin, breast, belly, vent, and inferior tail-coverts pure white; plumage plumbeous at base; scapulars and lesser wing-coverts margined with white; greater wing-coverts with a broad white tip; primaries surpassing the tip of the tail, blackish, slightly edged with whitish; exterior shaft white, shafts whitish on the middle of their length; rump blackish, plumage margined at tip with cinereous tinctured with rufous; tail-coverts white, submargins black; tail-feathers cinereous margined with white; two middle ones slightly longer, black margined with white; legs blackish. Adult male, length to tip of tail, seven inches; bill, seven-eighths of an inch."

This bird was shot in November near Engineer Cantonment. and Mr Say thought it was probably a variety of the very changeable *cinclus* (*Tringa alpina*) in its winter plumage. It is this very specimen that we have had represented of its full size in the annexed figure, in order that naturalists may judge if we are right in the course that we have chosen. Be it as it may, we are satisfied that *Tringa Schinzii* is a good species, well distinguished from *Tringa alpina* by its smaller size, and proportionally even shorter bill. The more exten- sively white upper tail-coverts are the best and most conspi- cuous mark; it is also to be observed that in the summer dress the ferruginous colour of the upper part is paler, the black spot of the breast more restricted and less pure, and the neck more broadly streaked. Both sexes are, moreover, perfectly alike in colour, which is never the case in the *alpina* in spring dress. It belongs to the subgenus *Tringa*, of which we have already treated, and it is common to both continents. In America it is found from far beyond the Mississippi to the Atlantic shores, and is rather common in autumn on the coasts of New Jersey, either in flocks by themselves, or mixing in company with other sandpipers, with which it has every habit in common.

The specimens that we shot in New Jersey measured seven inches in length and above fourteen in extent. The bill is very nearly, but not quite, an inch long, compressed and black from the base: the crown, neck above, and interscapulary region are of an ashy brown, much darker in the centre of each feather and lighter on their margins ; on the lower portion of their back this darker colour widening, predominates, and becomes black, so that the tips of the feathers only are of the general pale ashy colour ; the upper tail-coverts are white, blackish along the shaft and towards the margin of the outer vane : a whitish stripe runs from the very origin of the bill over each eye ; the cheeks, sides of the neck, and breast are whitish streaked with ashy dusky along the shaft of the feathers, giving these parts an obscurely lineated appearance ; the throat quite to the bill, and all the remaining under parts are white, the bottom of the plumage being plumbeous, and a few bands of that colour appearing across the lower flank-feathers. The wings are four inches and a quarter long, with the tertials and scapularies remarkably tapering and acuminate, shorter by a good inch than the two first quill-feathers : all the wing-coverts are of the colour of the body, but a little darker, each having a pale grey margin, the inner great coverts have a very pure white tip : the shafts of all the quill-feathers are pure white, at least for a good portion near the centre : the primaries are blackish ash : the secondaries paler and margined with whitish, the tertials are again blackish edged with pale greyish : the under surface of the wing is of a silvery grey ; the under wing-coverts white, marked with dusky. The tail is two and a quarter inches long : the four lateral feathers each side are very nearly equal in length, of a pale ash colour, margined and shafted with white : they become gradually darker as they are nearer the centre ; the fifth each side is blackish ash, a trifle longer than those already described, and has a very conspicuous pure white marginal tip on the inner web ; the two middle surpass the others by a quarter of an inch, are somewhat pointed, and entirely blackish. The feet are blackish ; the naked space above the heel half an inch ; the tarsus seven-eighths of an inch long, and much longer than the middle toe, the toes are cleft to the base ; the nails are blackish. As will easily be perceived, the specimen described is in the winter dress.

This sandpiper is well known to appear in a summer vesture analogous to that of *Tringa alpina* at the same season ; but we have never met with an American specimen in that state.

In the full-plumaged males the bill and feet are black ; irides brown : before the eye a small blackish patch surmounted by a white stripe dotted with blackish grey, head above, back, and wing-coverts bright rufous, the feathers with merely a black centre : colours not so bright as in *Tringa alpina* : wings above blackish grey with black shafts ; point of the primaries black, with white shafts : the ten middle tail-feathers as well as their upper coverts are blackish ; the lateral cinereous with their coverts white : the chin is white, the sides of the head and hind neck are of a ferruginous grey : throat white, longitudinally spotted with rufous grey ; the breast almost entirely of a jet-black colour, always interrupted by some insulated white feathers, and never so broadly black as in *Tringa alpina ;* all the remaining under parts are white, with a very few dusky streaks on the sides.

At one year of age the male is on the back of a less bright rufous spotted with black : on the breast the black consists merely of a spot, and is mixed with many white feathers. The female much resembles the male at the same age. The very young is above of a ferruginous colour varied with white, yellowish, and black ; all beneath white, streaked with dusky ferruginous on the throat.

They frequent marshy shores, and the borders of lakes and brackish waters. They are very social even in the breeding time, and are then by no means shy : during autumn they join company even with different birds, and become very wild. Their voice resembles that of *Tringa alpina,* but is more feeble. They feed on worms, aquatic insects, and similar food : build near marshes and lakes, among weeds : they lay four eggs, smaller and much less in diameter than those of *Tringa alpina,* of a yellowish grey spotted with olive or chestnut brown.

PIPING PLOVER. *(Charadrius melodus.)*

PLATE XXIV.—FIG. 3.

See *Wilson's American Ornithology*, Ring Plover, Charadrius hiatacula, vol. ii.
p. 360, pl. 37, fig. 3, for a reduced representation of the adult in spring dress,
and the history.—Charadrius melodus, *Ord, in the reprint of Wils. Orn.*
vii. p. 71, *and Gen. Ind. of the Water Birds, Suppl. Orn. Wils.* (ix.) p. ccxii.—
Nob. Obs. Wils. Orn. sp. 220; *Id. Cat. and Syn. Birds U. S.* sp. 217; *Id.
Specch. Comp. sp. Philad.*—Charadrius Okenii, *Wagler, Syst. Av.* i. *Charad.*
sp. 24.—Ringed Plover, var. B, *Lath. Gen. Hist.* ix. p. 327, sp. 12, var. B.
—*Philadelphia Museum.*

THE well-merited elevation of this bird to the rank of a
species fully vindicates our predecessor from the unjust censure of Temminck, who thought his figure of it intended for
the *Charadrius hiaticula.* The same censure is repeated and
aggravated by Mr Sabine, who probably thought it intended
for the *C. semipalmatus.* But if the figure is free from the
supposed fault of incorrectness, its extremely diminished size,
which renders it almost useless, requires that the bird should
now appear in this work in its full dimensions.

Not only is the true *C. hiaticula* of Europe not found on
the American continent, but the birds hitherto mistaken for
it constitute two very distinct and exclusively American
species, notwithstanding the awkward quotations in the new
edition of Cuvier's "Règne Animal," which, in this instance,
as in several others, is as far behind its age as the former was
in advance of it.

Although the never-too-much-lamented Wilson gave in his
fifth volume the present bird as a variety of which he intended
figuring the type in a future part of his work, when he came
to it in his seventh volume, he clearly and positively pointed
out the difference in markings, habits, migrations, and voice,
between the two which he then considered as distinct species:
he thus in reality established the species, and indeed so well,
that we cannot do better than refer to his conclusive reasonings.
The only essential point he omitted was to impose a name on

his species, which he undoubtedly would have done had he lived to publish himself the index to the water birds, as, in some instances, he supplied similar deficiencies for the land birds. Mr Ord has, however, filled this void by calling the bird *C. melodus*, which appropriate name we feel bound to adopt ; and the more so, as Mr Ord informs us that it would have been Wilson's own choice. Almost simultaneously with our endeavours in this country for permanently fixing the species, Dr Wagler in Europe, on his part, was also giving it a name, so that it is now furnished with two.

In the circumstance of its inner toe being cleft to the base, this bird approaches more closely to *C. hiaticula* of Europe than to *C. semipalmatus* ; but in colours it differs greatly from these so similar species, and the membrane that connects the outer toe is considerably smaller than in any. The synonyms of Wilson do not, of course, apply to this new species ; and, what is worse, though this is common to all writers upon the ring-plover, they do not belong to one and the same species.

Although, without doubt, related to the *Tringæ*, which are *Scolopacidæ*, the plovers belong to another family, that of the *Pressirostres* of Cuvier—which may be called *Charadridæ*— and through *Otis* and *Œdicnemus* these waders are connected somewhat with the gallinaceous birds. This natural family of ours—very different from the artificial one formed by so many authors for the three-toed waders indiscriminately, and adopted under the name *Charadriadæ* by the new English school, though professing to adhere to a natural arrangement —is well distinguished by its short (or moderately so) rather robust bill, the hind toe wanting, or when present, very short. It is composed of but eight genera, of which only three are found in North America, two aberrant, and the present, the only typical American, which is well distinguished by its bill, very short, rounded, obtuse, and somewhat turgid at tip. In order to exemplify how different from that of authors is this family, as we understand it, we may remark that the birds

forming it are scattered by Illiger through his *Campestres,*
Littorales, and *Limicolæ;* by Cuvier and Latreille divided
between their *Longirostres* and *Pressirostres;* by Vieillot
placed in *Pedionomi, Ægialites,* and *Helionomi;* in *Tachi-*
dromi and *Limose* by Ranzani and Savi; in *Charadriadæ* and
Scolopacidæ by Vigors, &c.

Our genus *Charadrius* has different limits from those of
perhaps any recent or former author, being more extensive
than in many, but more contracted than that of Wagler,
which comprehends all our typical *Charadridæ.* Linné, who
made it a sort of receptacle for nearly all three-toed waders,
has placed in *Tringa* some of our plovers that are furnished
with a rudiment of hind toe, and the same has been done by
Gmelin, Latham, Illiger, and even, though to a less extent,
by Cuvier. As long since restricted by the separation of
Himantopus and *Calidris,* which are not of the same family,
and of *Œdicnemus,* which truly is, it is much more natural ;
especially if with Wilson we unite with it, as Nature dictates,
those species that happen to possess the rudiment of a fourth
toe. Among the earlier writers, Brisson was the first who
assigned more natural limits to the genus, which he called
Pluviatis, and his two well-enough composed genera,
Pluvialis and *Vanellus,* include all our plovers. Cuvier,
Temminck, Vieillot, and Ranzani place the four-toed plovers
with the lapwings, *Vanellus.* Savi more recently has evinced
his good judgment by separating them at least from *Vanellus,*
if he does not unite them with *Charadrius,* which his pro-
fessedly artificial system did not allow.

I distinguish two subgenera in my extensive genus *Cha-*
radrius, regarding *Squatarola* of Cuvier and Savi as no more
than a section of my first subgenus, of so little importance do I
consider the anomaly of the hind toe, the sole characteristic
of that artificial group. These subgenera are—1. *Pluvialis,*
for the large mottled species without a collar, and with variegated plumage. Such are, amongst the three-toed, the European
and Asiatic *C. pluvialis* and *morinellus,* and the American

Virginicus (or *marmoratus*) ; and among the four-toed the Europeo-Asiatic bird *C. gregarius*, and the cosmopolite *C. Helveticus*. 2. *Ægialitis*, Boie, or the ring-plovers, which have a broad white collar around the neck. This is the more numerous in species, and the present belongs to it : it may form two sections, one for the semipalmated ring-plovers, whose toes are all connected at base by a membrane, and the other for this and the remaining ring-plovers, in which the inner toe is separated down to the base. As for the armed or spur-winged plovers, as well as the wattled species, all I have examined were perfectly similar to the armed and wattled lapwing, and they constitute in my arrangement a very natural subgenus under the name of *Hoplopterus*, which group, like *Pluvialis*, may be sectioned into those with three and those with four toes. This group of *Hoplopterus*, both by its tarsus and wings, takes place under my genus *Vanellus*, and differs subgenerically from the typical species merely by its longer legs, and hind toe less developed, or often wanting. *Pluvianus*, Vieillot, distinguished by a stouter bill, I never have examined, but have no doubt that it will find its place in my genus *Vanellus*, where it may be united to my three-toed *Hoplopteri*, or possibly become a subgenus by itself.

Both the three-toed and four-toed species that form my subgenus *Charadrius*, and are so easily known by their greater size and want of a collar, live in large damp meadows, or open and muddy champaign countries. They hardly ever alight on the beach, or even accidentally on river shores. During the nuptial season the males assume a brighter vesture. They do not breed in the temperate climates of Europe or North America, but only show themselves there in autumn and winter. Their flesh is exquisite food.

The ring-plovers, on the contrary, are shore birds in their habits, and may be known by their diminutive size and broad white collar. They frequent invariably the banks of rivers and sandy sea-beaches, and it is by accident if they are seen at a distance from their favourite element. Their plumage does not undergo extreme changes, and merely from darker to lighter. Several species breed in our climates, and their flesh is hardly esculent. Although not marked by any striking physical character, we regard the extensive group *Ægialitis* as a very natural one: it has numerous species in every part of our globe. The three European are modelled precisely after the same type as the present species, while the three other North American have each a strong distinctive character peculiar to itself; in the semipalmated it is the webbed toes, in the Wilson's the powerful and acute bill, and in the kildeer its large stature and oddly coloured wedge-shaped tail.

In all our plovers the bill is shorter than the head, rather slender, straight, cylindrical, depressed at base, obtuse and somewhat turgid at tip: the upper mandible is longitudinally furrowed two-thirds of its length, the lower is shorter: a remarkable character consists in the small opening of the bill, which is hardly cleft beyond the origin of the feathers. This peculiarity affords an excellent means of distinguishing them from the *Œdicnemi*, in which the gape extends to beneath the eye. The nostrils are basal, lateral, placed in the furrow, and covered by a membrane, leaving only a narrow longitudinal opening: the tongue is entire, obtusely lanceolate, channelled somewhat above, convex beneath. The head is large in proportion to the body, and the eyes large even for the head; the forehead is prominent, and the face wholly feathered. The feet are either three or four toed, with the hind toe exceedingly small and raised from the ground: the naked part of the tibia is moderate; the tarsi are longer than the middle toe, and reticulated; the toes scutellate, margined by a narrow squamulose membrane; the middle toe is longest, and connected to the outer, at least to the first joint, by a membrane: even in the species that have the inner toe cleft there are traces of the membrane, which is so much developed in the semipalmated ring-plover: the nails are compressed, curved, and acute. The wings are elongated, acute, and tuberculate: the first primary is longest, and after the second they decrease rapidly, thus presenting a most useful mark for discriminating between this and the kindred genus *Vanellus*, which has obtuse wings, the third primary being the longest, and the others decreasing gradually. The tail is more or less rounded, always composed of twelve feathers, rounded or lanceolate. The plumage of the under parts is soft, the feathers being numerous, wide, rather dense in the centre, with the barbs rather loose, and well furnished with down at base: the plumage of the upper parts is rather dense, and the feathers more or less rounded at the tips: the scapularies are long, at the tips attenuated and very flexible. In most of the species the males and females are alike, the young somewhat different from them. They moult generally twice in the year, when the colours of their plumage undergo some changes.

The plovers are all more or less gregarious in dispositions: their haunts are either meadows, as the mottled plovers, or the sea-shores, like the ring-plovers: they have a very remarkable habit of stirring the soil with their feet, to put in motion worms and aquatic insects, their exclusive food. They are more nocturnal than diurnal. They lay in the sand, about four large eggs. The young very soon after they are hatched follow the mother, and pick up the food which she with great care points out to them.

The piping plover is seven inches long, and fourteen in extent: the bill is bright yellow, slightly tinged with orange for half its length, thence black: the eyelids are bright yellow, and the irides dark brown. The plumage above generally, with the mere interruption of the ring on the neck, is of an extremely pale brownish or dusky, inclining strongly to whitish ash: the front, part of the head between the bill and eyes, and the whole inferior surface from the chin to the tip of the lower tail-coverts, and including the under wing-coverts and long axillary feathers, is pure white: the head and breast are ornamented, the former with a black crescent, that runs transversely between the eyes, and bounds the white forehead on one side, and the ash-coloured parts of the head on the other ; the latter by a curved band round its sides, forming the ring or half-collar round the neck, but narrow and almost interrupted before. The wings are four and three-quarter inches long, and reach when closed to the tip of the tail ; the wing-coverts are darker than the back feathers, and are all edged with white : the larger coverts are broadly terminated with white, constituting the band across the wings : the quill-feathers are dusky ; the secondaries are broadly white inside, with margins of the same : the primaries are blackish at the point, shafted and obliquely centred with white ; the four outer ones are blackish on their outer margins where the others are white. The tail is two and a half inches in length, nearly square at tip, being much less rounded than in the semipalmated species, white beneath for half its length, and blackish at tip ; the outer tail-feather is wholly white, the next is also white, and with a single spot of black, which on the third extends much more, and still more on the fourth and fifth, till the last is merely terminated with white, the middle ones being wholly dusky from the white of the base. The feet are greenish yellow tinged with orange, and the nails black.

Those authors who describe the autumnal plumage as much darker are still labouring under the erroneous opinion, which they had rejected, of this being the same with the *C. semipalmatus.* On the contrary, it is, if anything, still paler at that season, and considerably resembles that of the young birds, which are distinguished by the absence of the neck ring and sincipital crescent, and the bill being entirely blackish.

As will appear by referring to Wilson's two articles on the ring-plovers, this species is commonly met with during the whole summer along the sandy coasts of the United States, on the approach of winter retiring south : it lays in the month of July on the sandy beach, three or four eggs, very large for the bird, of an obscure clay colour, all sprinkled with numerous reddish spots. It runs rapidly, holding the wings half expanded ; and utters a very soft and mellow cry.

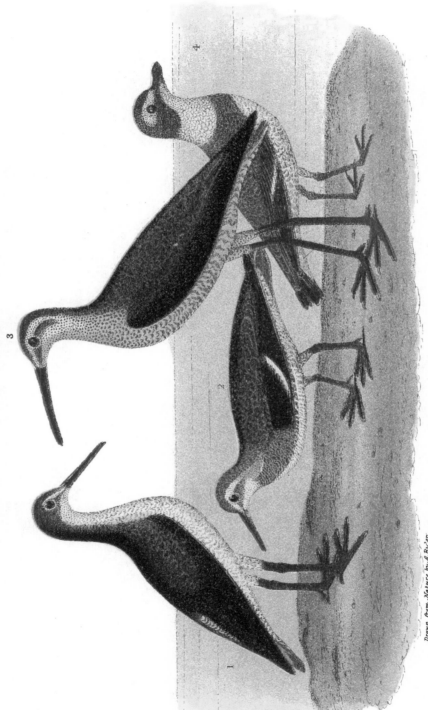

Drawn from Nature by A. Rider.

1. Wilson's Phalarope. Young.   2. Hyperborean Phalarope.   3. Long-legged Sandpiper.   4. Semipalmated Plover.

Phalaropus Wilsoni.   Phalaropus Hyperboreus.   Tripa Himantopus.   Charadrius Semipalmatus.

2 5

HYPERBOREAN PHALAROPE. (*Phalaropus hyperboreus.*)

PLATE XXV.—Fig. 2.

Tringa hyperborea, *Linn. Syst.* i. p. 249, sp. 9.—*Gmel. Syst.* i. p. 675, sp. 9.—*Retz, Faun. Suec.* p. 183, sp. 152.—*Mull. Prod. Zool. Dan.* sp. 196.—Tringa lobata, *Linn. Syst.* i. p. 249, sp. 8; *Id. Faun. Suec.* p. 64, sp. 179.—*Retz, Faun. Suec.* 152, young.—*Mull. Prod. Zool. Dan.* p. 195.—*Fahr. Faun. Grœn.* p. 109, sp. 75, adult and young, and history.—*Brunn. Orn. Bor.* p. 51, sp. 171, young—(*N.B.* Not of Gmel., who under this name had in view the *Ph. fulicarius*, though he unaccountably retained the Linnean phrase).—Tringa fusca, *Gmel. Syst.* i. p. 675, sp. 33, young.—Phalaropus hyperboreus, *Lath. Ind. Orn.* ii. p. 775, sp. 1.—*Muller,* sp. 196.—*Trans. Linn. Soc. Memoir, Birds of Greenland,* xii. p. 535.—*Temm. Man. Orn.* 1st ed. p. 457; *Id. Man. Orn.* 2d ed. ii. p. 709.—*Sabine, App. Franklin's Exp.* p. 690.—*Nob. Add. Orn. U. S. in Ann. Lyc. N. Y.* ii. p. 159; *Id. Cat. Syn. Birds U. S.* sp. 279; *Id. Specch. Comp.*—Phalaropus fuscus, *Briss. Orn.* vi. p. 15, sp. 3; *Id.* 8vo, ii. p. 363.—*Lath. Ind. Orn.* ii. p. 776, sp. 4.—*Linn. Trans.* xii. p. 535, young.—Phalaropus cinereus, *Briss. Orn.* vi. p. 15, sp. 2; *Id.* 8vo, ii. p. 362, adult.—*Nilss. Orn. Suec.* ii. p. 120, sp. 193.—*Meyer and Wolfe, Tasch. loc. cit.*—*Brehm, Lehrb. Eur. Vog.* ii. p. 665.—*Roux. Orn. Provence,* pl. 337, adult male.—Phalaropus Williamsii, *Haworth, Linn. Trans.* viii. p. 264.—Lobipes hyperboreus, *Selby and Jardine, Orn. Ill.* i. *Synops.* sp. i.—Larus fidipes alter nostras, *Ray, Syn. Av.* p. 132, A, 7.—*Will.* p. 270.—Tringa fusca rostro tenui, *Klein, Av.* p. 151, sp. 3.—Falaropo iperboreo, *Ranz. Elem. Zool.* iii. pt. viii. p. 283, sp. 1.—Le Coq d'Odin, *Anon. Icon. Rer. Nat.* ii. p. 8, pl. 20, adult.—Phalarope Cendré, *Buff. Ois.* viii. p. 224; *Id.* ed. 1784, ix. p. 124; *German Translation by Otto,* xxx. p. 111, cum figura.—*Roux. tab. cit.*—Phalarope de Siberie, *Buff. Pl. enl.* 766, male.—Phalarope Hyperboré, *Temm. loc. cit.*—*Vieill. Orn. Franc.* pl. 278, fig. *a,* summer dress; fig. *b,* winter.—Lobipède à hausse-col, *Cuv. Règn. Anim.* i. p. 532.—Cock Coot-footed Tringa, *Edw. Glean.* pl. 143, adult female.—Coot-footed Tringa, *Edw. Glean.* pl. 46, young.—Jonston's Small Cloven-footed Gull, *Will. Engl.* p. 355, § vii.—Red Phalarope, *Lath. Syn.* v. p. 70, sp. 1.—*Ubersetz (translation),* v. p. 289, sp. 1, tab. 94, male; *Id. Gen. Hist.* x. p. 1, sp. 1, and var. A and B (pl. 163)—(*N.B.* Var. c is *P. fulicarius,* taken from Wilson's work).—*Penn. Brit. Zool.* ii. p. 219, pl. 76; *Id.* ed. 1812, ii. p. 125, pl. 21.—*Arct. Zool.* ii. p. 494.—*Bewick,* ii. p. 139.—*Lewin,* v. pl. 193.—*Walck.* ii. pl. 127.—*Mont. Orn. Dict. Suppl. and App.*—Brown Phalarope, *Lath. Syn.* v. p. 274, sp. 4.—*Penn. Arct. Zool.* ii. sp. 214, young.—Red-necked Phalarope, *Bewick, Brit. Birds,* ii. p. 149.—Seeschnepfe, *Crantz, Hist. Grœnl.* p. 113.—Der Wasserstretter, *Schmidt, Vog.* p. 128, tab. 111.—Aschgrauer Wasserstretter, *Bechst. Nat. Deutschl.* iv. p. 372.—*Meyer and Wolfe, Tasch.* ii. p. 417.—Rothalrige Wasserstretter, *Wolfe and Meyer, Vog. Deutschl.* i. Heft. 15. fig. 1, adult male; fig. 2, young female; fig. 3, young male.—*Seligmann's Voy.* v. tab. 38.—Gemeine Wasserstretter, *Bechst. Nat. Deutschl.* ii. p. 317.—*Meyer, Vog. Deutschl.* i. *Heft.* 15, fig. 2 and 3, young at different ages.—*Naum. Vog. Nachts.* ii. p. 80, fig. 24, young.—Fisklita, *Act. Nidr.* iii. p. 575.—*Bam.* ii. p. 407.—Norduest Fugl, *Bom. Nat. Hist.* v. p. 599.—Nuorte-ladde, v. *Bieggush, Lieur. Finmark,* p. 290.—*My Collection,* adult and young.

By giving a representation of this phalarope, besides that we add a species to the American ornithology, we make good our promise of settling an important question. A glance at our figure of the hyperborean phalarope, here brought into comparison with the young Wilson's phalarope, will at once evince the incorrectness of Mr Ord's refined distinctions, and ultimate decision that they were the same bird. This comparison shows more conclusively than any argument to be found in our respective writers on this subject what are the real facts. We have previously observed, when illustrating the former species, that they even differ subgenerically, and that this one alone ought to form the genus *Lobipes* of Cuvier.

The *Lobipes* of Cuvier, since called by the recent English writers *Lobefoot,* and on which Vieillot imposed the name of *Phalaropus,* is formed in our opinion of this single species, notwithstanding that Cuvier and some English authors include the *P. Wilsonii* in it on account of its bill being similar. But the feet are too different to allow of such a reunion, being in this one precisely similar to those of the flat-billed species.

The bill of the lobefoot is moderate in length, slender, smooth, cylindrical throughout, and a little stoutish at base, subulate to the tip, with the point narrow and sharp: the upper mandible curves slightly upon the lower at tip, where they do not quite meet, as occurs in some *Totani:* the nostrils are not quite basal, as in the *Holopodius,* and are linear instead of the subovate form of the *Crymophilus* or true phalarope: the tongue is also filiform and acute, and by no means broad, fleshy, and obtuse, as in the same group. The tarsi are, however, longer than in this, though shorter and less compressed than in the *Holopodius:* the toes are likewise intermediate as to length between the two other groups: the middle one is connected with the inner to the first joint, and with the outer to the second; the edging membrane is broad, deeply scalloped, and finely pectinated: the hind toe is very short, only the nail touching the ground ; the wings are more elongated than in *Crymophilus :* the tail, on the contrary, is shorter, and the general form slender, in which respect, and some others also, they bear a resemblance to *Totanus.*

The hyberborean lobefoot, as represented in its summer, though not its perfect plumage, is seven and a half inches long, and fourteen and a quarter in extent. The bill is less than an inch long, black, exceeding slender, and with both mandibles remarkably acute, the upper being rather longer and somewhat inflected at tip. The irides are brown. The head, neck above, back, and wing-coverts, are very dark grey, which comes forward and round on the lower part of the neck, thus encircling the white throat : through the eye from the bill passes a broad dusky stripe to the hind head ; a rufous line arises behind the eye, which dilates into a large patch on each side of the neck, the two nearly joining at the back part : the sides of the neck and throat are white, the eyelids white ; the back and scapulars are of a darker colour than the adjoining parts, with large spots of ferruginous on the upper part of the back, occupying the outer side of the feathers : the rump and upper tail-coverts are banded dusky and white. The sides of the breast are dark cinereous, all the remaining lower parts are white, the base of the plumage being blackish ash, which rather predominates on the flanks, giving to these parts a very dark mixed appearance. The wings are four and a quarter inches long, and when closed reach precisely to the tip of the tail ; the under wing-coverts are varied with white and blackish ash ; the lesser and middle upper-coverts are dark blackish grey, the latter with a few white streaks at the tip of the outer one : the greater are almost blackish, and broadly pure white at the tips, which makes a conspicuous band of pure white across the wings : the primaries are blackish, slightly edged with paler, and with whitish shafts ; the secondaries are white at their base, and on the margin of their blackish tips, some of them being also white on their inner web, so that the white much predominates : the tertials are very long, and wholly blackish. The tail is little more than two inches long ; the feathers are blackish grey, edged with pale ferruginous at tip. The feet are of a greenish lead ; the naked space on the tibia nearly half an inch ; the tarsus little more than three-quarters of an inch, and precisely of the same length with the middle toe ; the hind toe no more than three-sixteenths of an inch.

In old and perfect specimens, especially old females, this sex being larger and much handsomer, the back, scapulars, and wing-coverts are of a very intense shining black, the anterior part of the back and scapulars being skirted with fulvous, and the wing-coverts edged near the tip with pure white ; the side and also the inferior portion of the neck are of a bright rufous ; the two middle tail-feathers are of the same deep black as the back, and the lateral ashy ones are edged with white. It will be remarked that the chief difference between the specimen figured and the quite perfect state resides in the ferruginous colouring of the sides of the neck, which does not meet on the breast, as it does quite broadly in adult birds : considerable variation takes place in this respect, which is entirely owing to the more or less advanced maturity of the birds.

The young before the summer moult are well distinguished by having the forehead, cheeks, throat, sides of the neck, and neck beneath pure white, as well as all the under parts, the neck and flanks being the only parts tinged with cinereous ; a slight yellowish tinge appears on the sides of the neck : the top of the head only, a band along the nucha, and a patch around the eyes are blackish grey slightly skirted with rufous ; the back and scapulars blackish, each feather broadly skirted with bright ferruginous ; the wing-coverts blackish, lesser margined with white, greater white at the tip; the inner part of the tarsus is yellow ; the exterior and the toes of a yellowish green.

During summer this bird resorts to lakes and fresh waters, though preferring at all times brackish water; in winter they betake themselves to the sea, and are even met with at great distances from land, floating among icebergs in the desolate seas of the North : they swim still better than the other phalaropes, and are met with farther at sea. This species is mostly seen in pairs, though sometimes in small flocks, and busily engaged in dipping their bill into the water after the minute and almost invisible animals of the ocean. They are also much on the wing, somewhat like the gulls and terns, and their cry resembles that of the greater tern.

Although the hyperborean phalarope is a very rare visitant in the United States, there being a few instances only of its being shot in Boston Bay and on Long Island, it breeds regularly at Hudson's Bay; arriving there annually in the beginning of June. In the middle of this month they lay three or four eggs on a dry spot among the grass : the nest is placed on a small hillock near a pond, and contains four very small pyriform eggs, resembling those of a snipe in shape, but much less, and of a deep olive colour, blotched with dusky, so thickly as nearly to obscure the ground colour. The young fly in August, and they all depart in September for less rigorous climes. In Greenland the species also arrives regularly in April and departs in September. This bird inhabits the Orkney and Shetland islands, as well as those of the Norwegian sea, in considerable numbers during summer, breeding there. It is very common in the marshes of Sanda and Westra, but especially Landa and North Ronaldsha, the two most northerly of the Orkney isles, in the breeding season, but leaves them in autumn for milder regions. Its favourite abode is the shores of lakes situated within the arctic circle : it is plentiful in the northern parts of Sweden, Russia, and Norway, as well as the northern coasts of Siberia, and between Asia and America, extending its irregular wanderings even to the Caspian Sea. In Iceland it is observed to come about the middle of May, and re-

main in flocks at sea ten miles from the shore, retiring early
in June to mountain ponds : remarkably faithful to each other,
both sexes are quarrelsome with strangers, and the males are
very pugnacious, fighting together, running to and fro on the
surface of the water, while the females are sitting.   The species
passes regularly along the north coasts of Scotland and the
continental coasts of the Baltic Sea.   It appears also, though
rarely, during spring and autumn in the southern Scandinavian
provinces.   In England it is very rare, and quite as accidental
as in the United States, though it has been casually observed
in Germany, France, and even on the great lakes of Switzerland : an individual was killed on the Lake of Geneva in
August 1806, the only one ever seen on that lake, where the
flat-billed phalarope is by no means so excessively rare : the
specimen alluded to was killed while swimming and picking
up small diptera from the surface of the water.   These
wanderers are always young birds ; but never within my
knowledge has an individual been known to stray into any
part of Italy.   The favourite food of this species is water
insects, especially diptera, that abound at the mouths of
rivers.   The old ones hover round their young when exposed
to any imminent danger, repeating *prip, prip,* and at the
commencement of August carry them out to sea, at the end
of that month being no longer to be found inland.   The
Greenlanders kill them with their arrows, and eat the flesh,
which being oily, suits their taste : they also keep the very
soft skin, making use of it to rub their eyes with, and thinking
it efficacious in curing a species of ophthalmia to which they
are subject.

Although the specific name of *lobata* was given first by
Linné to the present species before he bestowed upon it the
additional one of *hyperborea*, we have thought it proper to
retain the latter, which is also Linnean, because that of *lobata*
has been successively applied to each of the three species, and
by Latham exclusively appropriated to another, whilst the
present has never been so misapplied, and is long since unani-

mously consecrated to this species. By adopting the prior name of *lobata*, we should have been compelled to quote our own authority, and say *Ph. lobatus*, Nob., since *Ph. lobatus*, Lath., is the *Ph. fulicarius*, and *Ph. lobatus*, Ord, the *Ph. Wilsonii*.

LONG-LEGGED SANDPIPER. (*Tringa himantopus.*)

PLATE XXV.—Fig. 3.

Tringa himantopus, *Nob. in Ann. Lyc. New York,* ii. p. 157 ; *Id. Cat. and Syn. Birds U. S.* sp. 245 ; *Id. Specch. Comp. sp. Philad.—My Collection.*

The figure of this remarkable bird cannot fail to create a sensation among naturalists, and a careful examination may induce them to attach more importance to our subgenus *Hemipalama* than Baron Cuvier has done, and to admit that it is quite as distinct as his *Machetes*. That this has not already been done is no doubt because the real type, which is this species, was so little known. The *Tringa semipalmata* of Wilson, which we have united with it merely on account of its semipalmated toe, has no real affinity with it, but is similar to the other sandpipers, and we should never have thought of instituting a separate group for it alone, more than for the *Charadrius semipalmatus*.

The long-legged sandpiper is, in fact, one of those beings that, although intimately connected with several groups with which they have many things in common, yet possess peculiarities sufficient to insulate them completely from all that surround them. It is very remarkable for its anomalous characters. Though decidedly a *Tringa*, it connects, still more evidently than the other species with long subarched bills that have been placed in *Numenius* by German authors, this latter genus with its own, since to the other common traits of resemblance it unites the semipalmated toes ; so that, in fact, instead of placing it at the head of the *Tringæ*, it should rather be arranged last of the *Numenii*, were this not forbidden by the long and delicate legs and toes, as well as some other peculiarities easier to perceive than to express by words. As a species, in form, dimensions, and especially in plumage, this bird greatly resembles *Tringa subarquata* of Temminck (*Numenius Africanus*, Lath.), from which it is, however, clearly distinguished by its still longer and semipalmated feet, in which latter only it resembles *T. semipalmata*. It cannot for a moment be mistaken for any other *Tringa*, differing widely from all, and by a complication of anomalies resembling more in general garb and plumage a *Totanus* than a *Tringa*.

We are unable to say much of the habits of this curious sandpiper, further than that we met with it in the month of July 1826 near a small fresh-water pond at Long Branch. Being there in company with my friend Mr Cooper, we observed a flock flying about, at which I fired, and killed the one here represented. On first picking it up, I mistook it for a time for *T. subarquata*, a species very rare in the United States, though one of the most common in Italy, but was undeceived upon observing the web between the toes. This is the only specimen I have ever seen, though the gentleman just mentioned informs me that he has recently procured another, that was shot in the month of May on the south shore of Long Island.

This new species is nearly nine and a half inches long. The bill, much longer than the head, is decidedly subarched, and measures one inch and five-eighths, and is black. The general plumage is of the same grey colour usual in other sandpipers: the crown is dusky, mixed with whitish and blackish, and with a little bright rusty on the margins; a broad whitish line is above the eye; between the bill and eye dusky, a patch of rust colour on the auriculars: the neck above and on the sides is mixed with whitish; the back and scapulars black, the feathers tipped with dusky grey, and marked with pale rusty; the rump is plain dusky grey, and the upper tail-coverts white, regularly banded with black. The throat is whitish, obsoletely dotted with blackish; the whole under surface is then, including the tail-coverts, white, each feather being banded with blackish, and one of the bands terminal. The wings are five and a half inches long; all the coverts plain dusky with lighter margins; the under-coverts are marbled with blackish and whitish : the primaries are blackish, the first with a white shaft; the secondaries are pale dusky, edged with whitish. The tail is grey, even, and two inches long; the two middle feathers are acute, projecting beyond the others the length of their points; the outer on each side is also somewhat longer than the others : all are pale dusky with white shafts, the white spreading somewhat along the middle, but particularly at the base, where all the feathers but the middle ones are white, as well as the two outer also on the greater part of their inner vane. The feet are black, and the legs very long : the naked space on the tibia one inch and a quarter; the tarsus one and three-quarters long : the middle toe is very nearly one inch without the nail, and about as much over an inch including it : all the front toes are half-webbed, that is, with a membrane connecting them at base.

YOUNG SEMIPALMATED PLOVER. (*Charadrius semipalmatus.*)

PLATE XXV.—Fig. 4.

See *Wilson's American Ornithology*, Ring Plover, Charadrius (Tringa by a typographical error) hiaticula, vol. ii. p. 360 (Ord's ed. p. 69), pl. 59, fig. 3, for the adult in spring dress, and the history.—Charadrius semipalmatus, *Nob. Obs. Nom. Wils.* sp. 219 ; *Id. Cat. and Syn. Birds U. S.* sp. 216 ; *Id. Specch. Comp. sp. Philad.*—*Caup, Isis*, xii. 1825, p. 1375, t. 14 (the head and foot).—*Wagler, Syst. Av.* i. *Charadrius*, sp. 23.—*Philadelphia Museum.*

THE credit of first pointing out the curious though obscure character which distinguishes the present bird from its very near relative the *Ch. hiaticula* of Europe is due to Mr Ord ; and after verifying it in all our American specimens, we feel satisfied that the true *hiaticula* does not inhabit this continent, and those authors who have recorded it as American must have mistaken the present species for it : we might therefore

have swelled our limited list of synonyms with quotations
of all their American specimens described under this name.
The species was first established in our "Observations on the
Nomenclature of Wilson," and in our "Synopsis," and nearly
at the same time by Mr Caup also, on a single specimen in
the Museum of Darmstadt, whose origin was doubtful, but
the real one suspected.   By a fortunate coincidence, Mr Caup
and myself were led to select the same appropriate name for
our bird, which is the less extraordinary, as being suggested
by so material an anomaly in the characters; natural history
conducting us in this instance to the result of one of the most
exact sciences.

The distinctions between the three European species of
ring-plovers having been until lately but little understood, it
is not to be wondered at if those inhabiting these States were
not at once well established.   North America counts also three,
independently of the kildeer, and several others, not yet properly
determined, inhabit other parts of the world.

Being now regarded as a new and very distinct species, we
have not hesitated to reproduce of its natural size a bird that
Wilson has already represented reduced one half; but his
figure of the adult being remarkably good, we have thought it
best to give the young, with the subjoined description, referring
the reader for other particulars to the accurate account of our
predecessor.

The young semipalmated plover is seven inches long, and fourteen in extent: the bill is almost entirely black, being destitute
of orange, and with no more than a little dirty yellowish flesh
colour at the base of the under mandible.   The frontlet continued into the lora, and dilating broadly on the auriculars, is
of a darkish grey colour, somewhat tinged with brown: a frontal
band obscurely continued over the eyes is white ; there is no
sincipital black band : the top of the head is greyish brown
down to the neck, which colour unites, and forms a single mass
with the auriculars already described : the throat to the very
origin of the bill, and all the under parts, are pure white, with

the exception of a colour on the breast, which, as a continuation of the colour of the back, is of a brownish grey : the white encroaches somewhat upon the middle of this collar on the lower side, and extends in a broad ring all round the neck : after this collar, the whole upper parts of the body are brownish grey, precisely of the same hue as the top of the head, and like it have each feather slightly edged with pale. The wings are four and three-quarter inches long, exactly reaching the tip of the tail, the smaller and middle coverts and tertials are of the colour of the body ; the larger are darker, white at the tips, and they form a conspicuous band across ; the spurious wing and under wing-coverts are white, somewhat mixed with dingy : the quill-feathers are dark grey, blackish at their point and on their outer web : the shaft of all is white towards the middle, and the secondaries have, moreover, a white spot along it. The tail is two and a half inches long, slightly rounded ; the outer and shortest feather is white, with a small elongated spot towards the middle of its inner web ; the second each side has a much broader and darker one extending on both webs, dingy at base, and pure white on the shaft and at tip only : all the remaining ones are dusky at base, with a broad black space towards the point, and are terminated with white, less pure and less extended according as they are nearer to the true middle ones, which are merely edged with whitish. The feet are yellowish ; the tarsus is almost an inch long, and the middle toe three-quarters ; the outer is connected to the second joint with the middle one by a membrane ; and the inner is also connected with the middle, but no farther than the first joint.

In the adult, well described by Wilson, the bill is orange beyond the middle, black at the point : the margins of the eye-lids are orange : the irides are brown : the front, throat, neck broadly round, and all beneath pure white ; the head is of a grey colour, somewhat tinged with reddish : a broad sincipital band and a broad ring round the base of the neck jet-black : lora, continued through the eye into a broad patch dilating on the auriculars, blackish: the back and wing coverts are rufo-cinereous: the quills are blackish, the fifth, sixth, seventh, and eighth being white along the shaft: the secondaries are rufo-cinereous, white at their tips: the tail is blackish, and quite black towards the point; the outer tail-feather is white, the second, third, and fourth being also white at their tips.

In size, this species comes nearest *Charadrius curonicus* (*minor*) of Europe, but in colour and all else most resembles *C. hiaticula.*

On the coasts of New Jersey, this species arrives late in April, keeping then in flocks, and until late in May, when they depart in search of more northern climes   No instance is known of their breeding in the United States, but their flocks reappear periodically in September, protracting their stay till the last of October.   They run with rapidity, uttering a rather hissing short note, resembling the syllable *thyk, thyk.* It is a remarkable fact that these closely related species of ring-plovers, hardly cognisable at a distance by the eye, are at once detected by a practised ear, their note being so very different ; for who could mistake the hissing voice of the present for the soft and musical tones of the piping species, so happily compared by Wilson to a German flute ?   It is equally well known that the species of Europe differ also in this re-spect from each other, the true *hiaticula* having very nearly the same hissing voice as the semipalmated, whilst the *curo-nicus* has a very melancholy cry resembling *kirw ! kirw !*

PEALE'S EGRET HERON.   (*Ardea Pealii.*)

PLATE XXVI.—FIG. 1.

Ardea Pealii, *Nob. in Ann. Lyc. New York*, ii. p. 155.—*Id. Cat. Birds U. S. in Contr. Macl. Lyc.—Id. Syn. Birds U. S.—My Collection.*

AMONG the numerous and still badly known tribes of herons—a genus which, even as reduced according to the sounder views of modern authors, yet consists of about fifty species, spread pretty nearly in equal numbers over all parts of the world—a

Drawn from Nature by A.Rider.

1. Peal's Egret Heron. 3. Esquimaux Curlew. 2. Scolopaceous Courlan.
Ardea Pealii. Numenius Borealis. Aramus Scolopaceus

26

small group had been distinguished in common language, before it was recognised by naturalists, under the name of egret, and it may be admitted into the system as a secondary division of the subgenus *Ardea*, as this is distinguished from *Botaurus, Nycticorax*, &c. Their elegance of shape, long and slender bill, but especially their snowy whiteness, and the flowing train of plumes by which they are adorned in the perfect state, make them easily cognisable even at a distance, and seem fully to entitle them to such a distinction. But this very similarity, as one may well imagine, renders the several species—for there are several of them—liable to be easily confounded together. Besides their remarkable similarity of form, colours are wanting to discriminate them ; and we are reduced to those exhibited by the bills, lora, and feet, to the proportions of the bird and its respective members, and to the nature of the plumage of the crest and trains that ornament the adults. The privation of these ornaments in the young, and in the adults also when moulting, increases the difficulty, and has caused them to be taken until lately for distinct species. Fortunately this source of confusion has been removed, and the females have been ascertained to be similar to their males. The species of Europe and Northern Asia were therefore, upon good grounds, reduced to two, the great and the small, *Ardea alba* and *A. garzetta ;* but both formerly, and one even till now, were confounded with their two American analogues described by Wilson. In my "Observations on the Nomenclature" of that author, as well as my subsequent writings, without excepting my "Synopsis," I admitted the two North American species, and added as a third the bird now represented in our plate ; but I also erred in considering the large American species as the same with the large European : they are, in fact, no less distinct from each other, however closely related, than *Ardea candidissima* and *A. garzetta*. The name of *alba* belongs to the European, and that of *egretta* to the American ; although Illiger, Lichtenstein, and Temminck (?), not perceiving that it was the legitimate *egretta* of Gmelin and Latham, and having applied that name to the European *alba*, have given the American the new one of *A. leuce.*

Mr Ord, in the second edition of Wilson's " Ornithology," was therefore right in doubting the identity of the two species, and I was mistaken when I declared his doubts unfounded : but he ought not to have quoted as synonymous *A. egretta* of Temminck, &c. Indeed, I am unacquainted with a single instance in which, upon due examination, the rule will not hold good that no bird is common to both continents that does not inhabit during summer the high northern latitudes, and the *Ardea alba* and *A. egretta* are not winter birds, but, on the contrary, summer visitants of Europe and the United States, and do not even then range far to the north : the European, moreover, is chiefly found in the east, and hardly ever seen in the west of that continent. This alone ought to have led us to detect the discrepancy. In order to clear up this point before taking up the species which more immediately forms our subject, I think it proper to fix all the species of egrets of which I have a perfect knowledge. These are :—

1. *Ardea alba*, L. (*Ardea egretta*, Temm., *Ardea candida*, Briss.), which can easily be distinguished by its large stature combined with a small crest (which is wholly wanting in the American), a much longer bill and longer tarsi, and the fusco-corneous colour of the legs. It is well figured by Naumann, " Vog. Nachtr.," tab. 46, f. 91, and the young by Roux, " Ornithologie Provençale," pl. 314 (under the name of *egretta*). It inhabits Europe, especially the oriental parts, and is very common in the Caspian Sea, in Asiatic Turkey, &c.

2. The second species is *Ardea egretta*, Gmel., Lath. (*Ardea leuce*, Temm.), the one figured by Wilson, whose tall stature allows it to be confounded with the preceding, from which, however, it may be readily distinguished by its perfectly smooth head, its light orange and shorter bill, and black legs. It is found both in North and South America, being men-

tioned by D'Azara, and we have ourselves received it from Surinam.

3. The third is *Ardea flavirostris*, Temm., not yet figured, a smaller bird, with black legs also, at once known from its two above-mentioned close analogues—from the European by its yellow bill, from the American by its small crest. It is found in Southern Africa and the Australian islands.

4. The fourth egret in point of stature is the one we are treating of, well distinguished by its bill, which is flesh colour at base, besides the different texture of the ornamental feathers.

As a fifth species, we shall cite the *Ardea candidissima* of Wilson, which is the analogue of the *Ardea garzetta* of Europe, figured by Roux, Orn. Prov. pl. 315. Both these are alike in stature and dimensions, and differ only, as is well known, by the crest, which in the latter consists of but two or three elongated, narrow, subulate feathers; while in the American the crest is formed of numerous elongated pendulous feathers, with loose flowing barbs.

Specimens that we have received from Java, under the name of *Ardea nigripes*, Temm., we consider as the young of *A. garzetta*, and are confirmed in this opinion by the fact of young birds that we possess of the American *candidissima*, that stand precisely in the same relation to this species as the supposed *nigripes* does to the *garzetta*.*

The family of the *Herodii, Cultrirostres*, or *Ardeidæ*, especially when the group *Gruinæ* is withdrawn, and restricting it to our former *Ardeinæ*, is a highly natural one. It still comprises, it is true, many aberrant genera, birds of peculiar forms, and remarkable for their strange and oddly shaped bills, though still not so far different as to rank them more properly with any other class, and in their general structure, as well as their habits and dispositions, too much identified with these to justify their separation into an inde-

* I have lately been informed of the discovery of two new European species of egrets, one from Sardinia, the other from Moldavia, of which the names and characters are not yet given.

pendent family.   But the *Gruinœ*, of which the crane is the
type, bear a strong analogy, and even in many respects so
much affinity, to the gallinaceous birds, having shorter feet,
vegetable food, and even their habits being terrestrial, that
we think proper to unite them as a subdivision or subfamily
with the *Alectrides*.   The artificial character (which, as we
are not now treating of them, is all that need be mentioned)
by which they may be at once distinguished from the *Ardeidœ*
consists in having the hind toe short, and inserted so high up
as to be raised from the ground except merely at the tip ;
while in the *Ardeidœ* it is long, and bears with its whole length
on the ground, or nearly so.   But as, according to the axiom
of the great Linné, the character does not constitute the genus,
even if the most general and characteristic mark should fail
us, it is still no reason why the group is not natural which it
has hitherto been believed to represent.   A minute peculiarity
may furnish a most useful though artificial generic or specific
character, while an apparently important and evidently natural
one may be of no use for this purpose.   In our system, the
family *Ardeidœ* is composed of nine genera, of which none is
subdivided except *Ardea* itself, which with *Ciconia* are all
that are strictly typical.   Besides the more direct relations,
this family is connected with the *Rallidœ* by the curious
though anomalous courlan, also allied to the *Gruinœ* by its
feet, as well as to the *Scolopacidœ*.   But to these the genus
Eurypyga forms a very strongly marked and still better
passage.   At the same time the *Platalea*, which in its feet
shows the transition to *Phœnicopteridœ*, and by its curiously
flattened bill stands alone, is so similar in internal conformation,
and especially the sternal apparatus, to the genus *Ibis*, that
they ought in this respect to go together ; though *Tantalus*,
one of the *Ibidœ*, is constructed rather upon the osseous
plan of the *Ardeidœ !   Scopus, Anastomus, Canchroma*, and
even *Dromas* to a minor extent, each and all exhibits striking
anomalies in their bills, so that *Ardea* and *Ciconia* are the
only two typical genera with sharp-pointed bills of the whole

group. In order to comprehend all these forms of bills, it becomes necessary to restrict greatly the physical characters of the family, and we can merely observe, that in the *Ardeidæ* the bill, whatever be its form, is longer than the head, very robust, and almost always sharp, with cutting edges. The neck is long, the feet long, and always four-toed, the hind toe strong and well developed : the tarsus is longer than the middle toe, and toes and nails both are also long ; the wings are of moderate length, and obtuse : the tail is never long, nor otherwise remarkable, and consists of twelve, or only of ten feathers.

There is no marked external difference between the sexes, but the young vary greatly from the adults, and do not gain their complete plumage till their third year.

In habits and internal conformation these birds are all much more alike than in external. They have all a grave, deliberate, and well-poised gait : their flight is slow, though light and elevated, and they stretch back their legs like sticks in flying, even more so than other waders. They are faithfully monogamous in their loves: their nests are built with more art than those of aquatic birds generally, being placed in trees, thickets, aquatic grasses ; and some of the species, half domesticated, even nestle on housetops : the female incubates, while the male merely watches, and supplies her with food. Both unite in nursing and rearing their young, which remain in the nest until they are full fledged. The flesh of these waders is quite unpalatable.

The genus *Ardea*, when disembarrassed of the several species forced into it by ancient authors, is a very natural one, differing from the storks by having the inner toe cleft, whilst they have all the toes semipalmated at base : the storks also have the tarsi reticulated, and the middle toe-nail entire, whilst the herons have the former scutellated and the latter toothed like a saw, to assist in seizing and securing their slippery prey. A peculiarity of the herons, in which they not only differ from the storks, but from all other birds, is found in their anatomy :

they have but one cæcum, like quadrupeds, while other birds have two. The genus *Ardea* is admitted by all authors, though some modern writers have cut it up into several, which we employ as subgenera, or groups of still minor importance. Generally divided into three, and by Boie into five, they might with the same propriety be carried to seven or eight. We recognise no more than three, comprising eight secondary groups. The first, which we call more properly heron (*Ardea*), is well distinguished by its long and slender neck, all well clothed with shortish oppressed feathers, and by having a very large part of the tibia naked.

The second, called bittern (*Botaurus*), has the neck shortish, with loose, longish feathers, and the posterior more or less distichous and lanuginous : the naked part of the tibia is much limited.

In all the herons the bill is more or less longer than the head, cleft to beneath the eyes, straight, compressed, conic-elongate, acuminate, and very acute, higher than wide, and more or less robust. Both mandibles are near their base covered with a kind of very thin cere or membrane: the upper is scarcely longer than the lower mandible, and equal in height: it is longitudinally impressed on the sides with a straight furrow obliterated before : the upper ridge is therefore rather distinct and flat at base, terminated by the frontal feathers transversely placed ; towards the point the ridge is perfectly smooth, compressed, and slightly and gradually inclined at tip : the edges, nearly vertical, in some species are perfectly entire, in others obliquely and finely denticulated, in all emarginated at the extreme tip : the palate has in the middle a longitudinal sword-like process, perfectly straight, which towards the throat is more or less conspicuously doubled : the lower mandible has strong and flattened sides, more or less impressed towards the base ; it is sharply acute, with the edges drawn in, excessively sharp, quite straight, either entire or slightly serrated obliquely : the inferior ridge is slightly compressed, rather acute, and more or less ascending ; the mental angle is extended beyond the middle of the mandible, is exceedingly narrow, very acute, and feathered : the lora are naked, as well as a portion of the orbits. The nostrils, not quite basal, are placed in the furrow, and are linear, longitudinal, pervious, and above half closed by a naked membrane. The tongue is half the length of the bill, acute, very entire, narrow, membranous, and rather flattened. The body is much compressed. The feet are equilibrate, long, and four-toed : the tarsus is always longer than the middle toe, sometimes barely so, sometimes a great deal : in some species the tibia is almost entirely naked, whilst in others it is on the contrary nearly all feathered : the toes are elongated, slender, narrowly bordered by a membrane, all unequal ; the middle is connected to the outer one by a membrane that extends to the end of the first joint ; the inner toe, a little shorter than the outer, is merely furnished with a very minute basal membrane : the hind toe is long, half equal to the middle one, and all bearing on the ground, being inserted opposite to the inner toe : the nails are compressed, falcate, the hind one largest : the middle one is dilated on the inside into a pectinated sharp edge. The coverings of the tarsi are transversely clypeate, the upper and lower clypei being scutelliform, the opisotarsus and knee are covered with small hexagonal scales ; the toes are scutellated. These various forms of the scales are represented with inimitable accuracy by Mr Lawson in the plate of Peale's egret. The wings are broad, obtuse, tuberculated, the three outer primaries being longest, and the third hardly shorter than the two first. The tail is short and obtuse, and composed of ten or twelve feathers. The feathers of the lower neck before in the adult bird are pendulous, elongated, mostly acuminate, narrow, or ragged ; on the occiput and back they are in many species elongated sericeous, either linear, or laciniate-lacerated, seldom dense, oblong or rounded at the end ; the neck is bare at base on the sides, but concealed by a tuft of longish plumes originating at the shoulders : the neck-feathers in some species are short and closely pressed to the body ; in others they are softer, longer, especially on the sides, and woolly at base : the tail-feathers are always rounded at the end ; those of the lower parts of the body are longish, with the webs disjoined, and the barbs plumulose at base : the down is silky.

The females are like the males ; the young are different from the adults, only obtaining their full plumage after the third year. They moult annually. The adults are ornamented by long slender feathers, which they lose in moulting, and do not acquire again for some time, when they resemble the young.

These birds are remarkably dull : they inhabit marshes, or watch perched on trees near the water for their prey, which the conformation of their feet enables them to do with ease. They feed exclusively on animals, especially fishes and reptiles, but likewise large insects, and even small mammalia. They often stand motionless on the margins of ponds or marshes, concealed by the tall grass and weeds, with the neck so bent as to rest the head on the back, waiting patiently for their prey to pass within their reach, when they dart forward their sharp bill with inevitable aim ; but when tired of this, which is often unsuccessful, they overcome their natural indolence so far as to move slowly through the mud or water, stiring up as they walk by means of their long toes the frogs or fishes that may be lurking in such places. Timid and cowardly to a great degree, the smallest hawk will turn their flight, and often master them, though capable of inflicting a dangerous blow with their powerful beak. They build in companies in high trees, laying about four eggs. The parents are, to a proverb, tender of their offspring, and carefully provide for them during the long time that they require their assistance. Their voice is loud, hoarse, and monotonous, and heard chiefly at night, when most of them are in motion. Their flight is full of grace, and is performed with the neck bent backwards, and the head resting against the back.

The numerous species of this genus are dispersed over all climates and countries, excepting the very coldest. In no group does the size vary to the same extent as is exemplified in the American species by the gigantic *Ardea Herodias* and diminutive *Ardea exilis.*

The herons properly so called forming our subgenus *Ardea*, of which the group *Egretta* is a subdivision, have the bill much longer than the head, at base as broad, or even broader, than high, and quite straight. Their neck is very long, slender, and ornamented beneath with slender, elongated, pendant plumes : their flanks are thin, their legs very long, and have an extensive naked space above the heel.

They are more diurnal than nocturnal in their habits, are the tallest of the genus, and for the most part feed on fishes. There is scarcely a fish, however large, that a heron will not strike at and wound, even if unable to carry it off. They both seize them in shallow water by darting their bill, or in deep water by plunging it under as they pass on the wing : they are therefore extremely injurious to fishponds, which they devastate to an incredible extent, and consume so great a quantity that a single heron will destroy in a year several thousand large fishes, without taking into account the small fry which are their chief dependence. Even when gorged with prey, these greedy birds will sit meditating further mischief, with their long necks sunk between their shoulders, and their heads turned to one side, intently eyeing the pool ; and their extraordinary power of digestion soon enables them to recommence their task. But, like other lean and hungry gluttons, the heron is never satisfied ; his food avails him not, and he is generally an emaciated mass of skin and bones. They do not hide themselves in grassy places, nor attempt to escape danger by retreating to them, but, on the contrary, are careful to seek their prey where the weeds are not too high to prevent them from observing the approach of an enemy, to escape whom flight is their only resource. Highly social in their disposition, they travel, fish, and keep together in parties, and build on trees or hanging cliffs, hundreds in company, in retired haunts, where they may expect to enjoy perfect quiet

and security. Several of these retreats are celebrated both
in America and Europe. The naturalist whose courage and
perseverance enable him to penetrate the swamps, and a
thousand difficulties that surround one of these recesses, and
render them nearly inaccessible, is amply repaid by the astonishing spectacle he witnesses. He finds every branch, every
fork, the top of every bush, covered with the nests of these
birds; and the ear is stunned with the cries and flapping
of the wings of the alarmed multitude. The parents, and
such of the young as can fly, at once depart, their numbers
obscuring the sky: but their attachment for their offspring
overcoming their fears, the parents soon return to their defence,
and boldly attack any enemy, so that even the blows of sticks
or the report of the fatal gun has no terror for them. Their
nests are made with sticks, and lined with wool; but if they
find a nest already made, they do not take the pains to build
a new one. Their young are as voracious and hard to satisfy
as themselves.

The egret herons are entirely of a snowy whiteness, without
any coloured markings on the plumage whatever. We even ex-
clude from them the *Ardea russata*, that visits occasionally the
south of Europe, and possesses when adult in the greatest degree
the long flowing ornamental plumes. This, with the *Ralloides
speciosa* of Java, &c., we consider as forming a group equivalent
in rank to egret, and we apply to it Boie's name of *Buphus*.

Our second subgenus, *Botaurus*, including the bittern, night
herons, and other groups of authors, is characterised by the
bill being hardly longer than the head, much compressed, higher
than broad, with the upper mandible somewhat curved. Their
legs are comparatively short, and the naked space on the
tibia restricted: the neck is rather short, thickly and closely
covered with long, broad, and loose erectile feathers, and merely
downy above: their body is comparatively plump, even fleshy,
and sometimes good eating. They are chiefly nocturnal, and
haunt in marshy and sedgy places. Their food is principally
reptiles, insects, worms, fish-spawn, and they even eat vege-

tables, and are not by any means so destructive as the herons proper, nor so skilful at fishing. The birds of this subgenus never sit in open places, but, on the contrary, keep concealed amongst the highest reeds or grasses; and if an enemy approaches their retreat, they either squat on the ground, or escape between the reeds, and never resort to their slow heavily raised flight but in the last extremity. Instead of high trees, the bitterns place their nest in a sedgy margin or among the rushes; and instead of sticks and wool, they are contented with simpler materials, such as sedge, leaves of water-plants, or rushes; and they lay seven or eight eggs, twice the number of the true herons. The young do not require for so long a period the parental care, but, on the contrary, follow the mother after a few days. When excited, the bitterns have a curious mode of erecting their loose neck-feathers, causing it to appear very much enlarged. Although well defined as a group, these birds are connected with the true herons by means of intermediate species that might with propriety be placed in either. As an example of the intermediate species more allied to the herons, we might quote the beautiful *A. ralloides* of Southern Europe, which we look upon as the type of the group *Buphus.* Of those nearer to *Botaurus, A. virescens* is an example, with the form of the herons but the plumage of the bitterns: we establish it as the type of a natural though secondary group, to which we cannot do better than apply the name of *Herodias*, proposed by Boie. In the subgenus *Botaurus*, also, Nature has pointed out several small sections, of which nomenclators have eagerly availed themselves. As among the herons we have noticed the egrets, herons proper, *Herodias*, and *Buphus*, we may also indicate the *Nycticoraces* among the bitterns, which are distinguished by wearing in the adult state long tapering occipital feathers; and the *A. stellaris* of Europe, together with its close analogue, *A. minor* of Wilson, may be regarded as the types of a similar small group. Another group hardly distinct had been called *Crabier* by the French, but without any fixed character: we have divided these *Crabiers* into two groups, and made them regular by arranging them near the limits of our two subgenera : the larger striated species of bitterns have also been called *Onorés* (*Tigrisoma*, Sw.)

A third subgenus, which we first instituted, and called *Ardeola*, contains only three species, the smallest of the tribe, and closely allied in form, and even markings : one is the European *Ardea minuta*, the other the American *Ardea exilis*, and the third a still less, the New Holland *Ardea pusilla*. In these, the female differs somewhat from the male, and the young are different from both. The bill of these small herons is much the same as that of the true heron, being longer than the head, higher than broad at base, and with the upper mandible nearly straight : the neck likewise is elongated, and rather slender; but, as in the bitterns, it is merely downy above, and thickly covered on the remaining parts with long, loose, broad erectile feathers : the body is slender, and exceedingly compressed, like that of the rail, of which they remind one : the legs are comparatively short; but what strikes most, as a circumstance extraordinary in the waders, their tibiæ are completely feathered, as in the woodcock and the land birds : the membrane that unites the toes is, moreover, simply rudimental.

These birds, which are chiefly nocturnal, have much of the habits of the rails. They live and propagate in marshy grounds, hiding closely amongst the reeds, and running far and very fast in them rather than take wing. They feed on small fishes, reptiles, spawn, but more especially on water insects.

Returning to our egret, whose claims to be considered new have been set forth in the first page of this article, we have to state that it is dedicated to Mr Titian Peale, by whom it was first shot for us in Florida, as a just compliment to a naturalist to whom American zoology owes so much, and from whom so much may still be expected, retaining as he does all that zeal for science for which his family has been long conspicuous.

We regret not being able to relate any peculiarity in the habits of this bird, which besides Florida inhabits other analogous climates of America. It is never seen in the middle States, but appears not to be rare in Florida, for since the individual first brought by Mr Peale, we have observed it in almost all the collections of birds sent from that country.

Peale's egret heron is twenty-six inches long; the bill five inches, flesh colour for nearly three inches from the base, then black to the point; the lora and naked parts of the face are of the same flesh colour, but more delicate: the plumage is uniformly, and without exception, snowy white, as in all the egrets: the head, nearly from the origin of the bill down to the neck, is thickly and densely set with a large crest, formed of numerous compact, subulate feathers, more than three inches long; a bunch of these feathers, precisely of the same texture, and even longer, hangs down from the front part of the neck. The structure of these feathers most resembles that of the corresponding plumes of the *A. garzetta*, and is totally different from those of the *candidissima*. The long flowing plumes of the back are filiform or criniform rather than silky, being by no means delicate, and reach much beyond the tail, with their rays quite straight, and rather stiff, and by no means curled, nodding, or divaricate, as in the *candidissima*. The wings are thirteen inches long: the tail is four. The legs, including the toes and nails, are all black; the toes yellow beneath: the nakedness of the tibia extends more than three inches: the tarsus is full six inches long—that is, twice as long as the middle toe and nail: the hind toe without the nail measures more than an inch.

The young is distinguished by smaller proportions, a circumstance for which this group is more than usually remarkable, and by the absence of the ornamental feathers: we have, however, always observed, even in very young specimens, the tendency of the head-feathers to be long and pointed to a considerable extent, indicating the future crest.

SCOLOPACEOUS COURLAN. (*Aramus scolopaceus.*)

PLATE XXVI.—FIG. 2.

Ardea scolopacea, *Gmel. Syst. Nat.* i. p. 647, sp. 87.—*Lath. Ind. Orn.* ii. p. 701,
  sp. 89, a very bad description.—Aramus scolopaceus, *Vieill. Nouv. Dict.* viii.
  p. 300 ; *Id. Gal. Ois.* ii. p. 134, pl. 252.—*Nob. Ann. Lyc. New York,* ii. p.
  155 ; *Id. Specch. Comp. sp. Philad. ; Id. Cat. and Syn. Birds U. S.* sp. 237.
  —Aramus Carau, *Vieill. Nouv. Dict.* viii. p. 301.—Rallus Guarauna, *Ill.*
  (*mentio duntaxat*).—Rallus Gigas, *Licht. Berlin. Vog. Verz.* p. 79, sp. 815.—
  Rallus ardeoides, *Spix. Av. Brasil,* ii. pl. 91.—Rallus giganteus, *Nob. Add.
  Orn. U. S. in Journ. Ac. Nat. Sc. Philad.* v. p. 31.—Nothorodius Guarauna, *Wagler, Syst. Avium,* i. sp. 1.—*Goldfuss, Nat. Atlas. Aves,* pl. 239.—
  Courliri Courlan, *Vieill. loc. cit.*—Guarauna, *Marcgr. Brasil,* p. 204.—
  Courlan ou Courliri, *Buff. Ois.* vii. p. 442 ; *Id.* ed. 1783, viii. p. 266 ; *Id.
  Pl. enl.* 848.—Carau, *D'Azara, Voy.* iv. p. 223, sp. 366, an excellent descrip-
  tion.—Scolopaceous Heron, *Lath. Syn.* v. p. 102, sp. 79 ; *Id. Gen. Hist.* viii.
  p. 135, sp. 116.—*My Collection.*

HERE is a bird which, if any, might be considered as partaking of a double nature, some authors having regarded it as a
heron allied to the rails, and others as a rail somewhat analogous to the herons.   But notwithstanding these more striking
affinities, and many besides that shall be carefully pointed out—
for it is not contented with these—it fully deserves to constitute
a genus by itself.   After due consideration, therefore, we have
withdrawn it from the rails, where, unconsciously coinciding
in this with Spix, Illiger, and Lichtenstein, we at first arranged
it ; and finding the genus *Aramus* already proposed for it by
Vieillot, willing as we are to admit it to this rank, we do not
hesitate a moment to adopt his name, and although we must
acknowledge ourselves equally unable, with Dr Wagler, to
explain the meaning or etymology of the word, we do not
think this any reason why we should, with the German ornithologist, apply to this bird a new compound signifying
spurious heron.

It was supposed that South America might furnish us with
a second species of courlan, but it being now a well-ascertained fact that the *Carau* of D'Azara is the same as the
Guarauna of Marcgrave, the bird must stand alone in his

genus unless new discoveries shall supply him with a companion. This being settled, we shall proceed to give a minute description, that will therefore comprehend both its generic and specific characters.

Although there can be no doubt that our bird is the *Guarauna* of Marcgrave, it would be committing a great error to take it for the *Scolopax* (or *Numenius*) *guarauna* of systematical writers, that being a very different bird, a species of genuine *Ibis*, which they ought to place under their *Tantalus*, and which has nothing in common with our bird except a somewhat similar speckled appearance, the only source of all this confusion.

Instituting a genus for this bird does not, however, decide the question where it ought to be placed, for it may still be inquired, in what part of the system shall we arrange the genus ? The reader cannot fail to be surprised that we, who made a species of rail of the same bird, should place it as a genus in a very distant family. But this is the result of more mature reflection, and however apparently remote may appear to be at first sight the two families *Rallidæ* and *Ardeidæ*, we have already seen that the subgenus *Ardeola* claims some analogy with the former, and the *Aramus* forms a still better and closer link. It was principally on account of the greatly compressed form of its body that we called it a rail, and upon well examining the singular form of its bill, which is not observed in any other bird, every ornithologist will be satisfied of the propriety of the course we have finally adopted. We have no hesitation in placing it in the *Ardeidæ*, where it is eminently distinguished from all its fellow-genera by its toes, cleft to the base, and entirely separated. Together with *Eurypyga*, it aberrates somewhat towards the *Scolopacidæ*, whilst by the manner of insertion of its hind toe it tends a little towards the *Psophidæ*, subfamily *Gruinæ* (Cuvier even going so far as to make it a genuine *Grus*), and claims again a well-founded resemblance to the most typical form of the genus *Rallus*.

The scolopaceous courlan inhabits principally Cayenne, Brazil, and Paraguay, where it is rather common: it is numerous in the island of Cuba, and other warm parts of America. In the United States, Florida appears to be its most natural residence, and a few instances have occurred of its visiting the middle States. The courlan leads a solitary life, or at most keeps in pairs; night and day they cry out in a loud, sonorous, and resounding voice, *carau !* being in the full sense of the word a *crying-bird :* its chief food is mollusca and other aquatic animals, and even frogs, but not snakes nor fishes : when frightened, they move their tail. Like all solitary and reserved characters, this bird is remarkably shy : it carefully hides itself, but as soon as aware of being discovered, it starts rapidly to a great elevation, its flight being long continued : they walk also with great agility, but never willingly wade into the water: they alight on the very summit of trees : they build in the grass near stagnant water, concealing their nest with much art: they lay but two eggs : the young follow their parents soon after they are hatched, and are covered with blackish down, the throat only being whitish.

The specimen figured was a female, killed on the 5th of February by Mr Titian Peale, at Key Tavernier, on the Florida Reef. Mr Peale took it for the much-disputed crying-bird of Bartram. Mr Peale saw no other individual, but that we have described was brought by Mr F. Cozzens from Florida : one or two killed on the coast of New Jersey, near Long Branch, may be seen in the American Museum at New York. Mr Peale did not hear the bird utter any sound ; it was very unwilling to fly, and caused him some trouble to make it rise from the thick mangroves and other bushes where it kept. It appears to inhabit the low shores and swamps of the rivers and lakes of Florida, and perhaps Georgia, being merely a straggler north of this. Even there, we must conclude it to be rather a scarce species, as Mr Peale could never get information about it; and even upon showing it to the most experienced sportsmen, they declared themselves unacquainted with it, except a few who called it Indian-hen, as they probably would any other rare bird of its size. It runs through the grass exactly in the manner of the rails, compressing its narrow body to pass through a small hole, and very difficult to catch when wounded.

The scolopaceous courlan is two feet and three-fourths of an inch long, and three feet eight inches in extent. The bill, which has but a small gape, and by no means extending like that of the herons to beneath the eyes, measures four and three-quarter inches in length: of course it is longer than the head, and may be called much lengthened; it is slender, quite straight, much compressed, being more than thrice higher than broad, and of a corneous consistence: the upper mandible is of equal height almost throughout, slender; from the base to the middle it is compressed, and channelled each side with a deep furrow covered by a kind of cere-like membrane; from where the furrow ends it swells slightly on each side, being there quite smooth, and even appearing polished: there is no vestige of a notch, as in the herons, and the margins are perfectly entire: these margins, from the middle to the angle of the mouth, are revolute inside, and obtuse; towards the tip they are nearly vertical, and acute, forming throughout inside a straight medial channel; the upper ridge is somewhat depressed at base, then slightly inclined to the tip, being obtuse, and nowhere sharp: the lower mandible at base and beyond the middle is of nearly equal height, straightish in the middle; on the sides at base it is covered by a very thin membrane, and slightly furrowed lengthwise; from the middle to the point, it is as smooth and polished as the upper one, excessively compressed, with the ridge prominent, rather acute at tip, the margins are perpendicular, approximated, very entire; the bifurcation of the sides is very long, extending beyond the middle of the mandible; it is narrow, and the mental angle formed by it naked, acute, entering the corneous substance of the bill. The nostrils are placed rather distant

from the base, and in the lateral furrow they are entirely per-
forated, longitudinal, and somewhat elliptical: the tongue is
elastic, narrow, and acute. The bill is yellow at base, and
of a corneous blue-black at tip: the eyelids are yellow,
the iris brown; the legs pale lead-colour, and the nails
black.

The feet are elongated, and much of the tibia naked, the
bare space measuring three inches: the tarsus, four and a half
inches long, much exceeds the middle toe: the four toes are
slender, all cleft from the base, long, unequal, and compressed;
the inner is a little shorter than the outer, the middle longest,
measuring three inches without the nail; the hind toe is
rather more than one inch, and slender: it is inserted in an
unusual manner, opposite to the base of the inner toe, but
much higher, and with only the last joint, which is very
short, resting on the ground. The unfeathered part of the
tibia is covered behind with transverse scutella, the anterior
with large angulose scales; the tarsus behind has a double
longitudinal series of knobs, before it is covered with oblique
scutella; the cnemidia, that is, the lower part of the naked
tibia, are squamulose; the toes scutellate, and warty beneath:
the nails are moderate, arcuated, acute; the hind nail is rather
the smallest: the middle is the largest, and dilates internally
into a sharp edge, perfectly entire, and by no means pectinated,
any opinions or statements to the contrary notwithstanding.

The body is compressed, but fleshy: the neck cylindrical
and slender: the face and lora entirely feathered. When it is
stated that some specimens have these parts bare, it is because
the other *Guarauna*, which is an *Ibis*, has been confounded
with it. The tail is moderate, scarcely six inches long, plane,
broad, rounded, and composed of twelve broad feathers.

The wings are twelve and a half inches long, ample, and
rounded-obtuse: the first quill is moderately long, and equal
with the eighth, and by more than two inches shorter than
the second, which is equal to the sixth: it is peculiarly shaped,
narrower at base than at tip, where it is very blunt: the third

is the longest of all, being, however, but little longer than the fourth.

The feathers of the neck are short, and rather narrow; those of the body and wing-coverts are rounded on their margins, and soft and dense; the inferior are somewhat loose on their borders. There is no naked place on the sides of the breast, as in the herons. The general colour of the courlan is a deep chocolate-brown, or fuscous sooty hue, reigning all over the bird: the feathers are, however, paler on their margins, and there is on each from the base along the middle, including the shaft, with the exception of the tip, a large, broad, lanceolate, pure white spot. (In the *Ibis guarauna,* the white occupies the margin instead of the middle of the feathers.) This white spot is larger in proportion to the size of the feather, so that it is more conspicuous on the wing-coverts, both upper and under, especially as on the back, not reaching to the tip, it is mostly concealed by the overlapping of the feathers: on the larger coverts, however, it consists of a mere streak, as well as on a few of the lower tail-coverts and femorals. Generally speaking, however, these parts, as well as the rump, upper and lower tail-coverts, outer large wing-coverts, vent, all the quills, and tail-feathers, are un-spotted, and of a bright chocolate-brown, with even a greenish gloss, darker, and with purplish reflections on the quills and tail: on the contrary, on the head and neck all round, the brown colour is paler and duller, and as the feathers are on these parts much smaller, the more extended white longitu-dinal spots are more closely set, producing a thickly striated appearance. On the crown and cheeks the white is, more-over, neither so pure nor well defined, which, together with the much less intense ground colour, gives these parts a rufous grey look: the throat is entirely whitish.

The sexes present no difference, and the young soon put on the adult plumage.

ESQUIMAUX CURLEW.   (*Numenius borealis.*)

PLATE XXVI.—FIG. 3.

Numenius borealis, *Lath. Ind.* ii. p. 712, sp. 9 (not of Ord, which is *N. Hud-sonicus*).—*Nob. Obs. Wils. Orn. notes.—Id. Cat. and Syn. Birds U. S.* sp. 244.—*Id. Monogr. Num. in Osserv. Cuv. Regn. An.—Id. Sp. Comp. Rom.* sp. *Phil.* 187.—Scolopax borealis, *Forst. Phil. Trans.* lxii. p. 431 (not of Gmel., &c., which is *N. Hudsonicus*).—Numenius brevirostris, *Licht. Cat.* ii. *Vog.* p. 75, sp. 744.—*Temm. Pl. Col.* 381.—Numenius cinerius, Seaside Lesser Curlew, *Bartr. Trav.* p. 292.—Courlis Demi-bec, *Temm. loc. cit.—* Chorlito Champêtre? *D'Azara,* iv. p. 275, sp. 307.—Esquimaux Curlew, *Lath. Gen. Syn.* v. p. 125.—*Lath. Gen. Hist.* ix. p. 180, sp. 10.—*Forster, loc. cit.* (not of Pennant, which is *N. Hudsonicus*).—*American Museum at New York.*

IN Wilson's standard work are described but two species of curlew, and no more than this are given by Temminck in his very complete and excellent European Ornithology.   We have brought forward three North American and three European species, which, contrary to the generally received opinion, are all distinct from each other, and different in both continents, not one being found in Europe that is also an inhabitant of America.   These facts, independent of any reference to the almost interminable confusion pervading the works of preceding authors, will sufficiently justify us in repeating here, and stating with more details, what we have published in our Monography, in which, if no new species be introduced (and the list is already too long), we hope to have placed the old ones in a new and more advantageous light.

Perhaps no genus of birds has been less accurately studied, and notwithstanding that it is exceedingly natural, it has but very recently been restricted within its appropriate limits. The appellation it bears was first given by Brisson, yet he was far from assigning its true boundaries.   In addition to the curlews, he comprised in *Numenius* a few other birds (the *Tantali* of Linné), now forming the natural family of *Tantalidœ,* and divided into the genera *Tantalus* and *Ibis.*   The true *Numenii* had been much more philosophically classed by Linné in his extensive genus *Scolopax,* which, though not well formed, was still, with very few exceptions, entirely com-

posed of birds belonging to the natural family *Scolopacidæ*.
Under all circumstances, the union of *Numenius* with *Scolopax*
was far more natural than that with *Tantalidæ ;* and although
we make use of the name given by Brisson, the credit of
establishing it in its present acceptation is due to Latham, or
perhaps to Illiger, who freed it from extraneous species, and
we, with Temminck, Vieillot, and others, adopt it as we find
it. The species now regarded as *Numenii* form a very natural
group, being closely allied in manners, colours, and somewhat
even in size. Hence they have been continually mistaken for
each other, erroneously united, or wantonly multiplied, as will
be made amply apparent by the synonyms and scientific history
of each species.

All the species of curlews have the bill very long, slender,
feeble, much arched, slightly compressed, almost cylindrical,
hard and obtuse at tip, and entire: the upper mandible is
longest, furrowed for three-fourths of its length, rounded towards
the tip ; the lower a little shorter. The nostrils are basal,
lateral, longitudinal, linear, being placed in the furrow. The
tongue is very short, small, and acute. The face is attenuated,
and wholly feathered. The feet are rather elongated, slender,
bare above the heel ; the tarsi cylindrical, half longer than the
middle toe, with their integument reticulated : the three fore
toes are short, fimbriated, scutellated beneath, *all connected at
base by a short membrane*, extending to the first articulation ;
the hind toe is inserted high upon the tarsus, slender, short,
but longer than the phalanx of the fore toes, bearing on the
ground only at tip ; the claws are arcuate, rather short,
bluntish ; the cutting edge of the middle one being entire.

The wings are long, acute, falciform, with from twenty-eight
to thirty stiff quills: the first primary is longest ; the scapulars
are elongated. The tail, rather short, is somewhat rounded,
and of twelve feathers.

They moult once annually : the females perfectly resemble
the males in colour, and the young only differ, but can be known
at once, by their bill being much shorter and less bent.

Possessing numerous general features common to the waders of their family, and a few of those which distinguish the *Ibis* and *Tantali*, the curlews have nevertheless some peculiar traits of their own, more easy to perceive than to define. Their physiognomy may be thus described : They have a rather small head, with a remarkably long, slender, and arched beak, longish neck, and body deeper than broad, and apparently gibbous. The wings are long, the tail moderate, the feet rather slender, though not so much so as in the allied genera, and bare for a considerable space above the heel (commonly, but improperly, called the knee). The toes remarkably short and stout. The plumage of the curlews is composed of a rather thick covering of somewhat loose though silky feathers, abundantly furnished with down. The colours, consisting of a mixture of greyish brown, white, and blackish, are very dull, and hardly vary in the different species. The sexes are not distinguishable by difference of colour or stature : the female is perhaps a trifle smaller than the male. The young scarcely differ in plumage from the adults, but are well marked by their much shorter and straighter bill. They moult but once during the year, and late in the season. We have detected a clue to the species in the medial line of the crown, the colour of the rump, and of the under wing-coverts and long axillary feathers.

The curlews are mute, timid, shy, and wary. They frequent, and seek their food in, salt marshes, and along muddy coasts and inlets, where at low water they may be observed in company with other waders on the mud flats, or at high water roaming along the marshes. They but seldom alight on wet sands, and only when muddy shores are not to be found ; always preferring such on account of their flexible bill. They seldom desert the salt water, and are very rarely met with inland at a distance from the sea or large rivers : during summer, however, they often frequent dry fields in search of berries. They run swiftly, being much upon the ground : their flight is high, very rapid, and long sustained. The voice of the curlews is loud and whistling : when about to commence their great periodical journeys, they congregate in large flocks, rise to a great height, and extend themselves into a vast line : whilst thus travelling onward, they keep up an almost incessant whistling, carefully waiting for each other. These companies only separate during the breeding season. In captivity, though they may linger for weeks or months, they seem to perish at last from the continued operation of melancholy and want of proper food.

Their food is chiefly animal, and in a great degree marine. They prey indifferently upon worms, insects, mollusca, crustacea, and occasionally small fish, and are very dexterous in probing the mud with their long, soft, and slender bill, and pulling out of their holes small shellfish and crabs. In summer, however, they are very fond of berries, especially those of *Rubus trivialis,* or dewberries, and *Empetrum nigrum,* on which they soon fatten.

The spring is their season for breeding, and the northern regions the place they prefer for this purpose. They are monogamous, lay four or five pyriform eggs, which are deposited with little art on a few bits of reeds or grass, placed in the midst of tufts, or in small bushes, for shelter ; sometimes they are merely dropped in sandholes, or on wild open shores. Both sexes sit on the eggs ; but the young receive little attention from their parents, and almost as soon as hatched provide for themselves without requiring their assistance.

This genus, though by no means numerous in species, is not confined to any particular region of either continent, but is distributed everywhere along the shores, from the frozen regions of the north to those of the south pole, and they appear also in the torrid zone in winter. Their migrations may be traced from north to south according to the seasons. They pass the winter in our temperate regions, generally returning in May from the south, and in September from the north. In the economy of nature, these birds seem to be of some importance in preventing the superabundant multiplication of numerous marine animals, thus assisting to maintain the

equilibrium and preserve the harmony of the animal kingdom, as the flycatching birds serve to check the too great
increase of land insects. It is perhaps on this account that
they are so generally diffused. In relation to man they appear
to be of no less importance, since, without being delicious,
their flesh is very palatable, and even, when they have fed and
fattened on berries, tender and excellent meat : when their
nourishment has been derived from the sea, it is much inferior.
They are pursued both in Europe and America in various
ways, and brought in numbers to the city markets. In some
districts their eggs are much sought after, but those of other
aquatic birds are mixed with them, and offered for sale under
the same name.

Wherever the curlews may be classed by ornithologists,
their rank in the system of nature is at the head of the family
Limicolæ, which they connect with the *Falcati.* Their *linear*
place, therefore, is between the genera *Ibis* of the latter and
Tringa of their own family : species of the latter genus are
so closely related to them, as almost to fluctuate between the
two genera. There is a striking affinity on the one hand between some species of *Ibis* and *Numenius,* and on the other
between the smaller *Numenii* and *Tringæ* with slightly curved
bills, such as *Tringa subarquata,* and also those with semipalmated feet, but especially when they combine both these
characters, as our new *Tringa himantopus.* In their own very
natural family, the curlews are more immediately related to
Tringa and *Limosa,* both in aspect and manners. The genus
Scolopax we do not consider as approaching them within
several degrees.

Cuvier had attempted to divide this genus into two independent subgenera, but unsuccessfully, and they must be relinquished even as sections, inasmuch as the characters on which
they are based have no existence in nature, as he has since
virtually acknowledged by omitting all mention of the group
Phæopus in his new edition of the " Règne Animal." This is,
in fact, one of those very natural small genera which do not

admit even of well-based sections. If the species were numerous, we might perhaps divide them into those with white rumps, and those which have no white on that part, or into those showing the crown of the head marked with a central line, and those without this line. There being, however, but few species, we consider it more philosophical to view them as an undivided genus, beginning with the larger and ending with the smaller species : but at all events, the marks we have indicated (of the head and croup), together with those of the under wing-coverts and long axillary feathers, furnish us with what we have called the clue of the genus. For example, the *Numenius arquata* of Europe is distinguished by its head, not parted by the central line, its large size, long arched bill, white rump, white under wing-coverts and axillary feathers : its American analogue, whose still longer bill has gained for it the name of *longirostris*, has the croup of the same dark colour as the body, with the under wing-coverts, &c., rust coloured. The *phœopus* of Europe and *Hudsonicus* of North America, similar in colour and stature, and each ornamented with the medial coronal line, are in like manner distinguishable, the former by the white, the other by the dark-coloured croup ; and by the under-coverts, in the European white banded with black, whilst in the American they are banded with black and rusty.

The two smallest, the present American species and the *N. tenuirostris* of Europe, though less completely analogous, are nevertheless both destitute of the coronal line : the present has the rump dark, and the under wing-coverts banded with black and rusty ; while the *slender-billed* has them pure white, as well as the rump and ground of the tail-feathers. The diminutive size of the Esquimaux curlew will certainly prevent its being confounded with the gigantic *N. longirostris*, especially as its bill is remarkably short, and but little arcuated.

The reader will here have already remarked, we are confident, the curious fact, that all the European species of *Numenius* have white rumps and white under wing-coverts ; whilst the American all have the former uniform in colour with the remainder of the plumage, and the latter rust coloured.

The true Esquimaux curlew (we say the *true*, for it is neither the Esquimaux curlew of Wilson nor of the "Arctic Zoology") is one of the four species that are destitute of the medial coronal line. It is easily known from the large species by its diminutive size; from the small ones, by wanting the white rump; from all, by its very short bill.

It is but half the size of the species that has usurped its name of *short-billed*, being hardly fourteen inches in length, and twenty-four in breadth. The bill is no more than two and a half inches long, but little arched, remarkably slender, blackish, the lower mandible rufous at base: the head is pale, with longitudinal lines of brown: the forehead is deep brown, with pale spots: although there is no medial line, it is somewhat indicated by yellowish marks on that part: the eyebrows and chin are whitish; the neck, breast, belly, and vent are rufous white, the two first dashed with brown streaks and arrowheads, and a few slender streaks on the vent: the feathered parts of the thighs are rufous-white, spotted with brown; the sides under the wings, rufous, transversely fasciated with brown: the back is of a deep brown, the feathers margined with yellowish grey in a serrated manner, and the croup is uniform with the rest. The wings are long, reaching much beyond the tail; they are brown: the shafts of the prime quills are white; the secondaries and lesser coverts margined with grey: the lower coverts, as well as the long axillary feathers, are ferruginous banded with brown: the rump is brown, the feathers edged and spotted with whitish. The tail is short, brown ash, crossed with darker bands, and slightly edged with whitish. The legs are bluish black; the tarsus is one and three-quarter inches long. The female is perfectly similar to the male, except a very little inferiority in size.

This exclusively American bird is widely spread throughout both sections of the new continent, being traced from the fens of Hudson's Bay, in the extreme north, to the warm climates of Brazil, Monte Video, and Paraguay, a circumstance which, however recently observed or extraordinary, is often repeated with the waders that are peculiar to America. D'Azara informs us that in Paraguay this species makes its passage in the month of September, and keeps in the open champaigns, either wet or dry, and never on the borders of rivers or marshes; hence he calls it field curlew, *Chorlito champetre*.

At Hudson's Bay this curlew makes its appearance early in May, coming from the south, and going farther north, returning again to Albany Fort in August: it remains there till September, when it departs for the south. It is common in Maine and Nova Scotia during the months of October and November, and still more so at Newfoundland. We have received it from Maine, and from Prairie du Chien in Michigan, and have occasionally met with it also in the markets of New York and Philadelphia. In the middle States, however, it is by no means common, having escaped the industrious Wilson. This fact proves that our curlew is fond of extremely remote regions, without remaining for any length of time in the intervening countries between its winter and summer residences. They collect in small flocks of from ten to twenty; and when starting on the wing, utter a cry resembling *bibi:* this whistling note may be heard at a distance. The Esquimaux curlew lays four eggs, and keeps in flocks, composed of young and old together: they feed much on the berries of *Empetrum nigrum*, which imparts to their flesh a delicate flavour.

It has been the lot of all the species of curlews to be wantonly confounded with each other: only two were reckoned as European, and in them were merged as identical the three American. The *longirostris* was first definitively disunited from the *arquata* by Wilson. Vieillot unaccountably confounded as one two very different species, giving it more than one name, however. The *Hudsonicus*, though correctly described by Latham, was referred by all writers, including Temminck, to the European whimbrel, *N. phœopus*. The present one he forebore, through extreme caution, to unite also with it, observing that it might be a real species, or at least a constant variety. But when the bird actually fell into his hands, he called his specimens, which were from South America, *Numenius brevirostris*, not recognising in them the *N. borealis* of Latham.

Although we call this bird Esquimaux curlew, it would perhaps be better to condemn this name altogether, and give this one the really appropriate name of short-billed curlew, although this as well as the former appellation has been mis-applied. As for the legitimate scientific name, this also might be disputed. *Borealis* was first given by Gmelin to the Hudsonian curlew, but as he called them *Scolopax*, we have preferred retaining the appellation of Latham, who is admir-ably correct with respect to the curlews, being only wrong perhaps in the choice of the name, and certainly in the citation of Gmelin. As for Temminck, in declaring that the new species of Lichtenstein differs essentially from Latham's *N. borealis* (a fact which was doubted by the accurate German himself), he must have had in view our *N. Hudsonicus*, Lath., the *Scolopax borealis* of Gmelin.

We can form no opinion on the *N. rufiventris* of Vigors, a supposed new curlew from the north-west coast : the diagnosis is certainly inconclusive, not embracing the essential characters, and establishes no difference between it and *N. Hudsonicus*, of which it also has the size.

The *N. Madagascariensis* of Brisson forms a seventh species of *Numenius* peculiar to southern Africa and Oceanica, allied to the *arquata* and *longirostris* : it is figured on the Pl. Enl. 198, of Buffon. We do not know either *N. virgatus* or *N. lineatus* of Cuvier, but one of them, at all events, will have to be referred to the *Madagascariensis*.

1   2

1. *Florida Gallinule.*   2. *Yellow-breasted Rail.*

Gallinula galeata.   *Rallus noveboracensis.*

FLORIDA GALLINULE. *(Gallinula galeata.)*

PLATE XXVII.—Fig. 1.

Crex galeata, *Lichtenstein, Verzeich. Mus. Berlin*, p. 81, sp. 826.—Gallinula chloropus, *Nob. Cat. and Syn. Birds U. S.* sp. 275.—Fulica major pulla, fronte cera coccinea oblongo-quadrata glabra obducta, membrana digitorum, angustissima, *Browne, Nat. Hist. of Jam.* p. 479 (Red-faced Coote).—The Coot, *Sloane, Jamaica,* ii. p. 320, sp. 15.—*My Collection.*

IN all cases wherein we find two animals, however similar or apparently identical in other respects, but restricted within very far distant localities, between which no line of communication can be traced, and beyond which, as in the present case, they are not known to perform great periodical migrations, we may boldly assert that the individuals of the different countries belong to distinct species, having sprung from a different centre of creation, and not being descendants of the same original type. The few known exceptions to this excellent general rule are daily falling in with it, as they come under the closer observation of the more and more practised eye of the naturalist; and since the separation into different species of the gallinules that inhabit the different parts of the globe, there is reason to think that no exception whatever will be admitted to exist, and that all that remain are owing to the want of sufficiently minute comparison and examination. No birds, in fact, reappear in widely separated longitudes under forms and colours so similar as the gallinules, of which we are treating, and if all the species were found in the same country, they would hardly be looked upon even as individual varieties. Yet, upon the principle we have set forth, and which we do not fear to maintain, they have a right, and ought properly to be considered as real species. How different is the stand we now take, fortified by observations in the great field of nature, from that arbitrarily adopted by Buffon, who, on the contrary, saw everywhere the same species reproduced, but changed by climate, or I know not what, and, whenever he

could, referred every new bird he met with to the paltry crea-
tions of Europe !

But to come to facts, and without longer indulging in
theory, we shall merely state that the Florida gallinule differs specifically from the common gallinule of Europe no
less than the Java gallinule (*Gallinula ardosiaca*, Vieill.),
although the differences are almost imperceptible, so as to
justify those who have not hitherto distinguished between
them, among whom we are to be included ourselves. The
true *Gallinula chloropus* is spread over all Europe and the
temperate parts of Asia, and is also met with throughout the
continent of Africa from east to west and from north to south.
We have examined specimens from Egypt, others from Sene-
gambia and from the Cape of Good Hope. The size varies much,
even in specimens from the same country, but the *G. chloropus*
and *ardosiaca* have always the toes shorter than our American
analogue.   In fact, even in the largest specimen examined by
Lichtenstein, which was from Caffraria, and measured fourteen and a half inches, the middle toe without the nail was
only twenty-six lines long ; whilst in the Florida specimens
of the ordinary size of fourteen inches, the same toe measures
at least thirty-four lines. The tarsus, likewise, and the other
toes, are proportionally longer, and this forms the best discriminating mark. Another might also be drawn from the
frontal clypeus, but as this extends with age in the different
species, it may be deceptive. In full-grown birds, however,
it is proper to observe, that both the American and Javan
species differ from the common kind in having it much wider
and differently shaped: in the American it extends still
farther back, and is cut somewhat square behind, whilst the
Javan has it exactly rounded ; in the European it is much
less extended, narrow, and comparatively acute. In point
of form, markings, proportions of the primaries, and every
other particular we could think of, we have been unable to
find any distinction, however trifling, between the three
species.

The genus gallinule, restrained within its just limits,* is a small group composed of but five or six species, spread over all the warm and temperate climates of the globe, and exceedingly similar in form and colours: only one, that figured by Wilson, assumes the brilliant vesture of its near relations the *Porphyriones,* for which reason some authors have considered it as one of them. Together with the rails, the coots, and some others, it forms the natural family *Macrodactyli* (*Rallidæ*), and is more aquatic in its habits than many web-footed birds. Unlike the coots, however, the gallinules dislike salt or brackish water, and confine themselves to fresh, and to rivers and streams especially ; and they are solitary, or at most the hen is seen with her family, like the gallinaceous birds of that sex. Being chiefly nocturnal, the gallinules hide carefully by day among reeds and other aquatic plants; and even in a state of captivity they are so remarkable for this habit, that some which I kept in a yard would take advantage of every hiding-place to escape the eye of man. It was only at the approach of night that they would willingly display on the water their graceful evolutions, swimming in circles, and often striking the water with their tails. From time to time they would rest awhile, placing their necks on the reeds or large leaves of aquatic plants.

Not gifted by nature with the long wings of other waders, the water-hens, being anything but wanderers, obey both their conformation and natural disposition by not undertaking long periodical migrations, but are permanently resident in their native countries, merely removing from one station to another within certain provinces, and without roaming over the adjacent districts. They run with rapidity, fly badly, always in motion, and frequently carry their tail high, as represented in the plate, showing the white plumage of the vent, especially when running on the ground. They dive when frightened, but never after food. They feed on small fishes, insects, and

* The greater part of authors, and among them Latham and Temminck, improperly unite the short-billed rails with them.

some vegetables, picking them up as they swim.   They seldom leave the pond or river where they get their food and exercise, and are peculiarly attached to such as are bordered with sedge and bushes; and standing waters, green with vegetation, furnish them with abundant provision of animalcula and pond-weeds.   They lay twice or thrice in a season, building their nest upon low trees, stumps, and logs, with sticks and fibrous substances, rushes and weeds, or other coarse materials in great abundance, invariably placing it by the waterside. The eggs are very long, of a greenish white, spotted with rufous, and very pointed at the small end.   There are nine or ten in the first brood, the subsequent ones less and less numerous, and the mother never leaves the nest without carefully covering them with weeds.   The chicks are no sooner hatched than they swim with instinctive dexterity, pursuing their parent, and imitating all her motions.   Thus are two or three broods reared in a season, which, while under her care, she regularly after their evening's sport leads back to the nest, where she uses every exertion to make them warm, dry, and comfortable; but when grown up, and taught to provide for themselves, she turns them off.

The Florida gallinule, or water-hen, is fourteen inches long; the bill one and a quarter to the corner of the mouth, and one and an eighth to the posterior portion of the clypeus ; it is red, as well as the clypeus, with the point greenish.   This clypeus, or bare red membrane spreading over the forehead, is more than half an inch wide between the eyes, occupying a great portion of the head, and being posteriorly cut somewhat square or slightly cordate, the reverse of what is observed in the European, which is rather pointed at this place.   The whole plumage from the very base is of a dark plumbeous hue, or sooty black, the head and neck being a shade darker, and the lower portion lighter, and more tinged with bluish, so that they might be styled cinereous.   The mantle, that is, the whole back with the wing-coverts, is highly tinged with olivaceous ; the quills are blackish, and the tail deep black, much more than in the other allied species. The under tail-coverts are also deep black, with the lateral pure white: the white also lines the wings externally from all round the shoulder, almost, but not quite, to the tip of the outer quill, which is white on half the outer part of its narrow web: a few white longitudinal spots may likewise be seen on the under wing-coverts, and very large and conspicuous ones along the flanks, and a few whitish streaks mixed with the plumbeous on the belly. The wings are nearly seven inches long, and the tail more than three. The feet are greenish, with a red ring like a garter surrounding the tibia: the bare space on this is nearly three-quarters, and the tarsus two inches and three-eighths: the middle toe without the nail is more than two and a half, and the nail itself three-quarters: the lateral toes measure more than two, and the hind, one and an eighth. The sexes are precisely alike.

The little that is known of the habits of this gallinule does not allow us to doubt that it has all those of its close analogues. It is common in Florida and Jamaica on the streams and pools, and extends over a great portion of the southern continent of America: in the middle and northern United States it appears to be quite accidental, for although a few well-authenticated instances are known of its having been seen and shot, even as far as Albany in the State of New York, it has escaped the researches of Wilson as well as my own. It is by no means, therefore, a common bird, and is not known as inhabiting Arctic America, ranging much less to the north, even as a straggler, than its European analogue. Its voice is sonorous, resembling *ka, ka, ka.*

The genus *Gallinula* has the bill shorter than the head, rather stout, much higher than broad, tapering, compressed, straight, convex at the point: both mandibles are furrowed, the upper covers the margins of the lower, is inclined at the point, and spreads at base into a naked membrane occupying the forehead. This conformation, found also in the *Fulicæ*, to which Linné united them, more judiciously than they have since been united with the rails, in which the front is feathered, is in my opinion of considerable importance : the lower mandible is navicular : the tongue is moderate, compressed, entire. The legs have been described among the characters of the family, the anterior toes being in all extremely long, flattened beneath, and bordered by a narrow membrane, which circumstance alone distinguishes the gallinules from the coots, that have a broad membrane cut into festoons. The hind toe bears on the ground with several joints : the nails are compressed, subarched, and rather acute. The wings are convex rounded ; the first primary is shorter than the fifth, the second and third being longest. The tail is so short as hardly to appear from under the coverts. The females scarcely differ from the males, but the young are different from the adults. They moult annually.

The family *Macrodactyli*, or *Rallidæ*, when restricted to the five genera of which we compose it (one being *Fulica*, which nothing but blind caprice could separate from them), is surprisingly natural. The bill is short or of moderate length in the long-billed rails, hard, thick at the base, straight, compressed, entire, curved at the point, and sharp on the edges. The head is small, the neck well proportioned, the body slender and much compressed. The feet are moderate, rather robust, and without exception four-toed : the naked space on the tibia is rather limited ; the tarsus not longer, generally shorter, than the middle toe, and *scutellated :* the toes are three before and one behind, remarkably long (the most obvious trait of the family), slender, quite divided, and edged with a decurrent membrane : the hind toe is rather long, articulated almost on a level with the others, resting on the ground a good part of its length : the nails are slender, compressed, and acute. The wings rather short, wide, somewhat rounded, concave, and tuberculated ; the first primary is not much shorter than the second, the third or fourth being the longest. The tail is short, and of twelve feathers.

The female is smaller, but otherwise differs little from the other sex : the young often differ from the adults : even those that moult twice in a year do not change their colours in moulting.

All these birds have very similar habits : they are all solitary ; all fond of concealment and the immediate neighbourhood of water : they move nimbly about on marsh plants, walking on the softest mud, and even floating weeds, their characteristic long toes serving admirably the purpose of a broad base. Their food is small animals, seeds and vegetables. They are monogamous, and breed several times in the year : they build their nests on, or close to the water, some being even afloat, and therefore liable to be carried away in floods. The number of eggs varies from five to sixteen, and they are rounded : both sexes alternately sit upon them. The young run about under the parental care, and provide for themselves as soon as hatched ; they are remarkably brisk and lively, being born with a thick down of a beautiful velvet black colour, whatever else it may finally become. Those that migrate travel by night : owing to their short rounded wings, composed of flaccid feathers, their flight is slow and limited, and by no means rapid, so that they only have recourse to it in the last extremity, when it is performed with the legs hanging down in a way peculiar to themselves, and not stretched out as in the other waders, or drawn up to the belly as in the generality of birds. It is in running that they excel, and with their long compressed body they make their way so adroitly and swiftly amongst the grass or weeds, that their pursuers are left far behind. They also swim well, and even dive occasionally when there is necessity for it. Their flight is, however, rapid when elevated and fairly started. Their voice is strong but hoarse. Their flesh is well flavoured.

YELLOW-BREASTED RAIL.   (*Rallus noveboracensis.*)

PLATE XXVII.—FIG. 2.

Gallinula noveboracensis, *Lath. Ind.* ii. p. 771, sp. 16.—Fulica noveboracensis,
  Gmel. Syst. i. p. 701, sp. 15.—Rallus Ruficollis, *Vieill. Gal. Ois.* ii. p. 168,
  pl. 266 (a bad figure).—Rallus noveboracensis, *Nob. Cat. Birds U. S.; Id.*
  Syn. sp. 273 ; *Id. Sp. comp. sp. Phil.* 212.—Perdix Hudsonica? *Lath. Ind.*
  ii. p. 655, sp. 41.—Le Râle varié à gorge rousse, *Vieill. Nouv. Dict.* xviii. p.
  556.—Yellow-breasted Gallinule, *Lath. Syn.* iii. p. 262, sp. 15 ; *Id. Gen.*
  Hist. ix. p. 419, sp. 30.—*Penn. Arct. Zool.* ii. sp. 410.—Hudsonian Quail?
  Lath. Ind. Orn. Suppl. p. 224 ; *Id. Gen. Hist.* viii. p. 330, sp. 72.—*American*
  Museum at New York.

THE genus rail, and that of the gallinules, are so closely related, that many authors have either confounded them together,
or by their various definitions and acceptations made them to
interfere with each other.   Thus, for Latham, Temminck, and
others, the short-billed rails, among which ranks the present
species, are gallinules, although they want that obvious character upon which Linné founded his natural, though too much
extended, group *Fulica,* and which we also, with Vieillot and
others, adopt as its best representative character, namely, the
naked frontal clypeus.   The genus rail is therefore very comprehensive and numerous in species, which are spread over all
the globe, and may with propriety be divided into two subgenera or groups, the first of which will contain the long-billed
species, under the more restricted name of *Rallus,* containing
the true *Ralli* of all authors, whilst the name *Crex,* or rather
Porzana or *Ortygometra,* may be consecrated to the short-
billed rails, improperly ranked by authors with the gallinules.
I say rather *Porzana* or *Ortygometra,* because the name *Crex*
might be reserved for a secondary group, instituted for the
corncrake alone (*Rallus crex,* L.), a European bird, whose *dry-*
land habits, so different from those of its congeners, have,
with apparent propriety, induced Bechstein and others to elevate it to the rank of a full genus.   Its land habits are so
peculiar, resembling more those of gallinaceous birds than of

waders, that notwithstanding a perfect similarity of conformation, we do not hesitate to grant it the distinction of a
section for itself, especially as we are at last, after a minute
examination, able to assign it a character drawn from the
respective proportions of the toes and tarsus. This is, however,
the result of extraordinary pains. In the land-crake of Europe
(and probably in a few analogous foreign species), the middle toe
without the nail is shorter than the tarsus, whilst in the water-
crakes it is longer. The hind toe is also shorter and rather more
elevated from the ground. All the other rails and crakes are,
though much less aquatic than the gallinules and coots, always found in marshes, swamps, lakes and their reedy margins,
or in their vicinity, and they even swim occasionally, though
not habitually. The *Ortygometræ*, or crakes, are again subdivided by the modern English school into two groups, which
they elevate to the dignity of genera, under the names of
crake and *craker*, but to which they assign no character. At
least Dr Leach, the author of the genus *Zapornia*, did not, as
far as I know, characterise the group, nor is my good friend
at present able to point out the difference. However this
may be, the only species referred to it is the European *Rallus
pusillus*, whilst its close relative the *porzana*, and even the
R. Baillonii are left in *Ortygometra* with the *Rallus crex*,
which, with great inconsistency, the same writers omit to distinguish separately, as has been done by some Germans and
Italians. It will not be useless here to bear in mind that even
the two chief divisions of this natural genus pass so insensibly
into each other as to make it impossible to separate the connecting species, so that a great many Brazilian rails are arbitrarily placed in either subgenus, notwithstanding that the
extremes—which among the four North American species
may be exemplified by this, the yellow-breasted namely, and
the Virginia rail—are so widely different; and this furnishes
additional proof of the inexpediency of Latham's arrangement,
however it may have since been admired and imitated. Our
genus rail, which we maintain to be natural, though closely

related to *Gallinula*, and especially *Porphyrio*, is easily known at once from them all by the feathered front, common to all the species.

The bill, varying in length, which affords the means of distinguishing the two subgenera, is in all the rails more or less thick at base, generally straight, and always compressed: the upper mandible is furrowed each side, somewhat vaulted and curved at tip, its base extending upwards between the feathers of the front: the nostrils, placed in the furrow, are medial, oblong or longitudinal, open and pervious beneath, and covered at base by a membrane (by which conformation they differ essentially from the *Porphyrios*): the tongue is moderate, narrow, compressed, entire, acute, fibrous at tip: the forehead is feathered: the body very compressed and thin flanked. The naked space on the tibia is small, the tarsi subequal to the middle toe, somewhat compressed, so as to make up for the want of membrane in the analogy to the web-footed that other less aquatic wading birds exhibit. We are particular in remarking this, for the toes are entirely divided, and the decurrent membrane extremely narrow. The hind toe equals in length one phalanx of the middle, and is inserted a little higher than the others: the nails are short, compressed, curved, and acute. The first primary is shorter than the fifth; the second, third, and fourth being the longest. The tail is very short, the feathers flaccid, not appearing from beneath the coverts.

The female is generally, though not always, similar to the male, an exception being met with in one of the small European species. The young differ much from the adult. They moult twice a year.

The bill of the subgenus *Rallus* (true rails) may be thus described: Longer than the head, slender, straight, subequal throughout, compressed at base, cylindrical and obtuse at the point; upper mandible furrowed beyond the base; nostrils more basal, linear.

In the crakes, of which the present is an example, the bill

is shorter than the head, robust, much higher than broad at
base, tapering, compressed and acute at the point : upper
mandible furrowed at base only, a little curved at tip : the
lower is navicular : the nostrils exactly medial, oblong. Apparently the group is easy to define ; but as if nature took
delight in baffling our attempts at exactness, the species are
found to pass from one form to another by nice and insensible
degrees.

This rail, like all others, inhabits swamps, marshes, and
the reedy margins of ditches and lakes. By a singular coincidence, it was in the market of New York that, in the beginning of February 1826, I first met with this pretty species,
which appears to have escaped the industrious research of
Wilson, although found equally in Pennsylvania in winter,
where it is, however, very rare. We can hardly believe it is
to be found in the south or south-west, notwithstanding we
have been credibly informed of the circumstance. But we
have no hesitation in declaring it an arctic bird, for we do
not doubt that it is the Hudsonian quail of Latham, thus
miscalled by superficial observers on account of its general
resemblance in plumage and size to the true quail of Europe ;
besides which, we have received it ourselves from the extreme
northern limits of the American continent, and have information of its inhabiting near the most north-western lakes, such
as the Athabasca.

The crakes, as well as the true rails, lead a solitary life :
they are timid and shy, screening themselves from observation
amidst the tall reeds, so as hardly ever to be seen except when
surprised, which does not very often happen, and forced for
a moment to have recourse to their short wings. But they
prefer to evade dangers by their rapid movements among the
aquatic herbage, which the compressed form of their body
enables them to execute with the greatest facility, however
entangled the stalks or narrow the interstices. They also
swim and dive tolerably well, when compelled to take the
water, hiding all but the tip of the bill, but are by no means so

essentially aquatic as the gallinules, or their close relatives
the *Porphyriones.* They also breed in marshes, among weeds
and thickets, placing the nest near the water's edge, or, fasten-
ing it to the reeds, they build a floating habitation. In most
of the species (how it is in the present we do not know) the
eggs are about eight, generally seven or nine in number,
their colour being always of green, more or less tinged with
olive, and very oval in shape. Different in this from the
gallinules, they prefer stagnant to clear waters, and always
keep where the grass is high, and particularly avoid sand and
exposed shores. Notwithstanding their apparently limited
powers of flight, and a conformation similar to that of the
sedentary, unenterprising gallinules, they periodically undertake great journeys. They walk with agility and ease, raising
their head, elevating their feet, and jerking up their tail :
they alight sometimes on low branches, never on trees, except
to escape a very close chase. Of a nocturnal disposition, they
hide closely by day, seeking their food in the morning and
evening, or by moonlight, when they emerge from their
retreats. Their food is both animal and vegetable ; they
search eagerly after worms and snails, and are no less fond
of certain leaves and the seeds of marsh plants.

The following description is taken from a fine male, procured, as we have mentioned, in the neighbourhood of New
York in the winter.

Length, hardly six inches ; extent, about ten : bill six-eighths
of an inch long, exceedingly compressed, of a greenish-dusky
at base beneath on the margins of both mandibles, and the
ridge near the front dull yellowish orange ; irides dark drab ;
feet dirty flesh colour ; tarsus one inch ; middle toe an inch
and one-eighth long. Base of the whole plumage slate.
Head above chocolate-brown, the feathers being slightly
skirted with cinnamon-ferruginous, and on the hind part
minutely dotted at tip with white ; over each eye a broad stripe
of cinnamon-ferruginous, a chocolate spot between the bill and
eye inconspicuously continued beyond it ; the chocolate-brown

colour descends from the nucha to the back on the upper part of the neck in a broad stripe, the feathers of which are widely skirted with cinnamon-ferruginous, and crossed by two narrow white bands, one of which is terminal; those nearer to the neck, and the feathers of the rump, having only the terminal band; sides of the neck and whole under surface yellowish ferruginous, each feather being tipped with darker ferruginous, which gives a waved appearance to those parts, the waves being more intense on the lateral parts: throat and belly whitish, but passing insensibly into the general colour; flanks and thighs darker, with the two white transverse lines, as on the back. Wings when closed reaching to the tip of the tail; upper wing-coverts dark slate broadly margined with olive-ferruginous, and each with two white narrow spots representing the usual lines; margin and spots becoming by degrees inconspicuous towards the outer coverts; inferior wing-coverts and axillary feathers white; quill-feathers plain greyish, considerably lighter beneath, and with the shafts above darker; last of the primaries and first of the secondaries with two or three white dots very irregularly disposed, five or six nearest to the body white on a great part at tip, the last becoming, however, more generally greyish, and only mottled with white; tertials, or rather scapulars, blackish, very widely bordered each side with different shades of yellowish ferruginous, of which the palest is outside, and crossed by the two narrow white lines, having besides a rudiment of a third, equidistant; these scapulars form a whole with the wing-coverts and the feathers of the back, being of the same colour, only somewhat more brilliant. Tail very short, feathers blackish, each side ferruginous, with the two white lines, but interrupted, and neither at the tip; the tail is altogether concealed in its upper and lower coverts; the upper are of the same colour, but have only a terminal white band, whilst the inferior are black at base, and with a broad and vividly ferruginous tip.

This is the most brilliant specimen I have seen, and I must

declare that it had all the appearance of being adult.   Others did not, however, differ in anything except in having the colours duller and less decided ; nor did I notice any difference between the sexes, except a little in size, the female being smaller.   According to Vieillot, however, the plumage I have so minutely described could have been only that of the young bird : he states the adult male to be different in colour both from the adult female and the young ; but as the differences appear to consist more in the language of his imperfect descriptions than in anything else, we shall bestow no further notice upon them.

GENERAL INDEX

OF

ENGLISH AND LATIN NAMES.

The names printed in italics are the species introduced into the notes, not mentioned in the original ; those with the asterisk prefixed are synonyms.

THE END.

PRINTED BY BALLANTYNE, HANSON AND CO.
EDINBURGH AND LONDON